# Elastic Waves

## High Frequency Theory

# Monographs and Research Notes in Mathematics



**Variational-Hemivariational Inequalities with Applications**
*Mircea Sofonea, Stanislaw Migorski*

**Optimization and Differentiation**
*Simon Serovajsky*

**Willmore Energy and Willmore Conjecture**
*Magdalena D. Toda*

**Nonlinear Reaction-Diffusion-Convection**
Lie and Conditional Symmetry, Exact Solutions and Their Applications
*Roman Cherniha, Mykola Serov, Oleksii Pliukhin*

**Mathematical Modelling of Waves in Multi-Scale Structured Media**
*Alexander B. Movchan, Natasha V. Movchan, Ian S. Jones, Daniel J. Colquitt*

**Integration and Cubature Methods**
A Geomathematically Oriented Course
*Willi Freeden, Martin Gutting*

**Actions and Invariants of Algebraic Groups, Second Edition**
*Walter Ricardo Ferrer Santos, Alvaro Rittatore*

**Lineability**
The Search for Linearity in Mathematics
*Richard M. Aron, Luis Bernal-Gonzalez, Daniel M. Pellegrino, Juan B. Seoane Sepulveda*

**Difference Equations: Theory, Applications and Advanced Topics, Third Edition**
*Ronald E. Mickens*

*For more information about this series please visit: https://www.crcpress.com/Chapman--HallCRC-Monographs-and-Research-Notes-in-Mathematics/book-series/CRCMONRESNOT*

MONOGRAPHS AND RESEARCH NOTES IN MATHEMATICS

# Elastic Waves
## High Frequency Theory

Vassily M. Babich

St. Petersburg Department, Steklov Mathematical Institute, Russia
Mathematical Faculty, St. Petersburg State University, Russia

Aleksei P. Kiselev

St. Petersburg Department, Steklov Mathematical Institute, Russia
Physical Faculty, St. Petersburg State University, Russia
Institute for Problems in Mechanical Engineering, St. Petersburg, Russia

Translated from Russian by Irina A. So

CRC Press
Taylor & Francis Group
Boca Raton London New York

CRC Press is an imprint of the
Taylor & Francis Group, an **informa** business
A CHAPMAN & HALL BOOK

CRC Press
Taylor & Francis Group
6000 Broken Sound Parkway NW, Suite 300
Boca Raton, FL 33487-2742



Printed at CPI on sustainably sourced paper
Version Date: 20180228

International Standard Book Number-13: 978-1-1380-3306-1 (Hardback)



**Visit the Taylor & Francis Web site at**
**http://www.taylorandfrancis.com**

**and the CRC Press Web site at**
**http://www.crcpress.com**

# Contents

# Preface

This book describes mathematical foundations of the theory of high-frequency elastic waves, and it is natural that asymptotic methods run through it, with the large parameter being the frequency. The authors aimed not only at clear presentation of the grounds but also at doing so without drawing a veil over their mathematical charms. Luckily, mathematical charms are incompatible with unnecessary complexity and ornateness of mathematical techniques. Practically, all the text is accessible to the reader with standard physical or good engineering mathematical background. Whenever any "higher mathematics" is required, a necessary explanation is given.

We start by deriving equations of linear elastodynamics, which is followed by an account of the theory of volume and surface plane waves (in particular, including such a nontrivial question as the proof of existence of a Rayleigh wave in an anisotropic half-space). Solutions for point sources in unbounded media are presented, and related uniqueness theorems are outlined. A much detailed presentation of the ray method for volume waves in isotropic and anisotropic media is given. The ray method is understood as an asymptotic description of a high-frequency wave, which, roughly speaking, bears resemblance to a plane wave, the parameters of which smoothly vary from point to point. In our presentation, crucially important is the idea, traceable to N. A. Umov, of interpretation of energy as a fluid flowing with group velocity and obeying a continuity equation. The notion of local plane wave, which we introduce in this book, allows us to assign an exact mathematical meaning to this "energy fluid mechanics." Such a delicacy as interpretation of the Rytov law (which rules the rotation of the polarization of waves $S$) in terms of parallel translation in a Riemannian space is presented. With the help of the boundary layer approach, a description of the wave $S$ from a center of expansion in smoothly inhomogeneous media is given. An expression for the "nongeometrical" wave $S^*$ is found within the ray theory framework, and a ray theory of the Rayleigh wave propagation over a curved surface of an inhomogeneous, anisotropic elastic body is presented.

Such is a brief outline of the contents of the book, which is in large measure based on the research due to the authors.

We are indebted to several colleagues for advice and help. Those are first of all P. Chadwick, L. J. Fradkin, J. A. Hudson, I. V. Kamotsky, B. M. Kashtan, N. Ya. Kirpichnikova, P. V. Krauklis, O. V. Motygin, A. L. Shuvalov, A. M. Tagirdzhanov, I. D. Tsvankin, and T. B. Yanovskaya. Of special note is A. B. Plachenov's mammoths job of improving the text, which is far beyond the scope of an editor's duties.

This book is a cover-to-cover translation of a Russian monograph,[1] with few minor amendments. We are thankful to Irina A. So for a careful translation and helpful comments.

[1] *Бабич В. М., Киселев А. П.* Упругие волны. Высокочастотная теория. СПб: БХВ-Петербург, 2014.

# *Introduction*

The present book is devoted to the state-of-the-art theory of high-frequency elastic waves. This theory, on the one hand, is demanded by seismics, acoustics, and other applications. On the other hand, it is closely related to modern asymptotic methods of mathematical physics, Riemannian and Finsler geometries, variational and tensor calculus, functional analysis, etc. The theory of linear high-frequency elastic waves may now be regarded as more or less close to completion, and a systematic presentation of its fundamentals proves to be well-timed.

We deal mainly with waves that are locally, in the vicinity of each point, similar to plane waves, the parameters of which vary only slightly at a distance of order of the wavelength (such as, e.g., the classical spherical waves beyond a neighborhood of the source point). As a rule, by a *wave* we mean a solution of elastodynamics equations of such a kind. Waves propagate along rays the notion of which is commonly known from geometrical optics. The preceding admits refinement and, moreover, a precise mathematical statement, which is implemented in the ray method.

The behavior of the wavefield near singular points of a ray field is, in general, beyond our scope, with the exclusion of neighborhoods of point sources, which are considered in Chapters 3 and 6. Thus, we discuss mainly waves in the *case of a general position*, that is, "in all cases except for exclusionary ones," as V. I. Arnold used to say.

The asymptotic theory of elastic waves presented in this book owes much to research conducted by the authors and has two distinguishing features. The first is the introduction of a locally plane wave as a basic notion, and the second is the systematic employment of the calculus of variations. We solely deal with ideal elastodynamics.

Let us characterize the contents of the book in more detail.

Chapter 1 is an outline of the classical linear elastodynamics. Relevant equations and boundary conditions are derived from the natural assumption that each volume within elastic media can be treated as a mechanical system to which Hamilton's principle is applicable.

In Chapter 2, we address plane waves, which are the simplest and, at the same time, most important of the wave solutions. It splits into two parts.

The first part of Chapter 2 is devoted to volume plane waves in homogeneous media. It presents the classical, in its essence, theory of plane waves in isotropic and anisotropic media. Such significant notions as wave vector and phase and group velocities are introduced. The consideration of plane waves gives impetus to introducing the notion of local wave velocities in the case of inhomogeneous media, which allows us to clearly state a theorem on the domain of influence and a uniqueness theorem. These theorems are proved for general anisotropic media with compact inclusions and voids, under the assumption of a certain smoothness of a solution. Next we address reflection, transmission, and conversion of plane waves in the simple case of a plane traction-free surface of an isotropic half-space. A solution for an incident nonstationary plane wave of an arbitrary shape is given (which requires the consideration of a conjugate harmonic function if the total internal reflection occurs).

The second part of Chapter 2 concerns the plane surface waves. After presentation of the theory of classical Rayleigh and Love waves, we focus on waves in layered isotropic and anisotropic media. We establish several general statements on plane surface waves in such media, including the theorem on the group velocity. The chapter ends with a variational proof of the existence of the Rayleigh wave in a homogeneous isotropic half-space.

The major part of Chapter 3 relates to wavefields of point sources in boundless isotropic media. At the beginning, we recall the basics of the theory of generalized functions. Further, we separately deal with time-harmonic and nonstationary point sources. We consider boundary value problems describing small spherical emitters for which a center of expansion and a center of rotation are simple models (from the mathematical viewpoint). It is noted that, in the far-field area, wavefields of point sources resemble plane waves but demonstrate polarization anomalies. Similar anomalies will systematically arise further in the context of the ray method. Further, various versions of uniqueness theorems for outgoing waves are presented. In the case of general anisotropy, a theorem on the limiting absorption principle is proved. A sketch of the proof of the uniqueness theorem for nonstationary solutions of the Cauchy problem is given. As distinct to Chapter 2, the medium is assumed to be homogeneous and isotropic, but solutions are allowed to be generalized functions. For the isotropic case, a uniqueness theorem concerning the Jones conditions (which generalize the scalar Sommerfeld conditions) is stated.

The material presented in Chapter 4 is the keystone of the high-frequency theory. A detailed exposition of the classical results of the ray method for isotropic inhomogeneous media is given. Here, the theory of the eikonal equation (called *kinematics* in seismic research) is based on the traditional variational approach. The complex eikonal theory will be touched on briefly. The fundamental notion of *local plane wave* is introduced. This allows us to treat the wave propagation as a "rapid" process described by a local plane wave the parameters of which (such as the vector velocity, phase, and amplitude) vary comparatively slowly. We derive the fundamental *Umov equation* for the density of energy of a local plane wave. It can be interpreted as the continuity equation of *energy fluid*. The classical theory of this equation enables us to find leading-order approximate formulas for the intensities of waves $P$ and $S$. They involve "arbitrary" factors known as diffraction coefficients. To determine them, additional information is required concerning, to use the seismic nomenclature, the *dynamics* of the wave process. Also we describe higher-order terms of ray asymptotic expansions. In particular, we present explicit expressions for polarization anomalies. The problem on reflection of a wavefield, given by its ray expansion, from a curved traction-free surface is briefly considered.

Formulas of the ray method are interpreted in terms of Riemannian geometry the basics of which are presented therein. Finally, some explicit expressions for geometrical spreading, indispensable for calculation of the amplitudes, are derived. These are the general formula for homogeneous media and formulas related to reflection from a curved interface in the planar case.

Chapter 5 generalizes the ray method for volume waves to inhomogeneous anisotropic media. In contrast to the previous chapter, the eikonal equation (kinematics) is studied on the basis of the theory of characteristics. The higher-order theory is developed under the assumption that the related plane wave is nondegenerate. The importance of the Finsler geometry in the anisotropic case is demonstrated.

In Chapter 6, we determine diffraction coefficients for wavefields of point sources. It is shown that elementary consideration permits us to find the zeroth approximation for any point source in smoothly inhomogeneous isotropic media. Solutions of nonsimple problems of finding the wave $S$ from a center of expansion and the wave $P$ from a center of rotation in such media are presented. This is carried out by employing boundary layer techniques.

The objective of Chapter 7 is a ray description of the "nongeometrical" wave $S^*$ so much spoken about in the early 1980s. This wave is associated with the total internal reflection and arises in the supercritical area from a point source that is close to the interface and generates a wave $P$. We succeeded in finding a description of the wave $S^*$ in the context of the ray method thanks to the application of a trick based on the reciprocity principle.

The last, Chapter, 8 concerns the construction of ray formulas for the Rayleigh wave propagating along a smooth surface of a smoothly inhomogeneous anisotropic elastic body. An analog of Umov's equation is derived. Via its analysis, the leading-order amplitude of the Rayleigh wave is found. An explicit expression for the phase increment, which arises in propagating the wave and which is known as the Berry phase, is presented. A particular case of an isotropic body where the formulas simplify, still remaining very cumbersome, is considered. This chapter stands out for the number of analytic difficulties it overcomes.

The book ends with the Appendix providing a short account of the geometry of surfaces and tensor calculus required in Chapter 8.

The star $\star$ marks sections (mainly those heavily mathematized) that can be passed over at the first reading.

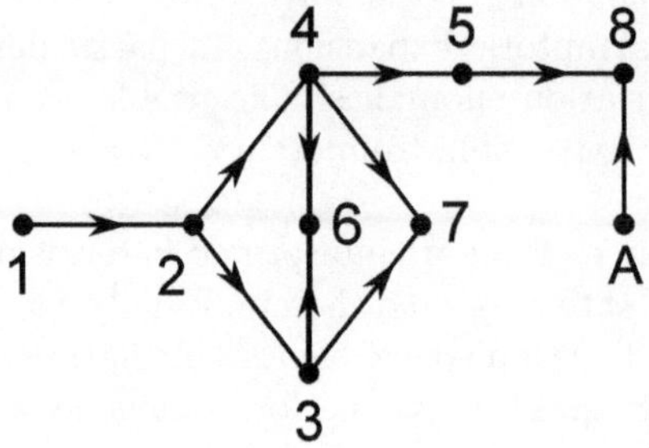


**Figure 1**: Relationship between chapters. A = Appendix

# *List of Basic Symbols*

$:=$ and $=:$ : equals by definition

$\mathbb{R}$ : real numbers

$\mathbb{R}^m$ : $m$-dimensional Euclidian space

Re, Im : real and imaginary parts

$*$ : complex conjugation

The summation from 1 to 3 with respect to repeated lower Latin subscripts and the summation from 1 to 2 with respect to repeated lower Greek subscripts are implied, unless otherwise stated. For example, $f_j g_j := f_1 g_1 + f_2 g_2 + f_3 g_3$, $h_\alpha k_\alpha := h_1 k_1 + h_2 k_2$

$\mathbf{u} \cdot \mathbf{v} := u_l v_l := \sum_{l=1}^{l=3} u_l v_l$ : contraction in $\mathbb{R}^3$

$(\mathbf{u}, \mathbf{v}) := \mathbf{u} \cdot \mathbf{v}^* = u_l v_l^*$ : complex scalar product (except for Chapter 8)

$\mathbf{u} \times \mathbf{v}$ : vector product

$\mathbf{e}_j$, $j = 1, 2, 3$ : unit coordinate vectors of the axes $x_1 = x$, $x_2 = y$, $x_3 = z$, forming a right triplet

$\boldsymbol{n}$ : unit normal to a surface

$\mathbf{n}$ : unit normal to a curve

$a$ and $b$ : velocities of waves $P$ and $S$

$(a_{ij}) = \|a_{ij}\|$ : matrix with entries $a_{ij}$

$\mathbf{I} = \|\delta_{ij}\|$ : identity matrix

$\mathfrak{L}$ : Lamé operator, $(\mathfrak{L}\mathbf{U})_i \equiv \mathfrak{L}_i(\mathbf{U}) := \partial_j(c_{ijkl}\partial_k \mathrm{U}_l)$

$\mathbf{L} = \mathfrak{L} - \rho\mathbf{I}\frac{\partial^2}{\partial t^2}$ : nonstationary Navier operator

$\boldsymbol{l} = \mathfrak{L} + \rho\omega^2\mathbf{I}$ : time-harmonic Navier operator

$\dot{} := \frac{\partial}{\partial t}$, $\ddot{} := \frac{\partial^2}{\partial t^2}$ : derivatives with respect to time

$\star$ : Sections that can be passed over at the first reading

# Chapter 1

## Basic Notions of Elastodynamics

This chapter is introductory; it presents classical notions of elasticity theory, which will continually appear in what follows.

### 1.1 Displacement, deformation, and stress

Displacement and stress are basic notions in elasticity theory.

#### 1.1.1 Displacement vector and strain tensor

Let $\mathbf{x}$ be a point in an *elastic medium* (an *elastic body*) and $x_1$, $x_2$, $x_3$ be its Cartesian coordinates, $\mathbf{x} = (x_1, x_2, x_3)$. When the medium is deformed, the coordinates of a material point that was in equilibrium state at $\mathbf{x}$ change and become $\mathbf{x}+\mathbf{u}= (x_1 + \mathrm{u}_1, x_2 + \mathrm{u}_2, x_3 + \mathrm{u}_3)$. The *displacement vector* $\mathbf{u}$ depends on $\mathbf{x}$ and time $t$, i.e., $\mathbf{u} = \mathbf{u}(\mathbf{x}, t)$. The dependence on $t$ will sometimes be omitted.

The process of deformation is characterized by changing the distance between near points $\mathbf{x}$ and $\mathbf{x} + d\mathbf{x} = (x_1 + dx_1, x_2 + dx_2, x_3 + dx_3)$ of the medium. In the presence of deformation, these points go over into $\mathbf{x} + \mathbf{u}(\mathbf{x})$ and $\mathbf{x} + d\mathbf{x} + \mathbf{u}(\mathbf{x} + d\mathbf{x})$, respectively (see Fig. 1.1).

Before deformation the squared distance between $\mathbf{x}$ and $\mathbf{x} + d\mathbf{x}$ was

$$d\sigma^2 = (d\mathbf{x})^2 = (dx_1)^2 + (dx_2)^2 + (dx_3)^2,$$

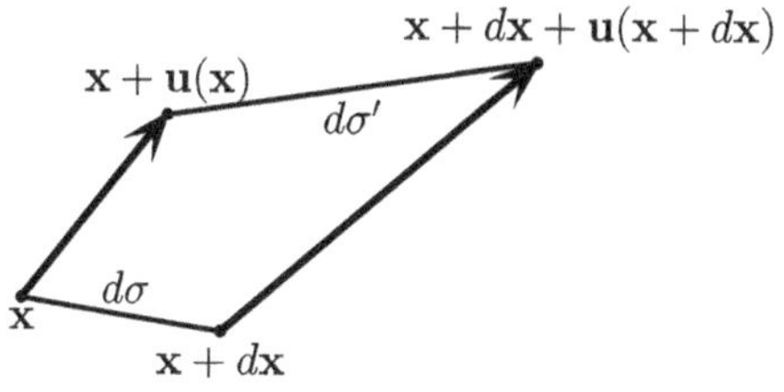


**Figure 1.1**: Deformation of a medium

and after deformation it becomes

$$d\sigma'^2 = (d\mathbf{x} + \mathbf{u}(\mathbf{x} + d\mathbf{x}) - \mathbf{u}(\mathbf{x}))^2.$$

The change in the squared distance is given, up to higher order terms in $dx_i$, as follows:

$$\begin{aligned} d\sigma'^2 - d\sigma^2 &= \sum_{i=1}^{3}\left(dx_i + \sum_{j=1}^{3}\frac{\partial \mathrm{u}_i}{\partial x_j}dx_j\right)^2 - \sum_{i=1}^{3}(dx_i)^2 \\ &= \sum_{i,j=1}^{3}\left(\frac{\partial \mathrm{u}_i}{\partial x_j} + \frac{\partial \mathrm{u}_j}{\partial x_i} + \sum_{l=1}^{3}\frac{\partial \mathrm{u}_l}{\partial x_j}\frac{\partial \mathrm{u}_l}{\partial x_i}\right)dx_i dx_j. \end{aligned}$$

Under the assumption that all $\frac{\partial \mathrm{u}_i}{\partial x_j}$ are small, we omit the terms of higher order $\sum_{l=1}^{l=3}\frac{\partial \mathrm{u}_l}{\partial x_j}\frac{\partial \mathrm{u}_l}{\partial x_i}$ (which corresponds to the assumption on the *geometric linearity* of a medium) and obtain:

$$d\sigma'^2 - d\sigma^2 = 2\varepsilon_{ij}(\mathbf{u})dx_i dx_j. \tag{1.1}$$

Henceforth, repeated lower Roman indices imply summation from 1 to 3, and

$$\varepsilon_{ij}(\mathbf{u}) = \frac{1}{2}\left(\frac{\partial \mathrm{u}_i}{\partial x_j} + \frac{\partial \mathrm{u}_j}{\partial x_i}\right), \quad i, j = 1, 2, 3. \tag{1.2}$$

Obviously,

$$\varepsilon_{ij}(\mathbf{u}) = \varepsilon_{ji}(\mathbf{u}). \tag{1.3}$$

The symmetric matrix $\boldsymbol{\varepsilon} = \|\varepsilon_{ij}(\mathbf{u})\|$ is called *the (Cauchy) strain tensor.*[1] Strain and stress tensors are probably the first tensors that appeared in the history of science.

### 1.1.2 Stress tensor

We now introduce stress, which is of a mechanical nature, as distinct from the geometrically defined displacement and deformation. Consider an oriented surface element $dS$, i.e., a surface element for which the direction of the normal $\boldsymbol{n}$ is specified. Let the element $dS$ lie within or on the boundary of an elastic body. Traditionally, the force acting on the element $dS$ from the side to which the normal $\boldsymbol{n}$ points is denoted by $\mathbf{t}^{\boldsymbol{n}}dS$ (see Fig. 1.2). Thus, $\mathbf{t}^{\boldsymbol{n}}$ is the surface density of this force. We note the obvious formula

$$\mathbf{t}^{-\boldsymbol{n}}(\mathbf{u}) = -\mathbf{t}^{\boldsymbol{n}}(\mathbf{u}), \tag{1.4}$$

which expresses the third Newton law.

[1]Such an interpretation of the tensor is possible as Cartesian coordinate systems are considered. A theory for tensors in arbitrary coordinate systems is provided in the Appendix.

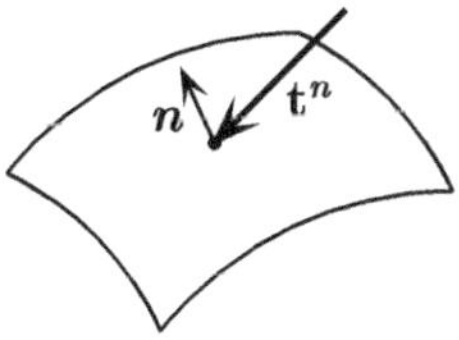


**Figure 1.2**: Force applied to a surface element

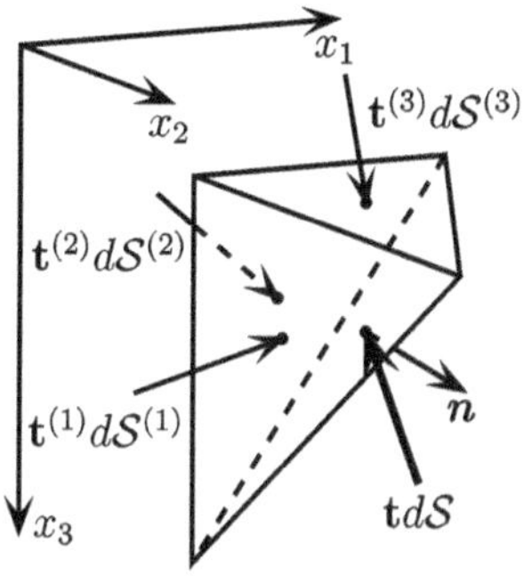


**Figure 1.3**: Surface forces applied to a tetrahedron

Figure 1.3 shows an infinitely small tetrahedron within a medium. According to the second Newton law, $m\mathbf{w} = \mathbf{\Phi}$, where $m$ is the mass of the tetrahedron, $\mathbf{w}$ is the acceleration of its center of gravity, and $\mathbf{\Phi}$ is the resultant force acting on the tetrahedron. The tetrahedron is subject to surface forces, which are external to it and act from the rest of the medium, as well as to volume forces (such as gravitation). Let the lengths of the edges of the tetrahedron be of order $O(h)$; then its volume is of order $O(h^3)$, and the area of its surface is of order $O(h^2)$. The volume and surface forces acting on the tetrahedron are of the same respective orders.

Let three faces of the tetrahedron be parallel to the coordinate planes and have areas $d\mathcal{S}^{(1)}, d\mathcal{S}^{(2)}$, and $d\mathcal{S}^{(3)}$, and let the outward normals to the faces be directed opposite to the unit coordinate vectors $\mathbf{e}_1$, $\mathbf{e}_2$, and $\mathbf{e}_3$. Denote the area of the fourth face of the tetrahedron by $d\mathcal{S}$ and the outward normal to it by $\boldsymbol{n} = (n_1, n_2, n_3)$ (see Fig. 1.3). We assume that the vector $\boldsymbol{n}$ is unit, i.e.,

$$|\boldsymbol{n}|^2 = n_1^2 + n_2^2 + n_3^2 = 1.$$

Comparison of the terms of order $O(h^2)$ in the second Newton law gives

$$-\mathbf{t}^{\mathbf{e}_1} d\mathcal{S}^{(1)} - \mathbf{t}^{\mathbf{e}_2} d\mathcal{S}^{(2)} - \mathbf{t}^{\mathbf{e}_3} d\mathcal{S}^{(3)} + \mathbf{t}^{n} d\mathcal{S} = 0.$$

Noting that

$$d\mathcal{S}^{(j)} = n_j d\mathcal{S}, \tag{1.5}$$

we have

$$\mathbf{t}^{\boldsymbol{n}} = \mathbf{t}^{\mathbf{e}_1} n_1 + \mathbf{t}^{\mathbf{e}_2} n_2 + \mathbf{t}^{\mathbf{e}_3} n_3 = \sum_{j=1}^{3} n_j \mathbf{t}^{\mathbf{e}_j}.$$

The quantities $\mathrm{t}_k^{\mathbf{e}_j}$ can be regarded as the entries of a matrix $\boldsymbol{\sigma} = \|\sigma_{kj}\|$, which we denote by

$$\sigma_{kj} = \mathrm{t}_k^{\mathbf{e}_j}. \tag{1.6}$$

The matrix $\boldsymbol{\sigma}$ is called a *stress tensor*. Obviously,

$$\mathrm{t}_k^{\boldsymbol{n}} = \mathrm{t}_k^{\boldsymbol{n}}(\mathbf{u}) = \sigma_{kj} n_j. \tag{1.7}$$

Elasticity theory is based on the assumption that stress is uniquely determined by deformation. The linear elasticity theory, which will only be considered here, implies a linear relation between them (this relation will be discussed in Section 1.3.2).

Now we recall fundamentals of the Lagrangian approach, which will be used later to derive elastodynamics equations. The Lagrangian approach is based on the derivation, by methods of the calculus of variations, of the main equations from the Hamilton principle, i.e., the condition under which the action functional is stationary.

---

## 1.2 Lagrangian approach to mechanical systems

The derivation of equations of motion will be based on the assumption that any volume chosen within an elastic medium can be regarded as a mechanical system, the motion of which can be described on the basis of the Hamilton principle. This assumption is natural and corresponds to the point of view accepted in theoretical physics that any fundamental physical law can be represented as the requirement that some functional be stationary (see, e.g., lectures by Feynman, published by Feynman, Leighton, and Sands 1966 [2]). It is a very strong assumption, because the Hamilton principle leads to equations having unique properties. Equations based on the Hamilton principle are related to self-adjointness and energy relations, in particular, to a fundamental relation such as the conservation law stated as the Umov – Poynting theorem.

First we recall some basic concepts of mechanics of systems with a finite number of degrees of freedom. Such a system can be described by using a Lagrangian $L(\boldsymbol{q}, \dot{\boldsymbol{q}})$.[2] In several important cases, the Lagrangian $L$ is the difference of kinetic energy $K(\dot{\boldsymbol{q}})$ and potential energy $W(\boldsymbol{q})$. Here, $\boldsymbol{q} = (q_1, q_2, \ldots, q_m)$ denotes generalized coordinates, $\dot{\boldsymbol{q}} := d\boldsymbol{q}/dt$, and $m$ is the

[2] In this connection, see, e.g., Landau and Lifshitz 2007 [7].

number of degrees of freedom. The linear theory assumes that $K$ and $W$ are quadratic forms with respect to $\dot{\boldsymbol{q}}$ and $\boldsymbol{q}$,

$$K = \frac{1}{2}\sum_{i,j=1}^{i,j=m} K_{ij}\dot{q}_i\dot{q}_j \quad \text{and} \quad W = \frac{1}{2}\sum_{i,j=1}^{i,j=m} W_{ij}q_i q_j,$$

where matrices $\|K_{ij}\|$ and $\|W_{ij}\|$ are positive definite.

The motion of a system in the absence of external forces is governed by the *Hamilton principle.* It states that during an arbitrary time interval $t_1 \leqslant t \leqslant t_2$ an isolated system behaves in such a manner that the *functional*[3] $\int_{t_1}^{t_2} Ldt$ is stationary, i.e., its variation vanishes,

$$\delta \int_{t_1}^{t_2} Ldt = 0. \tag{1.8}$$

Recall what is *a variation.* Let us give increments $\delta q_l$ to functions $q_l$, i.e., let us substitute in the integral (1.8) the sums $q_l + \delta q_l$ in place of $q_l$. Here, the variations of coordinates $\delta q_l$ are arbitrary smooth functions vanishing at instants of time $t = t_1$ and $t = t_2$,

$$\delta q_l|_{t=t_1} = \delta q_l|_{t=t_2} = 0, \quad l = 1, 2, \ldots, m. \tag{1.9}$$

In other problems, the assumptions on $\delta q_l$ may be different.

According to the Taylor formula,

$$\int_{t_1}^{t_2} L(\boldsymbol{q} + \delta\boldsymbol{q}, \dot{\boldsymbol{q}} + \delta\dot{\boldsymbol{q}})dt = \int_{t_1}^{t_2} L(\boldsymbol{q}, \dot{\boldsymbol{q}})dt + \int_{t_1}^{t_2} \sum_{l=1}^{l=m} \left(\frac{\partial L}{\partial q_l}\delta q_l + \frac{\partial L}{\partial \dot{q}_l}\delta \dot{q}_l\right) dt$$
$$+ O\left(\int_{t_1}^{t_2} \left((\delta\boldsymbol{q})^2 + (\delta\dot{\boldsymbol{q}})^2\right)\right) dt,$$

$(\delta\boldsymbol{q})^2 + (\delta\dot{\boldsymbol{q}})^2 = \sum_{l=1}^{l=m}(\delta q_l)^2 + \sum_{l=1}^{l=m}(\frac{d}{dt}\delta q_l)^2$. The *variation* of the functional (1.8) is defined by

$$\delta \int_{t_1}^{t_2} L(\boldsymbol{q}, \dot{\boldsymbol{q}})dt := \int_{t_1}^{t_2} \sum_{l=1}^{l=m} \left(\frac{\partial L}{\partial q_l}\delta q_l + \frac{\partial L}{\partial \dot{q}_l}\delta \dot{q}_l\right) dt, \tag{1.10}$$

where

$$\delta\dot{q}_l = \frac{d}{dt}\delta q_l.$$

This is a linear functional with respect to $\delta q_l$. We require that during the

[3]A functional is a mapping of a set of functions to a set of numbers. In this case, we define functions $q_i$ and obtain numbers $\int_{t_1}^{t_2} L\left(q_1(t), \ldots, q_m(t), \frac{dq_1(t)}{dt}, \ldots, \frac{dq_m(t)}{dt}\right) dt.$

motion of the system for any $\delta q_l$, for which (1.9) holds, the expression (1.10) vanish. Integration by parts yields

$$\delta \int_{t_1}^{t_2} L dt = \int_{t_1}^{t_2} \sum_{l=1}^{l=m} \left( \frac{\partial L}{\partial q_l} - \frac{d}{dt} \frac{\partial L}{\partial \dot{q}_l} \right) \delta q_l dt + \sum_{l=1}^{l=m} \frac{\partial L}{\partial \dot{q}_l} \delta q_l \Big|_{t_1}^{t_2},$$

and from (1.8) and (1.9) it follows that

$$\delta \int_{t_1}^{t_2} L dt = \int_{t_1}^{t_2} \sum_{l=1}^{l=m} \left( \frac{\partial L}{\partial q_l} - \frac{d}{dt} \frac{\partial L}{\partial \dot{q}_l} \right) \delta q_l dt = 0.$$

In view of the *Lagrange lemma*,[4] the arbitrariness of $\delta q_l$ implies the Euler equations

$$\frac{\partial L}{\partial q_l} - \frac{d}{dt} \frac{\partial L}{\partial \dot{q}_l} = 0, \quad l = 1, 2, \ldots, m. \tag{1.11}$$

For a nonisolated system, when the external forces are described by generalized forces $\mathscr{F}_j$, the equations of motion take the form

$$\frac{\partial L}{\partial q_l} - \frac{d}{dt} \frac{\partial L}{\partial \dot{q}_l} = -\mathscr{F}_l, \quad l = 1, 2, \ldots, m. \tag{1.12}$$

To include the external forces in the variational equation, we replace (1.8) by

$$\delta \int_{t_1}^{t_2} L dt + \int_{t_1}^{t_2} \sum_{l=1}^{l=m} \mathscr{F}_l \delta q_l dt = 0. \tag{1.13}$$

The derivation of (1.12) from (1.13) is completely analogous to that of (1.11). Expression (1.13) can be regarded as the main axiom of the Lagrangian mechanics (for linear systems with a finite number of degrees of freedom).

---

## 1.3 Elastodynamics equations

Equations describing the dynamics of elastic media can easily be obtained by applying the second Newton law to an infinitely small volume (as is presented in classical textbooks by Love 1944 [9], Sommerfeld 1950 [15], and others). We will use the Lagrangian approach, which will give us significant benefits afterwards. We will regard an elastic medium as a mechanical system with an infinite number of degrees of freedom. In Russian literature the suitable term "systems with distributed parameters" is accepted for such problems.

[4]Let $\psi(t)$ be a fixed function continuous at $t_1 \leqslant t \leqslant t_2$, and for an arbitrary smooth function $\eta(t)$ such that $\eta(t_1) = \eta(t_2) = 0$ the integral $\int_{t_1}^{t_2} \psi(t)\eta(t)dt$ vanishes. Then $\psi(t) \equiv 0$ in $[t_1, t_2]$.

A displacement vector $\mathbf{u}$ will be a counterpart of generalized coordinates; it is "numbered" by continuous variables $x_1$, $x_2$ and $x_3$ instead of $l = 1, 2, \ldots, m$, and summation over $l$ is replaced by integration over a volume occupied by a medium.

### 1.3.1 Kinetic and potential energies as quadratic functionals

Let us consider an arbitrary volume $\Omega$ chosen within a medium. Let the volume $\Omega$ be subject to external forces, both surface and volume ones. Let us introduce the *density of a Lagrangian* as the difference between the densities of kinetic and potential energies of the medium

$$\mathcal{L} = \mathcal{K} - \mathcal{W}. \tag{1.14}$$

*The density of kinetic energy* is defined as follows:

$$\mathcal{K} = \frac{1}{2}\rho\left(\frac{\partial \mathbf{u}}{\partial t}\right)^2 = \frac{1}{2}\rho\dot{\mathbf{u}}^2, \tag{1.15}$$

where $\rho = \rho(\mathbf{x})$ is the *volume density (of mass)*, which will always be assumed positive. We make a fundamental for *elasticity theory* assumption that the density of potential energy is determined at each point by the strain tensor $\boldsymbol{\varepsilon}(\mathbf{x})$,

$$\mathcal{W} = \mathcal{W}(\mathbf{x}, \boldsymbol{\varepsilon}(\mathbf{x})). \tag{1.16}$$

Since we confine ourselves to the linear elasticity theory, we assume that $\mathcal{W}$ is a positive definite quadratic form with respect to the components of the strain tensor $\varepsilon_{ij}$. The density of potential energy (1.16) is assumed to be a quadratic form with respect to $\boldsymbol{\varepsilon}$ (i.e., a homogeneous quadratic expression with respect to the components of $\boldsymbol{\varepsilon}$)

$$\mathcal{W} = \frac{1}{2}c_{ijkl}\varepsilon_{ij}\varepsilon_{kl}, \quad c_{ijkl} = c_{ijkl}(\mathbf{x}). \tag{1.17}$$

From (1.17) it easily follows that $c_{ijkl}$ is a fourth-order tensor. It is called an *elastic stiffness tensor*, and its components are called *elastic stiffnesses*.

### 1.3.2 Properties of elastic stiffnesses

By the symmetry of the strain tensor, without loss of generality the tensor $c_{ijkl}$ may be assumed to have the following symmetry properties:[5]

$$c_{ijkl} = c_{jikl} = c_{ijlk} = c_{klij}. \tag{1.18}$$

[5]If $c_{ijkl} \neq c_{jikl}$, the symmetry of $c_{ijkl}$ in the first pair of indices is achieved by replacing $c_{ijkl}$ and $c_{jikl}$ by $\frac{1}{2}(c_{ijkl} + c_{jikl})$, which does not change the quadratic form (1.17). If $c_{ijkl} \neq c_{klij}$, we replace $c_{ijkl}$ and $c_{klij}$ by $\frac{1}{2}(c_{ijkl} + c_{klij})$.

A fourth-order tensor satisfying the symmetry relations (1.18) is defined, as one can easily find out, by 21 parameters. Elastic stiffnesses are often represented as entries of symmetric $6 \times 6$ matrices $C_{pq}$, with the following correspondence between the pairs $ij$ and $kl$ and the numbers $p$ and $q$: $11 \leftrightarrow 1$, $22 \leftrightarrow 2$, $33 \leftrightarrow 3$, $23 = 32 \leftrightarrow 4$, $13 = 31 \leftrightarrow 5$, $12 = 21 \leftrightarrow 6$.

From the physical point of view, it is very important to assume that potential energy is a positive definite quadratic form with respect to the strain tensor, i.e.,

$$c_{ijkl}\varepsilon_{ij}\varepsilon_{kl} \geqslant \text{const} \sum_{i,j=1}^{i,j=3} (\varepsilon_{ij})^2 \tag{1.19}$$

for any symmetric real $\varepsilon_{ij}$ and some const $> 0$, wherein $\mathcal{W} = 0$ only if $\varepsilon_{ij} = 0$ for all $i$ and $j$. The above assumption implies an inequality for complex symmetric $\varepsilon_{ij}$:

$$c_{ijkl}\varepsilon_{ij}\varepsilon^*_{kl} \geqslant \text{const} \sum_{i,j=1}^{i,j=3} |\varepsilon_{ij}|^2, \tag{1.20}$$

which will be used many times in the sequel.

### 1.3.3 Derivation of elastodynamics equations

The Lagrangian for a volume $\Omega$ is defined by

$$L = K - W = \int_\Omega \mathcal{K} d\mathbf{x} - \int_\Omega \mathcal{W} d\mathbf{x}. \tag{1.21}$$

Here, $K$ is *the kinetic energy* of the volume $\Omega$, $W$ is its *potential energy*, and $d\mathbf{x}$ denotes a three-dimensional volume element.

Let us consider the dynamics of an arbitrary (not necessarily small) volume $\Omega$ in a time interval

$$t_1 \leqslant t \leqslant t_2.$$

By analogy with (1.13), consider the Hamilton principle in the form

$$\delta \int_{t_1}^{t_2} L dt + \int_{t_1}^{t_2} dt \int_\Omega \mathrm{F}_i \delta \mathrm{u}_i d\mathbf{x} + \int_{t_1}^{t_2} dt \int_{\partial\Omega} \mathrm{f}_i \delta \mathrm{u}_i dS = 0. \tag{1.22}$$

Here, $\mathbf{F} = (\mathrm{F}_1, \mathrm{F}_2, \mathrm{F}_3)$ is *the density of volume forces* (i.e., forces acting within $\Omega$), $\mathbf{f} = (\mathrm{f}_1, \mathrm{f}_2, \mathrm{f}_3)$ is *the density of surface forces* acting from the outside to the boundary $\partial\Omega$ of the volume $\Omega$, and $dS$ is a surface area element, i.e., $\mathbf{t}^{\boldsymbol{n}}(\mathbf{u}) = \mathbf{f}$. The external forces of both types are regarded as given. External forces (especially volume forces) are often referred to as *sources (of oscillations)*. Recall (see (1.7)) that

$$\sigma_{kj} n_j = \mathrm{f}_k. \tag{1.23}$$

In this section we constrain the variations by a condition analogous to (1.9)

$$\delta \mathbf{u}|_{t_1} = \delta \mathbf{u}|_{t_2} = 0. \tag{1.24}$$

We understand (1.22) as follows:

$$\int_{t_1}^{t_2} dt \int_{\Omega} \left( \rho \dot{\mathrm{u}}_i \delta \dot{\mathrm{u}}_i - \frac{\partial \mathcal{W}}{\partial \varepsilon_{ij}} \delta \varepsilon_{ij} \right) d\mathbf{x} + \int_{t_1}^{t_2} dt \int_{\Omega} \mathrm{F}_i \delta \mathrm{u}_i d\mathbf{x} + \int_{t_1}^{t_2} dt \int_{\partial\Omega} \mathrm{f}_i \delta \mathrm{u}_i dS = 0,$$

where

$$\delta \dot{\mathrm{u}}_i := \frac{\partial}{\partial t} \delta \mathrm{u}_i,$$

and

$$\delta \varepsilon_{ij} = \frac{1}{2} \delta \left( \frac{\partial \mathrm{u}_i}{\partial x_j} + \frac{\partial \mathrm{u}_j}{\partial x_i} \right) = \frac{1}{2} \left( \frac{\partial}{\partial x_j} \delta \mathrm{u}_i + \frac{\partial}{\partial x_i} \delta \mathrm{u}_j \right).$$

Let us apply the Ostrogradsky – Gauss *divergence theorem*,

$$\int_{\Omega} \frac{\partial}{\partial x_j} A_j d\mathbf{x} = \int_{\Omega} \operatorname{div} \boldsymbol{A} d\mathbf{x} = \int_{\partial\Omega} \boldsymbol{n} \cdot \boldsymbol{A} \ \ dS(\mathbf{x}) = \int_{\partial\Omega} A_j \cos(\widehat{nx_j}) dS, \quad (1.25)$$

where $\boldsymbol{A} = \boldsymbol{A}(\mathbf{x})$ is an arbitrary smooth vector, $\cos(\widehat{nx_j})$ are direction cosines of the outward normal $\boldsymbol{n}$ to the boundary $\partial\Omega$, and $dS$ is its surface element. Integrating by parts, with the symmetry of the displacement tensor $\boldsymbol{\varepsilon}_{ij}$ (1.3) taken into account, we obtain

$$\int_{t_1}^{t_2} dt \int_{\Omega} \left( \frac{\partial}{\partial x_i} \frac{\partial \mathcal{W}}{\partial \varepsilon_{ij}} - \rho \ddot{\mathrm{u}}_j + \mathrm{F}_j \right) \delta \mathrm{u}_i d\mathbf{x} + \int_{t_1}^{t_2} dt \int_{\partial\Omega} \left( \mathrm{f}_i - \frac{\partial \mathcal{W}}{\partial \varepsilon_{ij}} n_j \right) \delta \mathrm{u}_i dS = 0. \quad (1.26)$$

The arbitrariness of $\delta \mathrm{u}_i$, $i = 1, 2, 3$, results in

$$\frac{\partial}{\partial x_i} \frac{\partial \mathcal{W}}{\partial \varepsilon_{ij}} - \rho \ddot{\mathrm{u}}_j = -\mathrm{F}_j, \qquad \mathbf{x} \in \Omega, \ \ j = 1, 2, 3, \quad (1.27)$$

and

$$\frac{\partial \mathcal{W}}{\partial \varepsilon_{ij}} n_j = \mathrm{f}_i, \quad \mathbf{x} \in \partial\Omega, \ \ j = 1, 2, 3. \quad (1.28)$$

By comparing (1.28) and (1.23), we obtain

$$\left( \frac{\partial \mathcal{W}}{\partial \varepsilon_{ij}} - \sigma_{ij} \right) n_j = 0, \qquad \mathbf{x} \in \partial\Omega, \ \ i = 1, 2, 3.$$

The arbitrariness of $\Omega$ gives

$$\sigma_{ij} = \frac{\partial \mathcal{W}}{\partial \varepsilon_{ij}}, \qquad \mathbf{x} \in \Omega, \ \ i, j = 1, 2, 3. \quad (1.29)$$

From (1.29) and (1.17) it follows that

$$\sigma_{ij} = c_{ijkl} \varepsilon_{kl}. \quad (1.30)$$

A medium for which *Hooke's law* (1.30) holds is called *physically linear*. Expressions (1.18) and (1.30) result in the symmetry of $\boldsymbol{\sigma} = \|\sigma_{ij}\|$,

$$\sigma_{ij} = \sigma_{ji}, \quad i, j = 1, 2, 3. \quad (1.31)$$

We rewrite (1.27), (1.29), and (1.23) in the form

$$\rho \ddot{u}_j = \frac{\partial \sigma_{ij}}{\partial x_i} + F_j, \quad \mathbf{x} \in \Omega, \tag{1.32}$$

where

$$\mathbf{t}^n(\mathbf{u})|_{\partial\Omega} = \mathbf{f}, \tag{1.33}$$

which agrees with the definition of surface forces.

Equations (1.32) are the desired equations of motion of an elastic medium, expressing the second Newton law for its arbitrary volume. This system of three second-order equations is called *elastodynamics equations*, which is suitable, though little used in the Russian literature.

In the calculus of variations, boundary conditions that follow from the variational principle are called *natural boundary conditions*. An example of such a condition is given by (1.33).

### 1.3.4 Navier and Lamé operators

Elastodynamics equations (1.32) take the form

$$\mathbf{L}(\mathbf{u}) = -\mathbf{F}, \tag{1.34}$$

where the matrix differential *Navier operator* $\mathbf{L}$ has the components

$$L_i(\mathbf{u}) \equiv (\mathbf{L}\mathbf{u})_i = \frac{\partial}{\partial x_j}\left(c_{ijkl}\frac{\partial u_l}{\partial x_k}\right) - \rho\frac{\partial^2 u_i}{\partial t^2}. \tag{1.35}$$

This operator can also be written as

$$\mathbf{L} = \mathfrak{L} - \rho\mathbf{I}\frac{\partial^2}{\partial t^2}, \tag{1.36}$$

where $\mathbf{I}$ is the identity $3 \times 3$ matrix, and $\mathfrak{L}$ is the *Lamé operator*,

$$\mathfrak{L}_i(\mathbf{u}) \equiv (\mathfrak{L}\mathbf{u})_i = \frac{\partial}{\partial x_j}\left(c_{ijkl}\frac{\partial u_l}{\partial x_k}\right). \tag{1.37}$$

The expression for *the density of surface forces* (1.28) is written as

$$t_i^n(\mathbf{u}) = c_{ijkl} n_j \varepsilon_{kl}(\mathbf{u}) = c_{ijkl} n_j \frac{\partial u_k}{\partial x_l}. \tag{1.38}$$

## 1.4 Classical boundary conditions

Up to now, we dealt with an elementary volume in the medium of which the properties are continuously varying. We will focus on wave processes on the

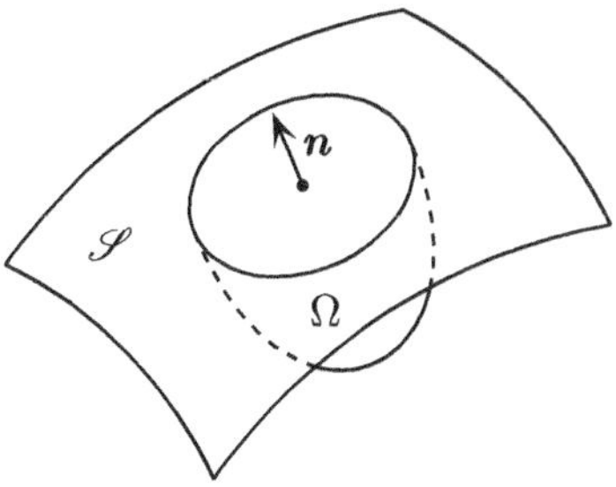


**Figure 1.4**: A volume adjacent to the boundary of a body

boundary of an elastic medium and on the interface between two physically different media. In the first place, several reasonable boundary conditions will be presented, and, in the second place, they will be discussed from the Lagrangian point of view.

### 1.4.1 List of boundary conditions

The physical boundary of an elastic medium is denoted by $\mathscr{S}$ (see Fig. 1.4). We now state two types of boundary conditions on the boundary $\mathscr{S}$ (assuming it to be sufficiently smooth).

#### Boundary of an elastic body

*1. The contact with an absolutely rigid body* along $\mathscr{S}$ is defined by the condition

$$\mathbf{u}|_{\mathscr{S}} = \boldsymbol{\varphi}, \tag{1.39}$$

where $\boldsymbol{\varphi} = \boldsymbol{\varphi}(\mathbf{x}, t)$ is a given vector describing the displacement of the surface under the impact from the external body. Equation (1.39) resembles the Dirichlet condition used in the theory of scalar waves. A special case of (1.39) is the condition of vanishing displacements on the boundary, i.e.,

$$\mathbf{u}|_{\mathscr{S}} = 0. \tag{1.40}$$

*2. The contact with an absolutely soft body* is defined by the condition

$$\mathbf{t}^n(\mathbf{u})|_{\mathscr{S}} = \mathbf{f}, \tag{1.41}$$

where $\boldsymbol{n}$ is the unit normal outward with respect to the medium under consideration, and the vector $\mathbf{f} = \mathbf{f}(\mathbf{x}, t)$ describing external surface forces is regarded as given (see (1.33)). An important special case of (1.41) is

$$\mathbf{t}^n(\mathbf{u})|_{\mathscr{S}} = 0, \tag{1.42}$$

and is called the case of a *traction-free boundary* or simply the case of a *free boundary.* This condition describes contact with vacuum. Conditions (1.41)–(1.42) resemble the Neumann condition.

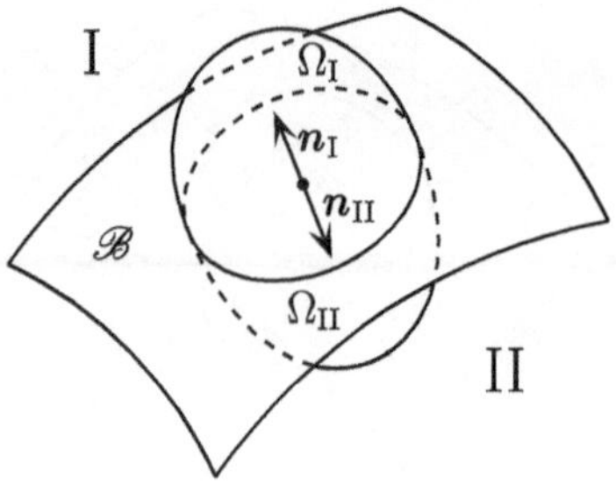


**Figure 1.5**: A volume divided by the interface $\mathscr{B}$

### Interface between two elastic bodies

Let $\mathscr{B}$ be an interface between two elastic bodies (see Fig. 1.5). Quantities relating to the bodies will be denoted by I and II. We state two types of boundary conditions on $\mathscr{B}$.

*1. Welded contact* is defined by

$$\mathbf{u}_{\mathrm{I}}|_{\mathscr{B}} = \mathbf{u}_{\mathrm{II}}|_{\mathscr{B}}, \quad \mathbf{t}^{\boldsymbol{n}}_{\mathrm{I}}|_{\mathscr{B}} = \mathbf{t}^{\boldsymbol{n}}_{\mathrm{II}}|_{\mathscr{B}} \tag{1.43}$$

or, in equivalent form, by

$$[\mathbf{u}]|_{\mathscr{B}} = 0, \qquad [\mathbf{t}^{\boldsymbol{n}}]|_{\mathscr{B}} = 0. \tag{1.44}$$

Here, $[\;]|_{\mathscr{B}}$ denotes the jump of a respective quantity on $\mathscr{B}$. As follows from Section 1.3, conditions (1.43) and (1.44) are automatically satisfied on any surface chosen within an elastic body.

*2.* In order to describe *slipping contact* (that is, contact with slippage) between I and II, we decompose displacements and stresses on the interface $\mathscr{B}$ into tangential ($\mathsf{T}$) and normal ($\mathsf{N}$), with respect to $\mathscr{B}$, components:

$$\begin{gathered} \mathbf{u} = \mathbf{u}^{\mathsf{T}} + \mathbf{u}^{\mathsf{N}}, \quad \mathbf{u}^{\mathsf{T}} \cdot \boldsymbol{n} = 0, \quad \mathbf{u}^{\mathsf{N}} \times \boldsymbol{n} = 0, \\ \mathbf{t}^{\boldsymbol{n}} = \mathbf{t}^{\boldsymbol{n}\mathsf{T}} + \mathbf{t}^{\boldsymbol{n}\mathsf{N}}, \quad \mathbf{t}^{\boldsymbol{n}\mathsf{T}} \cdot \boldsymbol{n} = 0, \quad \mathbf{t}^{\boldsymbol{n}\mathsf{N}} \times \boldsymbol{n} = 0, \end{gathered} \tag{1.45}$$

where $\boldsymbol{n}$ is a chosen unit normal to $\mathscr{B}$. *The conditions of slipping contact* take the form

$$\mathbf{u}^{\mathsf{N}}_{\mathrm{I}} = \mathbf{u}^{\mathsf{N}}_{\mathrm{II}}, \quad \mathbf{t}^{\boldsymbol{n}\mathsf{N}}_{\mathrm{I}} = \mathbf{t}^{\boldsymbol{n}\mathsf{N}}_{\mathrm{II}}, \quad \mathbf{t}^{\boldsymbol{n}\mathsf{T}}_{\mathrm{I}} = \mathbf{t}^{\boldsymbol{n}\mathsf{T}}_{\mathrm{II}} = 0, \tag{1.46}$$

where no condition is imposed on the jump of $\mathbf{u}^{\mathsf{T}}$. These conditions describe the contact of bodies slipping without friction along an infinitely thin layer of nonviscous liquid.

Let us proceed to discussion of the relationship between the above conditions and the Hamilton principle.

### 1.4.2 Hamilton principle and boundary conditions

#### Boundary of an elastic body

Condition (1.41) (in particular, (1.42)) appeared in considering the Hamilton principle (see (1.23)) and is therefore natural. At the same time, condition (1.39)–(1.40) is not natural.

#### Contact of two elastic bodies

Consider an interface $\mathscr{B}$ in an elastic body and apply the Hamilton principle to the volume $\Omega = \mathrm{I} + \mathrm{II}$ (see Fig. 1.5). Integrating by parts with the use of boundary conditions (1.28) on $\mathscr{B}$, we transform expression (1.26) into

$$\int_{t_1}^{t_2} dt \int_{\mathscr{B}} [\mathbf{t}^{n_{\mathrm{I}}}(\mathbf{u}_{\mathrm{I}}) \cdot \delta\mathbf{u}_{\mathrm{I}} + \mathbf{t}^{n_{\mathrm{II}}}(\mathbf{u}_{\mathrm{II}}) \cdot \delta\mathbf{u}_{\mathrm{II}}]\, dS = 0, \tag{1.47}$$

where $\boldsymbol{n}_{\mathrm{I}}$ is the unit normal to $\mathscr{B}$ that is outward for I, and $\boldsymbol{n}_{\mathrm{II}}$ is the unit normal to $\mathscr{B}$ that is outward for II, so that $\boldsymbol{n}_{\mathrm{I}} = -\boldsymbol{n}_{\mathrm{II}}$. Using (1.4), we obtain

$$\int_{t_1}^{t_2} dt \int_{\mathscr{B}} [\mathbf{t}^{n}(\mathbf{u}_{\mathrm{I}}) \cdot \delta\mathbf{u}_{\mathrm{I}} - \mathbf{t}^{n}(\mathbf{u}_{\mathrm{II}}) \cdot \delta\mathbf{u}_{\mathrm{II}}]\, dS = 0, \tag{1.48}$$

where $\boldsymbol{n} = \boldsymbol{n}_{\mathrm{I}}$.

*1.* First we assume that the media I and II are welded, i.e., displacements $\mathbf{u}_{\mathrm{I}}$ and $\mathbf{u}_{\mathrm{II}}$ on the interface $\mathscr{B}$ are equal, and accordingly $\delta\mathbf{u}_{\mathrm{I}} = \delta\mathbf{u}_{\mathrm{II}} := \delta\mathbf{u}$, where the vector $\delta\mathbf{u}$ is arbitrary. Then we have

$$\int_{t_1}^{t_2} dt \int_{\mathscr{B}} [\mathbf{t}^{n}(\mathbf{u}_{\mathrm{I}}) - \mathbf{t}^{n}(\mathbf{u}_{\mathrm{II}})] \cdot \delta\mathbf{u}\ \ dS = 0 \tag{1.49}$$

and arrive at the conditions of welded contact (1.44).

*2.* Now we allow slippage of the media I and II along $\mathscr{B}$ without intersection and without loss of contact between. This means that the tangential displacements $\mathbf{u}_{\mathrm{I}}^{\mathsf{T}}$ and $\mathbf{u}_{\mathrm{II}}^{\mathsf{T}}$ and their variations $\delta\mathbf{u}_{\mathrm{I}}^{\mathsf{T}}|_{\mathscr{B}}$ and $\delta\mathbf{u}_{\mathrm{II}}^{\mathsf{T}}|_{\mathscr{B}}$ on both sides of the interface are independent, whereas the normal displacements $\mathbf{u}_{\mathrm{I}}^{\mathsf{N}}$ and $\mathbf{u}_{\mathrm{II}}^{\mathsf{N}}$ coincide, as well as their variations $\delta\mathbf{u}_{\mathrm{I}}^{\mathsf{N}}$ and $\delta\mathbf{u}_{\mathrm{II}}^{\mathsf{N}}$. We can rewrite (1.48) in the form

$$\int_{t_1}^{t_2} dt \Big( \int_{\mathscr{B}} [\mathbf{t}^{n\mathsf{N}}(\mathbf{u}_{\mathrm{I}}) - \mathbf{t}^{n\mathsf{N}}(\mathbf{u}_{\mathrm{II}})] \cdot \delta\mathbf{u}^{\mathsf{N}} dS$$
$$+ \int_{\mathscr{B}} \mathbf{t}^{n\mathsf{T}}(\mathbf{u}_{\mathrm{I}}) \cdot \delta\mathbf{u}_{\mathrm{I}}^{\mathsf{T}} dS - \int_{\mathscr{B}} \mathbf{t}^{n\mathsf{T}}(\mathbf{u}_{\mathrm{II}}) \cdot \delta\mathbf{u}_{\mathrm{II}}^{\mathsf{T}} dS \Big) = 0.$$

The independence of variations $\delta\mathbf{u}^{\mathsf{N}}$, $\delta\mathbf{u}_{\mathrm{I}}^{\mathsf{T}}$, and $\delta\mathbf{u}_{\mathrm{II}}^{\mathsf{T}}$ results in the vanishing of all three integrals over $\mathscr{B}$. This yields conditions (1.46).

## 1.5 Isotropic medium

Consider a very special and, at the same time, very important class of elastic media. A medium is called *isotropic* at a point $\mathbf{x}$ if its elastic properties are identical in all directions. An elastic medium is isotropic if it is isotropic at all of its points. Otherwise, it is *anisotropic.*

We will show that the invariance of the density of potential energy with respect to rotation of coordinate axes around a point $\mathbf{x}$ implies that the elastic stiffness tensor for an isotropic medium is parameterized by two scalar quantities. Further we will derive inequalities for these quantities that ensure the positive definiteness of potential energy.

### 1.5.1 Consequences of the invariance of $\mathcal{W}$ with respect to rotations

Consider the characteristic polynomial for a matrix $\boldsymbol{\varepsilon}$, i.e., the polynomial $\det(\boldsymbol{\varepsilon} - \Lambda\mathbf{I})$ ($\mathbf{I}$ is the identity matrix) with respect to a numerical parameter $\Lambda$. The coefficients of this polynomial have a remarkable property to remain unchanged under rotation of a Cartesian coordinate system (see, e.g., Gelfand 1961 [5]) and are called the invariants of the matrix $\boldsymbol{\varepsilon}$.

In our case, the density of potential energy depends only on the invariants of $\boldsymbol{\varepsilon}$. Any $3 \times 3$ matrix

$$\boldsymbol{\varepsilon} = \begin{pmatrix} \varepsilon_{11} & \varepsilon_{12} & \varepsilon_{13} \\ \varepsilon_{21} & \varepsilon_{22} & \varepsilon_{23} \\ \varepsilon_{31} & \varepsilon_{32} & \varepsilon_{33} \end{pmatrix}$$

has the following invariants: its trace

$$I_1 = \mathrm{Tr}(\boldsymbol{\varepsilon}) = \varepsilon_{11} + \varepsilon_{22} + \varepsilon_{33}, \tag{1.50}$$

the quantity

$$I_2 = \begin{vmatrix} \varepsilon_{11} & \varepsilon_{12} \\ \varepsilon_{21} & \varepsilon_{22} \end{vmatrix} + \begin{vmatrix} \varepsilon_{22} & \varepsilon_{23} \\ \varepsilon_{32} & \varepsilon_{33} \end{vmatrix} + \begin{vmatrix} \varepsilon_{11} & \varepsilon_{13} \\ \varepsilon_{31} & \varepsilon_{33} \end{vmatrix}, \tag{1.51}$$

and the determinant $I_3 = \det(\boldsymbol{\varepsilon})$. Since $\mathcal{W}$ is assumed to be a second-degree homogeneous function of $\varepsilon_{ij}$, it must be a linear combination of $I_1^2$ and $I_2$. The traditional parameterization of $\mathcal{W}$ is as follows:

$$\mathcal{W}(\boldsymbol{\varepsilon}, \mathbf{x}) = \frac{1}{2}\left((\lambda + 2\mu)I_1^2 - 4\mu I_2\right). \tag{1.52}$$

The quantities $\lambda = \lambda(\mathbf{x})$ and $\mu = \mu(\mathbf{x})$ are called *Lamé (elastic) parameters.*

Formulas (1.29) and (1.52) give the representation of the stress tensor in the form

$$\sigma_{ij} = 2\mu\varepsilon_{ij}(\mathbf{u}) + \lambda\delta_{ij}\varepsilon_{kk}(\mathbf{u}). \tag{1.53}$$

Here, $\delta_{ij}$ is the Kronecker symbol,

$$\delta_{ij} = 1 \text{ if } i = j, \quad \text{and} \quad \delta_{ij} = 0 \text{ if } i \neq j. \tag{1.54}$$

The derivation of (1.53) is as follows. First consider the case of equal indices $i$ and $j$, for example, $i = j = 1$. The combination of (1.29), (1.50), (1.51), and (1.52) gives

$$\begin{aligned}\sigma_{11} &= \frac{\partial W}{\partial \varepsilon_{11}} = (\lambda + 2\mu)(\varepsilon_{11} + \varepsilon_{22} + \varepsilon_{33}) - 2\mu(\varepsilon_{22} + \varepsilon_{33}) \\ &= (\lambda + 2\mu)\varepsilon_{11} + \lambda(\varepsilon_{22} + \varepsilon_{33}),\end{aligned}$$

which coincides with (1.53) when $i = j = 1$. Now consider the case of different indices. When $i = 1$, $j = 2$, we have

$$\sigma_{12} = \frac{\partial W}{\partial \varepsilon_{12}} = 2\mu\varepsilon_{21} = 2\mu\varepsilon_{12},$$

which again agrees with (1.53). For other pairs of indices the calculation is similar.

Expression (1.53) can be written in the form

$$\sigma_{ij} = \mu\left(\frac{\partial \mathrm{u}_i}{\partial x_j} + \frac{\partial \mathrm{u}_j}{\partial x_i}\right) + \delta_{ij}\lambda \operatorname{div} \mathbf{u}. \tag{1.55}$$

From (1.18) and (1.53) we can obtain the following representation for the elastic stiffness tensor for isotropic media:

$$c_{ijkl} = \mu(\delta_{ik}\delta_{jl} + \delta_{il}\delta_{jk}) + \lambda\delta_{ij}\delta_{kl}. \tag{1.56}$$

The density of the surface force on the boundary with the outward normal $\boldsymbol{n}$ takes the form

$$\mathrm{t}_i^{\boldsymbol{n}}(\mathbf{u}) = \mu n_j\left(\frac{\partial \mathrm{u}_i}{\partial x_j} + \frac{\partial \mathrm{u}_j}{\partial x_i}\right) + \lambda n_i \operatorname{div} \mathbf{u} \tag{1.57}$$

or

$$\mathbf{t}^{\boldsymbol{n}}(\mathbf{u}) = \lambda \boldsymbol{n} \operatorname{div} \mathbf{u} + \mu[\boldsymbol{n} \times \operatorname{rot} \mathbf{u}] + 2\mu\frac{\partial \mathbf{u}}{\partial n}, \tag{1.58}$$

where $\partial/\partial n$ is the derivative along the outward normal.

The elastodynamics equations (1.34)–(1.35), with (1.53) taken into account, can also be written in the form

$$\mathbf{L}\mathbf{u} = \mathfrak{L}\mathbf{u} - \rho\ddot{\mathrm{u}} = -\mathbf{F}, \tag{1.59}$$

where

$$\begin{aligned}\mathfrak{L}\mathbf{u} := {} & (\lambda + 2\mu) \operatorname{grad} \operatorname{div} \mathbf{u} - \mu \operatorname{rot} \operatorname{rot} \mathbf{u} + \operatorname{grad} \lambda \operatorname{div} \mathbf{u} \\ & + [\operatorname{grad} \mu \times \operatorname{rot} \mathbf{u}] + 2(\operatorname{grad} \mu \cdot \operatorname{grad})\mathbf{u}.\end{aligned} \tag{1.60}$$

Here we use the notation

$$(\boldsymbol{p}\cdot\mathrm{grad})\mathbf{v} = p_j\frac{\partial\mathbf{v}}{\partial x_j} \tag{1.61}$$

and $\ddot{\mathbf{u}} := \frac{\partial^2\mathbf{u}}{\partial t^2}$. Equation (1.59) is significantly simplified in the case of an isotropic medium, i.e., in the case where $\lambda$ and $\mu$ do not depend on $\mathbf{x}$. Then we have

$$\mathbf{L}\mathbf{u} = (\lambda+2\mu)\,\mathrm{grad}\,\mathrm{div}\,\mathbf{u} - \mu\,\mathrm{rot}\,\mathrm{rot}\,\mathbf{u} - \rho\ddot{\mathbf{u}} = -\mathbf{F} \tag{1.62}$$

or

$$(\lambda+\mu)\,\mathrm{grad}\,\mathrm{div}\,\mathbf{u} + \mu\nabla^2\mathbf{u} - \rho\ddot{\mathbf{u}} = -\mathbf{F}, \tag{1.63}$$

where $\nabla^2 := \sum_{j=1}^3 \frac{\partial^2}{\partial x_j^2}$ is the *Laplace operator*.

### 1.5.2 Consequences of the positive definiteness of $\mathcal{W}$

From (1.52) and (1.3) it follows that

$$\begin{aligned}2\mathcal{W} &= 4\mu(\varepsilon_{12}^2+\varepsilon_{13}^2+\varepsilon_{23}^2) + (\lambda+2\mu)(\varepsilon_{11}^2+\varepsilon_{22}^2+\varepsilon_{33}^2)\\ &\quad + 2\lambda(\varepsilon_{11}\varepsilon_{22}+\varepsilon_{11}\varepsilon_{33}+\varepsilon_{22}\varepsilon_{23}).\end{aligned}$$

The latter expression can be regarded as the quadratic form with the $6\times 6$ matrix,

$$\begin{pmatrix} 4\mu & 0 & 0 & 0 & 0 & 0\\ 0 & 4\mu & 0 & 0 & 0 & 0\\ 0 & 0 & 4\mu & 0 & 0 & 0\\ 0 & 0 & 0 & \lambda+2\mu & \lambda & \lambda\\ 0 & 0 & 0 & \lambda & \lambda+2\mu & \lambda\\ 0 & 0 & 0 & \lambda & \lambda & \lambda+2\mu \end{pmatrix}, \tag{1.64}$$

acting on columns in the form $(\varepsilon_{12},\varepsilon_{13},\varepsilon_{23},\varepsilon_{11},\varepsilon_{22},\varepsilon_{33})^T$, where $^T$ denotes transposition. We apply the *Sylvester criterion*[6] to the matrix (1.64) and note that $D_1 = 4\mu > 0$, $D_2 = (4\mu)^2 > 0$, $D_3 = (4\mu)^3 > 0$, $D_4 = (4\mu)^3(\lambda+2\mu) > 0$, $D_5 = (4\mu)^4(\lambda+\mu) > 0$, $D_6 = 8(2\mu)^5(3\lambda+2\mu) > 0$. It is easy to observe that the inequalities $D_1 > 0$ and $D_6 > 0$ lead to the inequalities

$$\mu > 0 \tag{1.65}$$

and

$$3\lambda + 2\mu > 0, \tag{1.66}$$

---

[6] A symmetric real matrix $\mathbf{A} = \begin{pmatrix} A_{11} & \cdots & A_{1n}\\ \vdots & \ddots & \vdots\\ A_{n1} & \cdots & A_{nn}\end{pmatrix}$ is positive definite, that is, $(\mathbf{A}\boldsymbol{h},\boldsymbol{h}) > 0$ for any nonzero vector $\boldsymbol{h}$ if and only if $D_1 := A_{11} > 0$, $D_2 := \det\begin{pmatrix} A_{11} & A_{12}\\ A_{21} & A_{22}\end{pmatrix} > 0, \ldots$, $D_n := \det(\mathbf{A}) > 0$ (see, e.g., Gantmakher 1959 [4]).

which yield the remaining ones. Inequalities (1.65) and (1.66) are necessary and sufficient conditions for the positive definiteness of potential energy. Note the following remarkable consequence of (1.65) and (1.66):

$$\lambda + \mu > 0. \tag{1.67}$$

Conditions (1.65) and (1.66) are to be understood in the sequel wherever an isotropic medium is meant.

---

## 1.6 Energy balance

The equation of energy balance in an elastic medium can be deduced directly from the Hamilton principle as a consequence of the invariance of the Lagrangian with respect to time. Instead, we will obtain it as a simple consequence of the elastodynamics equations.

Let us return to the general case of anisotropy. Multiplying (1.32) by $\dot{\mathrm{u}}_j := \frac{\partial \mathrm{u}_j}{\partial t}$ and summing over $j$, we obtain

$$\frac{\partial}{\partial t}\left[\frac{1}{2}\rho\dot{\mathbf{u}}^2 + \frac{1}{2}\sigma_{ij}\frac{\partial \mathrm{u}_i}{\partial x_j}\right] - \frac{\partial}{\partial x_j}(\sigma_{ij}\dot{\mathrm{u}}_i) = \mathrm{F}_j\dot{\mathrm{u}}_j.$$

We note that the latter expression in the square brackets can be transformed into

$$\sigma_{ij}\frac{\partial \mathrm{u}_i}{\partial x_j} = \frac{\sigma_{ij}}{2}\left(\frac{\partial \mathrm{u}_i}{\partial x_j} + \frac{\partial \mathrm{u}_j}{\partial x_i}\right) = \sigma_{ij}\varepsilon_{ij} = 2\mathcal{W},$$

whence the *energy balance equation* easily follows:

$$\frac{\partial \mathcal{E}}{\partial t} + \operatorname{div}\mathbf{S} = \mathbf{F}\cdot\frac{\partial \mathbf{u}}{\partial t}. \tag{1.68}$$

Here,

$$\mathcal{E} = \mathcal{K} + \mathcal{W} \tag{1.69}$$

is *the density of energy*, and the vector $\mathbf{S}$ defined by the expression

$$\mathrm{S}_j = -\sigma_{jp}\frac{\partial \mathrm{u}_p}{\partial t} \tag{1.70}$$

is called *the vector of the density of energy flow*, or the *Umov vector*.

For interpretation of equation (1.68) we integrate both its parts over an arbitrary volume $\Omega$, taking into account the fact that on $\partial\Omega$

$$\mathbf{S}\cdot\boldsymbol{n} = -\sigma_{ij}(\mathbf{u})n_j\dot{\mathrm{u}}_i = -\mathbf{t}^n(\mathbf{u})\cdot\dot{\mathbf{u}},$$

and obtain

$$\frac{\partial}{\partial t}\int_\Omega \mathcal{E}\,d\mathbf{x} = \int_\Omega \mathbf{F}\cdot\dot{\mathbf{u}}\,d\mathbf{x} + \int_{\partial\Omega}\mathbf{t}^n(\mathbf{u})\cdot\dot{\mathbf{u}}\,dS. \tag{1.71}$$

Equation (1.71) shows that the increment of energy in a time unit in a volume $\Omega$ equals the work of volume and surface forces in the time unit. The expression $-\mathbf{S}\cdot\boldsymbol{n}\,dS = -\mathbf{t}^{\boldsymbol{n}}(\mathbf{u})\cdot\dot{\mathbf{u}}\,dS$ has a clear interpretation — it is the work of external surface forces on a surface element $dS$.

## 1.7 Time-harmonic solutions

Now we introduce complex solutions, which are a handy technical tool for analyzing linear wave processes.

### 1.7.1 Basic notions

Of great importance is the class of solutions of the elastodynamics equations having a special dependence on time of the form

$$\mathbf{u}(\mathbf{x}, t) = \boldsymbol{u}(\mathbf{x}; \omega)e^{-i\omega t}. \tag{1.72}$$

These solutions are called *time-harmonic.* The parameter $\omega$ is called the *frequency.*[7] The frequency is generally assumed positive. In formula (1.72), some authors write $e^{+i\omega t}$ instead of $e^{-i\omega t}$, and their respective formulas are complex conjugate to ours. The time-independent vector $\boldsymbol{u}$ is typically called the displacement, and its time dependence is often omitted. It is also convenient to consider complex volume and surface forces with the same time dependence

$$\mathbf{F}(\mathbf{x}, t) = \boldsymbol{F}(\mathbf{x}; \omega)e^{-i\omega t}, \quad \mathbf{f}(\mathbf{x}, t) = \boldsymbol{f}(\mathbf{x}; \omega)e^{-i\omega t}. \tag{1.73}$$

We will use for time-harmonic quantities $\mathbf{u}$, $\mathbf{F}$, $\mathbf{f}$, ... the corresponding notation $\boldsymbol{u}$, $\boldsymbol{F}$, $\boldsymbol{f}$, ...

Saying that a complex vector in the form (1.72) satisfies the elastodynamics equations (1.32), we mean that the vectors with real components $\operatorname{Re}\boldsymbol{u} = (\operatorname{Re}\mathrm{u}_1, \operatorname{Re}\mathrm{u}_2, \operatorname{Re}\mathrm{u}_3)$ and $\operatorname{Im}\boldsymbol{u} = (\operatorname{Im}\mathrm{u}_1, \operatorname{Im}\mathrm{u}_2, \operatorname{Im}\mathrm{u}_3)$ satisfy the equations.

Of course, complex solutions of equations (1.32) should not be understood as physical displacements as such. The traditional interpretation of a complex displacement vector consists of assigning the corresponding physical meaning to its real part

$$\operatorname{Re}\mathbf{u} = \frac{1}{2}(\mathbf{u} + \mathbf{u}^*) = \frac{1}{2}\big(\boldsymbol{u}e^{-i\omega t} + \boldsymbol{u}^* e^{i\omega t}\big). \tag{1.74}$$

Henceforth, the asterisk * stands for complex conjugation. Alternatively, a physical meaning could be assigned to the imaginary part; we choose the real part just for the sake of definiteness.

[7] It is more accurate to call $\omega$ the *circular frequency* but we use the shorter name "frequency."

The physical displacement, strain, and stress are naturally expressible via their respective complex quantities, for example,

$$\varepsilon_{ij}(\mathrm{Re}\,\mathbf{u}) = \frac{1}{2}[\varepsilon_{ij}(\mathbf{u}) + \varepsilon_{ij}(\mathbf{u}^*)] = \frac{1}{2}\left[\varepsilon_{ij}\left(\boldsymbol{u}e^{-i\omega t}\right) + \varepsilon_{ij}\left(\boldsymbol{u}^* e^{i\omega t}\right)\right]. \tag{1.75}$$

The corresponding density of kinetic energy and density of potential energy are as follows:

$$\mathcal{K} = \frac{\rho}{2}\,\mathrm{Re}\,\dot{\mathbf{u}} \cdot \mathrm{Re}\,\dot{\mathbf{u}} = \frac{\rho\omega^2}{8}\left(2\boldsymbol{u}\cdot\boldsymbol{u}^* + \boldsymbol{u}^*\cdot\boldsymbol{u}^* e^{2i\omega t} + \boldsymbol{u}\cdot\boldsymbol{u}e^{-2i\omega t}\right) \tag{1.76}$$

and

$$\begin{aligned}\mathcal{W} &= \frac{1}{2}\sigma_{jk}(\mathrm{Re}\,\mathbf{u})\varepsilon_{jk}(\mathrm{Re}\,\mathbf{u}) \\ &= \frac{1}{8}\left(\sigma_{jk}^*\varepsilon_{jk} + \sigma_{jk}\varepsilon_{jk}^* + \sigma_{jk}\varepsilon_{jk}e^{2i\omega t} + \sigma_{jk}^*\varepsilon_{jk}^* e^{2i\omega t}\right).\end{aligned} \tag{1.77}$$

Here,

$$\varepsilon_{jk} = \varepsilon_{jk}(\boldsymbol{u}), \quad \varepsilon_{jk}^* = \varepsilon_{jk}(\boldsymbol{u}^*), \quad \sigma_{jk} = \sigma_{jk}(\boldsymbol{u}), \quad \sigma_{jk}^* = \sigma_{jk}(\boldsymbol{u}^*). \tag{1.78}$$

Note that the first two terms on the right-hand side of (1.77) are equal.

The physical meaning of volume and surface forces is assigned to the expressions

$$\mathrm{Re}\,\mathbf{F} = \frac{1}{2}\left(\boldsymbol{F}e^{-i\omega t} + \boldsymbol{F}^* e^{i\omega t}\right) \tag{1.79}$$

and

$$\mathrm{Re}\,\mathbf{f} = \frac{1}{2}\left(\boldsymbol{f}e^{-i\omega t} + \boldsymbol{f}^* e^{i\omega t}\right). \tag{1.80}$$

For example, the work of volume forces with respect to a complex displacement $\mathbf{u} = \boldsymbol{u}e^{-i\omega t}$ equals

$$\frac{1}{4}\left(\boldsymbol{u}^*\cdot\boldsymbol{F} + \boldsymbol{u}\cdot\boldsymbol{F}^* + \boldsymbol{u}\cdot\boldsymbol{F}e^{-2i\omega t} + \boldsymbol{u}^*\cdot\boldsymbol{F}^* e^{2i\omega t}\right). \tag{1.81}$$

The elastodynamics equations in the harmonic case take the form

$$\boldsymbol{l}(\boldsymbol{u}) \equiv \boldsymbol{l}\boldsymbol{u} = -\boldsymbol{F}, \tag{1.82}$$

$$\boldsymbol{l} = \mathfrak{L} + \rho\omega^2\mathbf{I}, \quad (\boldsymbol{l}\boldsymbol{u})_j \equiv l_j(\boldsymbol{u}) = \frac{\partial\sigma_{jp}(\boldsymbol{u})}{\partial x_p} + \rho\omega^2 u_j. \tag{1.83}$$

### 1.7.2 Time-averaging

We introduce here the operation of averaging harmonic quantities over time. Let $\phi(\mathbf{x}, t)$ be an arbitrary function (e.g., the density of energy or a component of a displacement vector). We use the notation

$$\overline{\phi(\mathbf{x}, t)} := \frac{1}{T}\int_{t_1}^{t_1+T} \phi(\mathbf{x}, t)dt = \frac{1}{T}\int_0^T \phi(\mathbf{x}, t)dt, \tag{1.84}$$

where

$$T = \frac{2\pi}{\omega} \tag{1.85}$$

is the *period of oscillation*, corresponding to the frequency $\omega$. Expression (1.84) is called the *time average of a function* $\phi$. Obviously,

$$\overline{e^{-2i\omega t}} = 0 \quad \text{and} \quad \overline{e^{+2i\omega t}} = 0. \tag{1.86}$$

Let us derive a useful formula for averaging complex quantities. We assume that $\alpha$ and $\beta$ do not depend on time. Then we have

$$\overline{\mathrm{Re}(\alpha e^{-i\omega t})\,\mathrm{Re}(\beta e^{-i\omega t})} = \frac{1}{2}\,\mathrm{Re}(\alpha\beta^*) = \frac{1}{2}\,\mathrm{Re}(\alpha^*\beta). \tag{1.87}$$

Indeed, assuming that $\alpha = Ae^{i\Phi}$, $\beta = Be^{i\Psi}$, where $A$, $B$, $\Phi$, and $\Psi$ are real, we obtain

$$\mathrm{Re}(\alpha e^{-i\omega t}) = A\cos(\Phi - \omega t), \quad \mathrm{Re}(\beta e^{-i\omega t}) = B\cos(\Psi - \omega t).$$

Further,

$$\overline{\mathrm{Re}(\alpha e^{-i\omega t})\,\mathrm{Re}(\beta e^{-i\omega t})} = \frac{AB}{T}\int_0^T \cos(\Phi - \omega t)\cos(\Psi - \omega t)dt$$

$$= \frac{AB}{2T}\int_0^T [\cos(\Phi - \Psi) + \cos(\Phi + \Psi - 2\omega t)]dt = \frac{1}{2}\,\mathrm{Re}(\alpha\beta^*) = \frac{1}{2}\,\mathrm{Re}(\alpha^*\beta).$$

Similarly, if $\boldsymbol{\alpha}$ and $\boldsymbol{\beta}$ are time-independent vectors, then

$$\overline{\mathrm{Re}(\boldsymbol{\alpha} e^{-i\omega t})\cdot\mathrm{Re}(\boldsymbol{\beta} e^{-i\omega t})} = \frac{1}{2}\,\mathrm{Re}(\boldsymbol{\alpha}\cdot\boldsymbol{\beta}^*) = \frac{1}{2}\,\mathrm{Re}(\boldsymbol{\alpha}^*\cdot\boldsymbol{\beta}). \tag{1.88}$$

For time averages of the density of a Lagrangian and of the density of energy flow, we obtain

$$\overline{\mathcal{L}} = \overline{\mathcal{K}} - \overline{\mathcal{W}}, \tag{1.89}$$

where

$$\overline{\mathcal{K}} = \frac{\omega^2}{4}\rho\boldsymbol{u}\cdot\boldsymbol{u}^* = \frac{\omega^2}{4}\rho(\boldsymbol{u},\boldsymbol{u}), \qquad \overline{\mathcal{W}} = \frac{1}{8}(\sigma_{kl}^*\varepsilon_{kl} + \sigma_{kl}\varepsilon_{kl}^*). \tag{1.90}$$

We also give a formula for the averaged density of energy flow:

$$\overline{\mathcal{S}_j} = -\overline{\mathrm{Re}[\sigma_{jl}(\mathbf{u})]\mathrm{Re}[\dot{\mathrm{u}}_l]} = \frac{1}{2}\mathrm{Re}[\sigma_{jl}(\mathrm{u})(-i\omega\mathrm{u}_l)^*] = -\frac{\omega}{2}\,\mathrm{Im}\left[\sigma_{jp}\mathrm{u}_p^*\right]. \tag{1.91}$$

---

## 1.8 Reciprocity principle

The equations known as reciprocity principles are in a close relation with the Lagrangian nature of elastodynamics equations.

We start by establishing a simple but important relation. Let $\mathbf{u}$ and $\mathbf{v}$ be displacement vectors. From (1.17) and (1.18) we easily obtain the chain of relations

$$\sigma_{ij}(\mathbf{u})\varepsilon_{ij}(\mathbf{v}) = c_{ijkl}\varepsilon_{kl}(\mathbf{u})\varepsilon_{ij}(\mathbf{v}) = c_{ijkl}\varepsilon_{kl}(\mathbf{v})\varepsilon_{ij}(\mathbf{u}) = \sigma_{ij}(\mathbf{v})\varepsilon_{ij}(\mathbf{u}).$$

The resulting equation

$$\sigma_{ij}(\mathbf{u})\varepsilon_{ij}(\mathbf{v}) = \sigma_{ij}(\mathbf{v})\varepsilon_{ij}(\mathbf{u}) \tag{1.92}$$

is called *Betti's reciprocity law.*

### 1.8.1 The time-harmonic case

In the remaining part of this section, except for formulas (1.97) and (1.98), we will constrain the discussion to time-harmonic processes (1.72) resulting from time-harmonic external forces and omit the time dependence $e^{-i\omega t}$.

Let $\boldsymbol{u}(\boldsymbol{x})$ and $\boldsymbol{u}'(\boldsymbol{x})$ be arbitrary smooth vectors. Consider the integral

$$\mathsf{J} = \int_\Omega \boldsymbol{u}' \cdot \boldsymbol{l}(\boldsymbol{u}) d\mathbf{x} = \int_\Omega \frac{\partial}{\partial x_i}\left(u'_j \sigma_{ij}(\boldsymbol{u})\right) d\mathbf{x} - \int_\Omega \frac{\partial u'_j}{\partial x_i}\sigma_{ij}(\boldsymbol{u}) d\mathbf{x} + \int_\Omega \rho\omega^2 \boldsymbol{u}\cdot\boldsymbol{u}' d\mathbf{x}$$

(the definition of the operator $\boldsymbol{l}$ is given in (1.83)). Applying the divergence theorem (1.25) to the vector with components $A_j = u'_i \sigma_{ij}(\boldsymbol{u})$, we get

$$\mathsf{J} = \int_{\partial\Omega} u'_i \sigma_{ij}(\boldsymbol{u}) \cos(\widehat{nx_j}) d\mathbf{x} - \int_\Omega \left\{\sigma_{ij}(\boldsymbol{u})\frac{\partial u'_i}{\partial x_j} - \rho\omega^2 u_m u'_m\right\} d\mathbf{x}.$$

After swapping $\boldsymbol{u}$ and $\boldsymbol{u}'$, we have

$$\int_\Omega \boldsymbol{u}\cdot\boldsymbol{l}(\boldsymbol{u}') d\mathbf{x} = \int_{\partial\Omega} u_i \sigma_{ij}(\boldsymbol{u}') \cos(\widehat{nx_j}) d\mathbf{x} - \int_\Omega \left\{\sigma_{ij}(\boldsymbol{u}')\frac{\partial u_i}{\partial x_j} - \rho\omega^2 u_m u'_m\right\} d\mathbf{x},$$

and eventually, with the use of (1.92), we find

$$\int_\Omega \left\{\boldsymbol{u}\cdot\boldsymbol{l}(\boldsymbol{u}') - \boldsymbol{u}'\cdot\boldsymbol{l}(\boldsymbol{u})\right\} d\mathbf{x} = \int_{\partial\Omega} \left\{\boldsymbol{u}\cdot\mathbf{t}^{\boldsymbol{n}}(\boldsymbol{u}') - \boldsymbol{u}'\cdot\mathbf{t}^{\boldsymbol{n}}(\boldsymbol{u})\right\} dS. \tag{1.93}$$

This result is similar to the classical Green formula for the Helmholtz operator

$$\int_\Omega \left\{u(\nabla^2 + k^2)u' - u'(\nabla^2 + k^2)u\right\} d\mathbf{x} = \int_{\partial\Omega}\left(u\frac{\partial u'}{\partial n} - u'\frac{\partial u}{\partial n}\right) dS, \tag{1.94}$$

$k = \text{const}$. The operators $\boldsymbol{l}$ and $\mathbf{t}^{\boldsymbol{n}}$ play in (1.93) the same role as $\nabla^2 + k^2$ and the operator of differentiation along the outward normal $\frac{\partial}{\partial n}$ to the boundary in (1.94), respectively.

Let $\boldsymbol{u}$ and $\boldsymbol{u}'$ be solutions of equations $\boldsymbol{l}(\boldsymbol{u}) = \boldsymbol{F}$ and $\boldsymbol{l}(\boldsymbol{u}') = \boldsymbol{F}'$ for the same medium, each satisfying some boundary conditions from those described in Section 1.4. From (1.93) we have

$$\int_\Omega \left\{\boldsymbol{u}\cdot\boldsymbol{F}' - \boldsymbol{u}'\cdot\boldsymbol{F}\right\} d\mathbf{x} = \int_{\partial\Omega} \left\{\boldsymbol{u}\cdot\mathbf{t}^{\boldsymbol{n}}(\boldsymbol{u}') - \boldsymbol{u}'\cdot\mathbf{t}^{\boldsymbol{n}}(\boldsymbol{u})\right\} dS. \tag{1.95}$$

If the boundary conditions for $\boldsymbol{u}$ and $\boldsymbol{u}'$ are the same, for example,

$$\mathbf{t}^{\boldsymbol{n}}(\boldsymbol{u})|_{\partial\Omega} = \mathbf{t}^{\boldsymbol{n}}(\boldsymbol{u}')|_{\partial\Omega} = 0$$

or

$$\boldsymbol{u}|_{\partial\Omega} = \boldsymbol{u}'|_{\partial\Omega} = 0,$$

then the *reciprocity principle* holds:

$$\int_\Omega \boldsymbol{u} \cdot \boldsymbol{l}(\boldsymbol{u}') d\mathbf{x} = \int_\Omega \boldsymbol{u}' \cdot \boldsymbol{l}(\boldsymbol{u}) d\mathbf{x}. \tag{1.96}$$

This formula expresses the property of formal self-adjointness of the elastodynamics equations.

The reciprocity principle can be extended to infinite domains.

### 1.8.2 Non-time-harmonic case

For the non-time-harmonic case we derive a counterpart of formula (1.93). Let $\mathbf{u}(\mathbf{x}, t)$ and $\mathbf{u}'(\mathbf{x}, t)$ be smooth vectors in a space-time domain $\mathbf{X} := (\mathbf{x}, t) \in \mathrm{D} \subset \mathbb{R}^3 \times \mathbb{R}^1$ the boundary of which and the outward normal to which will be denoted by $\partial\mathrm{D}$ and $\boldsymbol{\nu}$. Then

$$\int_{\mathrm{D}} \{\mathbf{u}' \cdot \mathbf{L}(\mathbf{u}) - \mathbf{u} \cdot \mathbf{L}(\mathbf{u}')\}\, d\mathbf{X} = \int_{\partial\mathrm{D}} \{\mathbf{u}' \cdot \mathbf{p}(\mathbf{u}) - \mathbf{u} \cdot \mathbf{p}(\mathbf{u}')\}\, d\mathrm{S}, \tag{1.97}$$

where $d\mathbf{X}$ and $d\mathrm{S}$ denote a volume element D and a surface element $\partial\mathrm{D}$, respectively, an operator $\mathbf{p}(\mathbf{u})$ is defined as follows:

$$\mathrm{p}_i(\mathbf{u}) = \left\{ \sigma_{ij}(\mathbf{u}) \cos(\widehat{\nu, x_j}) - \rho \frac{\partial \mathrm{u}_i}{\partial t} \cos(\widehat{\nu, t}) \right\}\Bigg|_{(\mathbf{x},t)\in\partial\mathrm{D}}, \tag{1.98}$$

and the operator $\mathbf{L}$ is introduced in (1.35). Formula (1.97) is known as the *Green – Volterra formula.*

---

## 1.9 ⋆ Comments to Chapter 1

Fundamentals of elasticity theory are presented in detail in the books by Love 1944 [9], Sommerfeld 1950 [15], Nowacki 1975 [13], and several others. Equations of elasticity theory are traceable to the paper by Navier 1827 [12], who considered an isotropic medium for the case $\lambda = \mu$, in the modern notation. The elastodynamics equations are called *Navier equations* (whereas elastostatics equations, which are similar to the Laplace equation, are called Lamé equations). The methodical value of the variational approach to analytic mechanics and elasticity theory are well demonstrated by the textbooks by Landau and Lifshitz 1986, 2007 [7, 8]. Parameterizations of the elastic stiffness tensor for anisotropic media possessing other particular symmetries are considered, e.g., by Petrashen' 1980 [14] and, in more detail, by Musgrave 1970 [11]. Boundary conditions other than the classical ones described in Section 1.4 are discussed by Martin 1992 [10]. Averaging over a period is often used in the theory of harmonic waves (see, e.g., Whitham 1974 [16]). Detailed discussion of reciprocity principles is given by Achenbach 2003 [1].

Equation (1.68) is somewhat similar to the Poynting theorem in electrodynamics, where the Umov vector which is temporarily denoted by $\mathbf{S}^U$ is a counterpart of the Poynting vector $\mathbf{S}^P$. Let $\Omega$ be an arbitrarily chosen volume within a medium, $\partial\Omega$ its boundary, and $\boldsymbol{n}$ its outward normal. In each case, the quantity $\int_{\partial\Omega}(\mathbf{S}^{U,P}, \boldsymbol{n})\,dS\,\delta t$ can be understood as the amount of energy flowing out during $\delta t$ through the boundary of the volume $\Omega$. The vector $(\mathbf{S}^U, \boldsymbol{n})\,dS\,\delta t$ has a clear local interpretation — this is the work of forces acting on the surface $dS$ during $\delta t$. The vector $(\mathbf{S}^P, \boldsymbol{n})\,dS\,\delta t$ has no local interpretation, because the integral in question does not change when adding to $\mathbf{S}^P$ any solenoidal vector (see, e.g., Fock 1964 [3], Jones 1986 [6]).

---

## References to Chapter 1


[1] Achenbach, J. D. 2003. *Reciprocity in elastodynamics.* Cambridge: Cambridge University Press.

[2] Feynman, R. P., Leighton, R. B. and Sands, M. 1966. *The Feynman lectures on physics. Mainly electromagnetism and matter.* Reading, MA: Addison Wesley. *Фейнман Р., Лейтон Р., Сэндс М.* Фейнмановские лекции по физике. Т. 6: Электродинамика. Эдиториал УРСС, 2004.

[3] Fock, V. A. 1964. *The theory of space, time and gravitation.* Pergamon: Oxford. *Фок В. А.* Теория пространства, времени и тяготения. М.: ЛКИ, 2007.

[4] Gantmacher, F. R. 1959. *The theory of matrices.* New York: Chelsea Publishing Company. *Гантмахер Ф. Р.* Теория матриц. М.: Наука, 1966.

[5] Gelfand, I. M. 1961. *Lectures on linear algebra.* New York: Dover. *Гельфанд И. М.* Лекции по линейной алгебре. М.: МЦНМО, 2007.

[6] Jones, D. S. 1986. *Acoustic and electromagnetic waves.* Oxford–New York: Clarendon Press.

[7] Landau, L. D. and Lifshitz, E. M. 2007. *Mechanics.* Amsterdam: Elsevier. *Ландау Л. Д., Лифшиц Е. М.* Механика. М.: Физматлит, 2012.

[8] Landau, L. D. and Lifshitz, E. M. 1986. *Theory of elasticity.* Oxford: Pergamon Press. *Ландау Л. Д., Лифшиц Е. М.* Теория упругости. М.: Физматлит, 2003.

[9] Love, A. E. H. 1944. *A treatise on the mathematical theory of elasticity.* Dover. *Ляв А.* Математическая теория упругости. М.–Л.: ОНТИ НКПТ СССР, 1935.

[10] Martin, P. A. 1992. Boundary integral equations for the scattering of elastic waves by elastic inclusions with thin interface layers. *J. Nondestruct. Evaluation.* 1:167–74.

[11] Musgrave, M. J. P. L. 1970. *Crystal acoustics.* San Francisco: Holden-Day.

[12] Navier, C. 1827. Mémoire sur les lois de l'équilibre et du mouvement des corps solides élastiques. *Mém. de l'Acad. Roy. Sci.* 7:375–93.

[13] Nowacki, W. 1975. *Theory of elasticity.* Moscow: Mir. [in Russian]. *Новацкий В.* Теория упругости. М.: Мир, 1975.

[14] Petrashen', G. I. 1980. *Propagation of waves in anisotropic media.* Leningrad: Nauka [in Russian]. *Петрашень Г. И.* Распространение волн в анизотропных средах. Л.: Наука, 1980.

[15] Sommerfeld, A. 1950. *Mechanics of deformable bodies.* New York: Academic Press. *Зоммерфельд А.* Механика деформируемых сред. М.: Изд-во иностр. лит., 1954.

[16] Whitham, G. B. 1974. *Linear and nonlinear waves.* New York: Wiley. *Уизем Дж.* Линейные и нелинейные волны. М.: Мир, 1977.

# Chapter 2

## Plane Waves

Plane waves are the best-known solutions describing linear wave processes. They have already been known to D'Alembert and Euler. In considering plane waves, important notions of the theory of wave phenomena naturally appear; in particular, the notions of phase and group velocities, slowness, wavenumber, etc. Generalization of these notions marks out the solutions of the elastodynamics equations that are called *waves*.

### 2.1 Plane-wave ansatz

A plane wave is the simplest solution of the elastodynamics equations in the absence of external forces (see (1.35))

$$\mathbf{Lu} \equiv \mathbf{L}(\mathbf{u}) = 0. \tag{2.1}$$

Let us consider a *homogeneous medium*, that is, elastic stiffnesses and density are constant (do not depend on $\mathbf{x}$), for which

$$\mathrm{L}_q(\mathbf{u}) \equiv (\mathbf{Lu})_q = c_{qjml}\frac{\partial^2 \mathrm{u}_l}{\partial x_j \partial x_m} - \rho\frac{\partial^2 \mathrm{u}_q}{\partial t^2}. \tag{2.2}$$

For this case, a *plane wave* or a *plane-wave ansatz*[1] is a solution in the form

$$\mathbf{u} = \mathbf{h} f(\mathbf{k}\cdot\mathbf{x} + \mathrm{k}_0 t), \tag{2.3}$$

where $\mathbf{k} = (\mathrm{k}_1, \mathrm{k}_2, \mathrm{k}_3)$ and $\mathbf{h} = (\mathrm{h}_1, \mathrm{h}_2, \mathrm{h}_3) \neq 0$ are constant vectors, $f$ is an arbitrary function of one variable, and $\mathrm{k}_0 \neq 0$ is a real constant. The vector $\mathbf{k}$ is real, unless the opposite is stated. It is accepted to call $f$ a *waveform*, its argument $\theta = \mathbf{k}\cdot\mathbf{x} + \mathrm{k}_0 t$ is a *phase*, and the vector $\mathbf{h}$ is a *vector amplitude*. If $f$ and $\mathbf{h}$ in formula (2.3) are chosen so that the vector $\mathbf{u}$ is complex, then the physical meaning is assigned to its real part $\mathrm{Re}\,\mathbf{u}$.

Solutions of the elastodynamics equations for a homogeneous medium described by (2.3) with real vectors $\mathbf{k}$ are called *volume homogeneous plane*

[1] *Ansatz*, a term of German origin, is widely used as a synonym for the expression "the analytic structure of a solution."

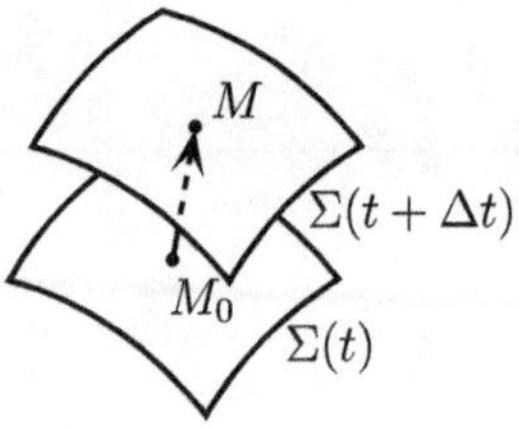


**Figure 2.1**: Moving surface at close instants of time

*waves*, to observe the terminology accuracy. In the sequel, *inhomogeneous volume plane waves*[2] with complex $\mathbf{k}$ will appear. In Sections 2.7–2.10, we will consider solutions of a special kind for inhomogeneous (so-called layered) media, which are named *surface waves.* We will often call solutions of the form (2.3) with real vectors $\mathbf{k}$ just *plane waves*.

We will pay much attention to *time-harmonic plane waves*, where $\mathrm{k}_0 = -\omega$, and

$$f(\theta) = e^{i\theta},$$

i.e.,

$$\mathbf{u}(\mathbf{x}, t) = \boldsymbol{h}e^{i(\boldsymbol{k}\cdot\mathbf{x}-\omega t)} = \boldsymbol{u}e^{-i\omega t}, \quad \boldsymbol{u} = \boldsymbol{h}e^{i\boldsymbol{k}\cdot\mathbf{x}}, \tag{2.4}$$

admitting complex $\boldsymbol{h}$. First we will assume that the *wave vector* $\boldsymbol{k} = \mathbf{k}|_{\mathrm{k}_0=-\omega}$ is real, and the *frequency* $\omega$ is positive. Waves with real $\boldsymbol{k}$ are called homogeneous.

The requirement that expression (2.4) be a solution of the elastodynamics equations (2.1) (if $\boldsymbol{h} \neq 0$) leads to a relation between $\omega$ and $\boldsymbol{k}$, which is traditionally written in the form

$$\omega = H(\boldsymbol{k}) \tag{2.5}$$

and is called a *dispersion equation*, or a *dispersion relation.* The notion of dispersion equation is of utmost importance. It will appear again when discussing the ray method in Chapters 4, 5, and 8. An intrinsic property of fundamental notions is to appear in different forms.

For plane waves of the form (2.3), considered in Sections 2.3 and 2.4, the function $H(\boldsymbol{k})$ is *first-degree homogeneous* with respect to $k_1$, $k_2$, $k_3$, i.e.,

$$H(\mathrm{C}\boldsymbol{k}) = \mathrm{C}H(\boldsymbol{k}) \quad \text{for any constant} \quad \mathrm{C} > 0. \tag{2.6}$$

If (2.6) holds, a wave is said to be *nondispersive.* The velocity of such waves (in this chapter the notion of velocity will be specified) does not depend on frequency. If the homogeneity condition (2.6) does not hold (as, for example, in Section 2.7.2), then *dispersion* is present, and a wave is called *dispersive.* The major part of this chapter relates to nondispersive waves. Examples of dispersive waves are considered in Sections 2.7.2, 2.8, and 2.9.

[2]Rather a strange name, though widely used.

The notions introduced above for an unbounded homogeneous, though, generally speaking, anisotropic, medium will be further generalized in various manners. An essential part of this book relates to the description of the asymptotic *ray method*, which generalizes plane waves.

## 2.2 Phase velocity

### 2.2.1 Normal velocity of a moving surface

We begin with derivation of a simple but an important auxiliary formula. Consider a smooth surface $\Sigma(t) \subset \mathbb{R}^3$ moving in time. The surface is defined by the equation

$$g(\mathbf{x}, t) = C, \quad C = \text{const}. \tag{2.7}$$

Here, $g$ is real and

$$\operatorname{grad}_{\mathbf{x}} g \neq 0. \tag{2.8}$$

Positions of the surface at two close instants $t$ and $t + \Delta t$ are shown in Fig. 2.1. Let us choose a point $M_0$ on the surface $\Sigma(t)$ and raise a perpendicular to $\Sigma(t)$ at $M_0$, the perpendicular pointing in the direction in which the surface moves. The perpendicular meets $\Sigma(t + \Delta t)$ at a point $M$.

We call the vector

$$\boldsymbol{v} = \lim_{\Delta t \to 0} \frac{\overrightarrow{M_0 M}}{\Delta t} = \lim_{\Delta t \to 0} \left( \frac{\Delta x_1}{\Delta t}, \frac{\Delta x_2}{\Delta t}, \frac{\Delta x_3}{\Delta t} \right) \tag{2.9}$$

the *normal velocity vector* of the surface (2.7). Applying the Taylor formula to (2.7), we have

$$C = g(\mathbf{x} + \Delta \mathbf{x}, t + \Delta t) = g(\mathbf{x}, t) + \operatorname{grad}_{\mathbf{x}} g \cdot \Delta \mathbf{x} + \frac{\partial g}{\partial t} \Delta t + \dots,$$

where dots denote the terms of higher order. By subtracting (2.7) from the latter, then dividing it by $\Delta t$ and taking the limit as $\Delta t \to 0$, we obtain

$$\frac{\partial g}{\partial t} + \boldsymbol{v} \cdot \operatorname{grad}_{\mathbf{x}} g = 0.$$

Taking into account the fact that $\boldsymbol{v}$ is parallel to $\operatorname{grad}_{\mathbf{x}} g$, we eventually arrive at the following expression for the normal velocity vector:

$$\boldsymbol{v} = \frac{(\boldsymbol{v} \cdot \operatorname{grad}_{\mathbf{x}} g)}{|\operatorname{grad}_{\mathbf{x}} g|} \frac{\operatorname{grad}_{\mathbf{x}} g}{|\operatorname{grad}_{\mathbf{x}} g|} = -\frac{\frac{\partial g}{\partial t}}{|\operatorname{grad}_{\mathbf{x}} g|} \frac{\operatorname{grad}_{\mathbf{x}} g}{|\operatorname{grad}_{\mathbf{x}} g|}. \tag{2.10}$$

The *scalar normal velocity of a surface* is defined as follows:

$$c = \frac{\left| \frac{\partial g}{\partial t} \right|}{|\operatorname{grad}_{\mathbf{x}} g|}. \tag{2.11}$$

### 2.2.2 Phase velocity and slowness

The velocity of a wave can be defined in different ways, and its values and directions will, generally speaking, be different. For different purposes different definitions are suitable.

The *phase velocity* of a plane wave (2.3) is defined as the normal velocity of its front, i.e., the surface of a constant phase, which is the plane

$$\mathrm{k}_m x_m + \mathrm{k}_0 t = \text{const}. \tag{2.12}$$

This velocity is defined as follows:

$$\boldsymbol{v}^{ph} = -\frac{\mathrm{k}_0 \mathbf{k}}{\mathbf{k}\cdot\mathbf{k}}. \tag{2.13}$$

The vector $\boldsymbol{v}^{ph}$ points in the direction of movement of the wavefront. The unit vector of this direction is denoted by

$$\mathbf{s} = -\frac{\mathbf{k}}{|\mathbf{k}|}\frac{\mathrm{k}_0}{|\mathrm{k}_0|} = -\frac{\mathbf{k}}{\sqrt{\mathbf{k}\cdot\mathbf{k}}}\frac{\mathrm{k}_0}{|\mathrm{k}_0|}, \tag{2.14}$$

and thus

$$\boldsymbol{v}^{ph} = \frac{|\mathrm{k}_0|}{|\mathbf{k}|}\mathbf{s}. \tag{2.15}$$

For a time-harmonic wave (2.4) we have

$$\mathbf{k} = (\mathrm{k}_1, \mathrm{k}_2, \mathrm{k}_3) = (k_1, k_2, k_3) = \boldsymbol{k}, \quad \mathrm{k}_0 = -\omega. \tag{2.16}$$

Expression (2.13) takes the form

$$\boldsymbol{v}^{ph} = \omega\frac{\boldsymbol{k}}{\boldsymbol{k}\cdot\boldsymbol{k}} = \omega\frac{\mathbf{s}}{|\boldsymbol{k}|}. \tag{2.17}$$

The absolute value of the phase velocity, which is also called the *phase velocity*, is denoted by

$$c := |\boldsymbol{v}^{ph}| = \frac{\omega}{|\boldsymbol{k}|}. \tag{2.18}$$

The *wave vector* $\boldsymbol{k}$ of a plane wave has the same direction as the phase velocity vector. The *direction of wave propagation* typically means the direction of the phase velocity of the wave.

Let $\mathbf{s}$ be a unit vector in the direction of the phase velocity. Obviously,

$$\boldsymbol{v}^{ph} = c\mathbf{s} = c\frac{\boldsymbol{k}}{|\boldsymbol{k}|}. \tag{2.19}$$

The vector

$$\boldsymbol{p} = \frac{\mathbf{s}}{c} = \frac{\boldsymbol{k}}{c|\boldsymbol{k}|} = \frac{\boldsymbol{k}}{\omega} \tag{2.20}$$

is called a *slowness vector*, and the scalar $\frac{1}{c}$ inverse to the phase velocity is called *slowness.* The slowness of a wave has the following kinematic meaning: it is a time interval needed for the wave to pass a unit path in this direction. The notion of slowness is no less important than that of phase velocity, though it is not commonly used.

In the absence of dispersion (see (2.6)), dividing both parts of dispersion equation (2.5) by $\omega$ and taking into account relation (2.20), we obtain

$$H(\boldsymbol{p}) = 1, \tag{2.21}$$

which can be rewritten as

$$H^2(\boldsymbol{p}) = 1. \tag{2.22}$$

Formula (2.22) is equivalent to the dispersion equation (2.5), which will be met many times henceforth.

Using (2.20) and (2.18), we arrive at yet another form of the dispersion equation for a nondispersive wave:

$$c(\mathrm{s}) = \frac{\omega}{|\boldsymbol{k}|} = \frac{H(\boldsymbol{k})}{|\boldsymbol{k}|} = H(\mathrm{s}), \quad \mathrm{s} = \frac{\boldsymbol{k}}{|\boldsymbol{k}|}. \tag{2.23}$$

Thus, in the absence of dispersion the velocity of a wave may depend on the direction of the wave, but not on its frequency.

---

## 2.3 Plane waves in unbounded isotropic media

### 2.3.1 Eigenvalue problem

Equation (2.1) for an isotropic medium can be written in the form

$$\mathbf{Lu} = (\lambda + \mu)\,\mathrm{grad}\,\mathrm{div}\,\mathbf{u} + \mu\nabla^2\mathbf{u} - \rho\ddot{\mathbf{u}} = 0. \tag{2.24}$$

For a plane wave (2.3), obviously,

$$\mathrm{grad}\,\mathrm{div}\,\mathbf{u} = \mathbf{k}(\mathbf{k}\cdot\mathbf{h})f''(\theta), \quad \nabla^2\mathbf{u} = \mathbf{k}^2\mathbf{h}f''(\theta), \quad \ddot{\mathrm{u}} = \mathrm{k}_0^2\mathbf{h}f''(\theta),$$

where prime denotes differentiation of a function with respect to its argument, and $\mathbf{k}^2 = (\mathbf{k}{\cdot}\mathbf{k}) = \mathrm{k}_j\mathrm{k}_j$. Substituting (2.3) into (2.1) and assuming that $f'' \neq 0$, we cancel by $f''(\theta)$ and obtain

$$\mathfrak{N}\mathbf{h} = 0, \tag{2.25}$$

where $\mathfrak{N}$ is the matrix with entries

$$\mathfrak{N}_{ij} = \mathfrak{N}_{ij}(\mathbf{k}, \mathrm{k}_0) = (\lambda + \mu)\mathrm{k}_i\mathrm{k}_j + \delta_{ij}(\mu\mathbf{k}^2 - \rho\mathrm{k}_0^2). \tag{2.26}$$

It is useful to regard equation (2.25) as an eigenvalue problem

$$\mathbf{\Gamma h} = \rho \mathrm{k}_0^2 \mathbf{h} \;\Leftrightarrow\; (\lambda+\mu)(\mathbf{h}\cdot\mathbf{k})\mathbf{k} + \mu \mathbf{k}^2 \mathbf{h} = \rho \mathrm{k}_0^2 \mathbf{h} \tag{2.27}$$

for the matrix $\mathbf{\Gamma}$ with entries

$$\Gamma_{ij} = \Gamma_{ij}(\mathbf{k}) = (\lambda+\mu)\mathrm{k}_i \mathrm{k}_j + \delta_{ij}\mu \mathbf{k}^2. \tag{2.28}$$

The matrix $\mathbf{\Gamma}$ is obviously symmetric and, when $\lambda + \frac{2}{3}\mu > 0$, $\mu > 0$ and $\mathbf{k} \neq 0$, as follows from (1.65), (1.67), is positive definite, so its eigenvalues are positive. We will find them explicitly.

### 2.3.2 Wave $P$

There are only two possibilities: either $\mathbf{h}$ is parallel to $\mathbf{k}$ or $\mathbf{h}$ is not parallel to $\mathbf{k}$. We start with the first one. Let

$$\mathbf{h} \parallel \mathbf{k}, \tag{2.29}$$

then since $\mathbf{k} \neq 0$, $\mathbf{h} \neq 0$,

$$\mathbf{h} = \phi \mathbf{k} \tag{2.30}$$

for some scalar $\phi \neq 0$. Substituting (2.30) in (2.27), we obtain

$$a^2 \mathbf{k}^2 = \mathrm{k}_0^2, \tag{2.31}$$

where

$$a = \sqrt{\frac{\lambda + 2\mu}{\rho}}. \tag{2.32}$$

As follows from (2.13)–(2.19), $a$ is the phase velocity of the wavefront (2.12) (as we will see, it coincides with the value of the group velocity) and is called simply *the velocity of waves P*. Let $\mathrm{k}_0 < 0$ (which agrees with the relation $\mathrm{k}_0 = -\omega$ and the positiveness of $\omega$), then the vectors of phase velocity and slowness are defined, respectively, as follows:

$$\boldsymbol{v}^{ph} = \frac{\mathbf{k}}{|\mathbf{k}|} a = a\mathbf{s}, \quad \boldsymbol{p} = \frac{\mathbf{s}}{a} = \frac{1}{a}\frac{\mathbf{k}}{|\mathbf{k}|}, \tag{2.33}$$

where $\mathbf{s} = \mathbf{k}/|\mathbf{k}|$ is the unit vector directed along $\mathbf{k}$.

A solution described by formulas (2.30)–(2.32) is called *a plane wave P*. Obviously, the displacement vector is parallel to the direction of wave propagation. Straight lines parallel to $\mathbf{s}$ and oriented along $\mathbf{s}$ are called *rays* of a plane wave $P$.

### 2.3.3 Wave $S$

Now we suppose that $\mathbf{h}$ is not parallel to $\mathbf{k}$. Then from (2.27) we have

$$\mathbf{h} \perp \mathbf{k}, \quad \text{i.e.,} \quad \mathbf{h} \cdot \mathbf{k} = 0. \tag{2.34}$$

From (2.34) and (2.27) it follows that

$$b^2 \mathbf{k}^2 = \mathrm{k}_0^2, \tag{2.35}$$

where $b$ is a constant defined by

$$b = \sqrt{\frac{\mu}{\rho}}. \tag{2.36}$$

Solution (2.34)–(2.36) is called a *plane wave $S$*, and the constant $b$ is called *the velocity of waves $S$*. Expressions for a phase velocity vector and a slowness vector are

$$\boldsymbol{v}^{ph} = \frac{\mathbf{k}}{|\mathbf{k}|} b = b\mathbf{s}, \quad \boldsymbol{p} = \frac{\mathbf{s}}{b} = \frac{1}{b} \frac{\mathbf{k}}{|\mathbf{k}|}. \tag{2.37}$$

Obviously,

$$\boldsymbol{v}^{ph} \parallel \mathbf{k}, \tag{2.38}$$

similarly to waves $P$. The displacement vector is perpendicular to the direction of propagation and parallel to the wavefront plane. Since the eigensubspace of the matrix $\mathbf{\Gamma}$, which corresponds to waves $S$, is two-dimensional, they are not, generally speaking, linearly polarized (see further Section 2.3.5). Straight lines parallel to $\mathbf{k}$ and oriented along $\mathbf{k}$ are called *rays* of a plane wave $S$.

From (1.67) it follows that

$$a > b, \tag{2.39}$$

therefore a wave $P$ propagates faster than a wave $S$. Seismologists call a wave that comes earlier the primary one, and a wave that comes later the secondary one, from which the notation $P$ and $S$ originates.

A wave $P$ is sometimes called *longitudinal*, because the displacement therein is directed along the direction of propagation, or a *pressure wave*, whereas a wave $S$ is called a *transverse* or *shear* one. This terminology does not fully agree with the properties of the natural generalization of the plane waves $P$ and $S$, which will be considered in Chapter 4.

### 2.3.4 Time-harmonic waves $P$ and $S$

Let us consider the harmonic case (2.4).

For waves $P$, the previous consideration gives $\boldsymbol{h} \parallel \boldsymbol{k}$, i.e., the representation (2.30) takes the form

$$\boldsymbol{h} = \phi \boldsymbol{k}, \tag{2.40}$$

where $\phi$ is a constant, possibly complex, $|\boldsymbol{k}| = \sqrt{\sum_{j=1}^{3} k_j^2}$, and the relation between $\omega$ and $\boldsymbol{k}$ is defined by the dispersion equation

$$\omega = a|\boldsymbol{k}|. \tag{2.41}$$

It can also be written in the form

$$\boldsymbol{k} = \frac{\omega}{a}\mathbf{s}, \tag{2.42}$$

where $\mathbf{s}$ is a unit vector in the direction of propagation of the wavefront.

For waves $S$, accordingly, $\boldsymbol{h} \perp \boldsymbol{k}$, i.e.,

$$\boldsymbol{h} \cdot \boldsymbol{k} = h_j k_j = 0, \tag{2.43}$$

and the dispersion equation has the form

$$\omega = b|\boldsymbol{k}|. \tag{2.44}$$

It can be written as

$$\boldsymbol{k} = \frac{\omega}{b}\mathbf{s}. \tag{2.45}$$

We note that in both cases (2.41) and (2.44) the function $H(\boldsymbol{k})$ (see (2.5)), which is defined respectively as $a|\boldsymbol{k}|$ or $b|\boldsymbol{k}|$, is first-degree homogeneous with respect to $\boldsymbol{k}$, and dispersion is absent. Obviously, the velocities $a$ and $b$ do not depend on frequency.

### 2.3.5 Polarization of time-harmonic waves $P$ and $S$

Let us fix a point $\mathbf{x}$. A curve described by the vector $\mathbf{u} = \mathbf{u}(\mathbf{x}, t) =: \mathbf{u}(t)$ is the trajectory of motion of a particle.

#### Linear polarization of a wave $P$

Let us discuss trajectories depicted by a medium particle in a harmonic wave $P$, without assuming the reality of the amplitude vector $\boldsymbol{h}$ (2.40). We assume that $\boldsymbol{k}$ is real, i.e., the *wave is homogeneous.* In the general case in (2.40) $\phi = Ae^{i\psi}$, where $A$ and $\psi = \psi(\mathbf{x})$ are real. Therefore the physical displacement is defined as follows:

$$\operatorname{Re}\mathbf{u}(\mathbf{x}, t) = A\boldsymbol{k}\cos(\psi - \omega t). \tag{2.46}$$

A particle moves according to a harmonic law along the vector $\boldsymbol{k}$, i.e., a wave is *linearly polarized.*

#### Elliptic, linear, and circular polarization of a wave $S$

***Preliminary transformation of the expression for* u.** Let us fix a point $\mathbf{x}$ and regard the trajectory $\operatorname{Re}\mathbf{u}(\mathbf{x}, t)$ of motion of a particle in a harmonic wave $S$ as a function of time. Decompose the complex amplitude into the real and imaginary parts,

$$\boldsymbol{h} = \boldsymbol{P} + i\boldsymbol{Q}, \quad \boldsymbol{h} \neq 0, \tag{2.47}$$

where $\boldsymbol{P} = \operatorname{Re}\boldsymbol{h}$ and $\boldsymbol{Q} = \operatorname{Im}\boldsymbol{h}$ are both real and orthogonal to the direction of propagation $\boldsymbol{k}$,

$$\boldsymbol{P} \cdot \boldsymbol{k} = \boldsymbol{Q} \cdot \boldsymbol{k} = 0. \tag{2.48}$$

However, $\boldsymbol{P}$ and $\boldsymbol{Q}$ are not necessarily mutually orthogonal.

Let us show that $\boldsymbol{h}$ can be represented in the form

$$\boldsymbol{h} = e^{i\psi}\left(\boldsymbol{P}' + i\boldsymbol{Q}'\right) \tag{2.49}$$

with some real $\psi$ and real vectors $\boldsymbol{P}'$ and $\boldsymbol{Q}'$, which are mutually orthogonal and orthogonal to $\boldsymbol{k}$, that is,

$$\boldsymbol{P}' \cdot \boldsymbol{Q}' = 0, \tag{2.50}$$

and

$$\boldsymbol{P}' \cdot \boldsymbol{k} = \boldsymbol{Q}' \cdot \boldsymbol{k} = 0. \tag{2.51}$$

***Determination of*** $\psi$**.** Relations (2.49)–(2.51) allow one to find $\psi$. Formulas (2.47) and (2.49) lead to

$$\boldsymbol{P}' = \boldsymbol{P}\cos\psi + \boldsymbol{Q}\sin\psi, \quad \boldsymbol{Q}' = -\boldsymbol{P}\sin\psi + \boldsymbol{Q}\cos\psi. \tag{2.52}$$

Relations (2.51) hold for any $\psi$, which follows from (2.48). Let us find out for which $\psi$ condition (2.50) is satisfied. By (2.52), relation (2.50) is equivalent to

$$2\boldsymbol{P} \cdot \boldsymbol{Q}\cos 2\psi - (\boldsymbol{P}^2 - \boldsymbol{Q}^2)\sin 2\psi = 0, \tag{2.53}$$

from which $\psi$ can be found.

For further consideration it is crucial whether the following quantity differs from zero:

$$\mathsf{S} := \sqrt{(2\boldsymbol{P} \cdot \boldsymbol{Q})^2 + (\boldsymbol{P}^2 - \boldsymbol{Q}^2)^2}. \tag{2.54}$$

If

$$\mathsf{S} > 0, \tag{2.55}$$

we can divide relation (2.53) by $\mathsf{S}$ and determine the angle $\gamma$, $0 \leqslant \gamma < 2\pi$, from the conditions

$$\cos\gamma = -\frac{1}{\mathsf{S}}(\boldsymbol{P}^2 - \boldsymbol{Q}^2), \quad \sin\gamma = -\frac{2}{\mathsf{S}}\boldsymbol{P} \cdot \boldsymbol{Q}. \tag{2.56}$$

Then relation (2.53) is written in the form

$$\sin 2\psi\cos\gamma - \cos 2\psi\sin\gamma \equiv \sin(2\psi - \gamma) = 0,$$

from which it follows that we can put

$$\psi = \frac{\gamma}{2}. \tag{2.57}$$

In the case where $\mathsf{S} = 0$, obviously, $\boldsymbol{P} \cdot \boldsymbol{Q} = 0$ and $\boldsymbol{P}^2 - \boldsymbol{Q}^2 = 0$, and any real number can be taken as $\psi$.

***Physical displacement.*** When $\psi$ is found, $\boldsymbol{P}'$ and $\boldsymbol{Q}'$ are defined by formulas (2.52). Let us introduce in the wavefront plane $\boldsymbol{k} \cdot \mathbf{x} - \omega t = \text{const}$, to which the displacement in a wave $S$ is parallel, the Cartesian coordinates $(x', y')$, taking $\frac{\boldsymbol{P}'}{|\boldsymbol{P}'|}$ and $\frac{\boldsymbol{Q}'}{|\boldsymbol{Q}'|}$ as the unit coordinate vectors of $x'$- and $y'$-axes, respectively. If it appears that $\boldsymbol{P}' = 0$, then we take $\left[\frac{\boldsymbol{Q}'}{|\boldsymbol{Q}'|} \times \frac{\boldsymbol{k}}{|\boldsymbol{k}|}\right]$ as the unit coordinate vector of the axis $x'$. In the case where $\boldsymbol{Q}' = 0$ we act in a similar way. $\boldsymbol{P}'$ and $\boldsymbol{Q}'$ cannot be equal to zero simultaneously, because $\boldsymbol{h} \neq 0$ (see (2.47)).

For projections $(X', Y')$ of the vector $\operatorname{Re} \boldsymbol{u}$ to the axes $(x', y')$, we get

$$X' = |\boldsymbol{P}'| \cos(\Theta - \omega t), \quad Y' = -|\boldsymbol{Q}'| \sin(\Theta - \omega t), \\ \Theta = \boldsymbol{k} \cdot \mathbf{x} + \psi. \tag{2.58}$$

Thus, the physical displacement takes the form

$$\operatorname{Re} \mathbf{u} = \operatorname{Re}\Big[(\boldsymbol{P}' + i\boldsymbol{Q}')e^{i(\Theta - \omega t)}\Big] = \boldsymbol{P}' \cos(\Theta - \omega t) - \boldsymbol{Q}' \sin(\Theta - \omega t). \tag{2.59}$$

***Elliptic, linear, and circular polarization.*** Three cases are to be considered.

*1*. In the case of a general position specified by (2.55) we have $\boldsymbol{P}' \neq 0$ and $\boldsymbol{Q}' \neq 0$. Then formulas (2.59) describe the motion of a point along an ellipse with half-axles having the lengths $|\boldsymbol{P}'|$ and $|\boldsymbol{Q}'|$ and directed along $\boldsymbol{P}'$ and $\boldsymbol{Q}'$, respectively. The ellipse does not degenerate into a circle because $|\boldsymbol{P}'| \neq |\boldsymbol{Q}'|$.[3] Therefore formula (2.58) describes the motion of a point along a nondegenerate *polarization ellipse* with nonequal half-axles $|\boldsymbol{P}'| \neq |\boldsymbol{Q}'|$. In this case, the wave is *elliptically polarized.*

*2*. Inequality (2.55) holds, where one of the vectors $\boldsymbol{P}'$ and $\boldsymbol{Q}'$ vanishes. As noted above, the other is not equal to zero. Formulas (2.58) describe the motion of a point either along the vector $\boldsymbol{P}'$, i.e., within the segment $[-|\boldsymbol{P}'|, |\boldsymbol{P}'|]$ of the axis $x'$ if $\boldsymbol{Q}' = 0$, or within the segment $[-|\boldsymbol{Q}'|, |\boldsymbol{Q}'|]$ of the axis $y'$ directed along $\boldsymbol{Q}'$ if $\boldsymbol{P}' = 0$. In this case, the wave is *linearly polarized.*

[3]Indeed, let us assume the opposite, that is, $\boldsymbol{P}'^2 - \boldsymbol{Q}'^2 = 0$. Formulas (2.52) then imply that

$$\boldsymbol{P}'^2 - \boldsymbol{Q}'^2 = 2\boldsymbol{P} \cdot \boldsymbol{Q} \sin 2\psi - (\boldsymbol{P}^2 - \boldsymbol{Q}^2) \cos 2\psi = 0. \tag{2.60}$$

Equations (2.53) and (2.60) can be regarded as a system of linear algebraic equations for $2\boldsymbol{P} \cdot \boldsymbol{Q}$ and $\boldsymbol{P}^2 - \boldsymbol{Q}^2$. Its determinant is equal to $\sin^2 2\psi + \cos^2 2\psi = 1 \neq 0$. Thus, $2\boldsymbol{P} \cdot \boldsymbol{Q} = 0$, $\boldsymbol{P}^2 - \boldsymbol{Q}^2 = 0$, whence $\mathsf{S} = 0$, which contradicts the assumption (2.55).

*3*. A degenerate case where $\mathsf{S} = 0$ and therefore the relations $\boldsymbol{P} \cdot \boldsymbol{Q} = 0$ and $\boldsymbol{P}^2 - \boldsymbol{Q}^2 = 0$ simultaneously hold. Then from (2.51) it follows that for any $\psi$ we have $\boldsymbol{P}' \perp \boldsymbol{Q}'$ and $\boldsymbol{P}'^2 = \boldsymbol{Q}'^2 > 0$, which implies that $|\boldsymbol{P}'| = |\boldsymbol{Q}'|$. A point $(X', Y')$, see (2.58), uniformly moves along a circle of radius $|\boldsymbol{P}'| = |\boldsymbol{Q}'|$. This is the case of *circular polarization*.

### 2.3.6 Group velocity

In the above discussion of plane waves $P$ and $S$, important notions of the theory of wave processes, namely, that of group velocity naturally appeared.

We saw that $\mathrm{k}_0^2$ (for harmonic waves it is $\omega^2$) is an eigenvalue of a positive definite self-adjoint operator of a very simple structure. In more complicated cases, which will be considered further, $\omega^2$ proves to be an eigenvalue of a more complicated but again positive definite self-adjoint operator. In these cases, $\omega$ is a function of the wave vector. The *group velocity* is defined as

$$\boldsymbol{v}^{gr} := \mathrm{grad}_{\boldsymbol{k}}\, H(\boldsymbol{k}) = \mathbf{e}_j \frac{\partial H(\boldsymbol{k})}{\partial k_j} = \left( \frac{\partial H}{\partial k_1}, \frac{\partial H}{\partial k_2}, \frac{\partial H}{\partial k_3} \right) \tag{2.61}$$

(it is assumed that $\omega = H(\boldsymbol{k})$ is the corresponding dispersion relation (2.5)). In the sequel, we will find out a relation between the group velocity and the propagation of energy.

For the group velocity of the wave $P$ from (2.41) we obtain

$$\boldsymbol{v}^{gr} = \mathrm{grad}_{\boldsymbol{k}} H(\boldsymbol{k}) = \mathrm{grad}_{\boldsymbol{k}}(a|\boldsymbol{k}|) = a\mathbf{s}, \tag{2.62}$$

which coincides with its phase velocity. Similarly, for the wave $S$ from (2.44) it follows that

$$\boldsymbol{v}^{gr} = b\boldsymbol{s}, \tag{2.63}$$

and its group velocity coincides with its phase velocity.

The coincidence of the phase and group velocities is a specific feature of isotropic media without dispersion. The isotropy of a medium means that all propagation directions are equivalent, so that $\omega = H(\boldsymbol{k}) \equiv H(|\boldsymbol{k}|)$ depends only on the modulus of $\boldsymbol{k}$. The absence of dispersion is equivalent to the first-degree homogeneity of the function $H(\boldsymbol{k})$ (see (2.6)), i.e., $H(\boldsymbol{k}) = H(|\boldsymbol{k}|) = \mathrm{const}|\boldsymbol{k}|$. The dispersion equations (2.41) and (2.44) for waves $P$ and $S$ have exactly this form.

### 2.3.7 Energy relations for time-harmonic waves

With examples of plane waves $P$ and $S$, we will illustrate two important general facts associated with time-averaged quantities. The first fact touches on densities of potential and kinetic energies, and the second one concerns the relationship between the density of energy, the energy flow, and the group velocity.

## Virial theorem

In the most general case of harmonic plane wave (2.4), formula (1.90) gives for the averaged kinetic energy the expression

$$\overline{\mathcal{K}} = \frac{1}{4}\rho\omega^2|\boldsymbol{h}|^2. \tag{2.64}$$

For waves $P$ from (1.55), (2.4), and (2.40) we find that $\varepsilon_{jl}(\mathbf{u}) = ik_jk_l\varphi e^{i\psi}$, $\sigma_{jl}(\mathbf{u}) = i(2\mu k_jk_l + \lambda\delta_{jl}\boldsymbol{k}^2)\varphi e^{i\psi}$, where

$$\psi := \boldsymbol{k}\cdot\mathbf{x} - \omega t.$$

Hence, with (1.87) taken into account, it follows that

$$\overline{\mathcal{W}} = \frac{1}{2}\overline{\mathrm{Re}[\varepsilon_{jl}]\ \mathrm{Re}[\sigma_{jl}]} = \frac{1}{4}(\lambda + 2\mu)\boldsymbol{k}^4\ |\varphi|^2 = \frac{1}{4}\rho a^2\boldsymbol{k}^2|\boldsymbol{h}|^2. \tag{2.65}$$

Using (2.42), we get

$$\overline{\mathcal{K}} = \overline{\mathcal{W}}. \tag{2.66}$$

Counterparts of relation (2.66) exist for many types of waves. In accordance with the terminology accepted in mechanics of systems with a finite number of degrees of freedom (see, e.g., Landau and Lifshitz 2007 [33]), we will call relation (2.66) the *virial theorem.*

We proceed to waves $S$. Let us calculate $\overline{\mathcal{W}}$. Obviously, $\varepsilon_{jl}(\mathbf{u}) = \frac{1}{2}i(h_jk_l + h_lk_j)e^{i\psi}$. With the use of (2.34), we derive that $\sigma_{jl}(\mathbf{u}) = i[\mu(h_jk_l + h_lk_j) + \delta_{jl}\lambda h_sk_s]e^{i\psi} = i\mu(h_jk_l + h_lk_j)e^{i\psi}$, from which

$$\overline{\mathcal{W}} = \frac{1}{2}\overline{\mathrm{Re}[\varepsilon_{jl}]\ \mathrm{Re}[\sigma_{jl}]} = \frac{1}{4}\rho b^2\boldsymbol{k}^2|\boldsymbol{h}|^2.$$

Comparing the latter with (2.64) and recalling (2.45), we obtain the virial theorem (2.66) for a wave $S$.

In the sequel, it will be convenient to regard equations of the form (2.66) as dispersion relations.

## Group velocity theorem

Now we average an energy density vector over a period (see (1.91)). For waves $P$

$$\overline{S_j} = \frac{\omega(\lambda + 2\mu)}{2}|\boldsymbol{h}|^2k_j,$$

or

$$\overline{\boldsymbol{S}} = \frac{\omega(\lambda + 2\mu)}{2}|\boldsymbol{h}|^2\boldsymbol{k} = \frac{\omega^2\rho a}{2}|\boldsymbol{h}|^2\mathbf{s}, \tag{2.67}$$

and therefore

$$\overline{\boldsymbol{S}} = \overline{\mathcal{E}}\ \boldsymbol{v}^{gr}. \tag{2.68}$$

Similarly, for waves $S$ we obtain

$$\overline{\boldsymbol{S}} = \frac{\omega\mu}{2}|\boldsymbol{h}|^2\boldsymbol{k} = \frac{\omega^2\rho b}{2}|\boldsymbol{h}|^2\mathbf{s}, \tag{2.69}$$

from which relation (2.68) follows. It is known as *the group velocity theorem.* Counterparts of relation (2.68) exist for waves of diverse nature and will appear many times henceforth.

Formula (2.68) agrees with the idea of N. A. Umov who considered energy to be like fluid "flowing in an elastic medium with elastic waves" (see the original work by Umov 1874, 1950 [54, 55] and historical research by Gulo 1977 [24]).

### 2.3.8 ⋆ Potentials

In the case of a homogeneous isotropic elastic medium, it is convenient to use some special representations of solutions, which are described below.

#### Nonstationary case

It may be useful to seek solutions of the nonstationary equations of isotropic elastodynamics (2.24) in the form

$$\mathbf{u} = \operatorname{grad}\Phi, \quad \Phi = \Phi(\mathbf{x}, t), \tag{2.70}$$

i.e., $u_j = \partial_j\Phi$. The function $\Phi$ is called a *scalar potential.* Since $\operatorname{rot}\operatorname{grad} = 0$, $\operatorname{div}\operatorname{grad} = \partial_j\partial_j = \nabla^2$, we have

$$\operatorname{grad}\left((\lambda + 2\mu)\nabla^2\Phi - \rho\ddot{\Phi}\right) = (\lambda + 2\mu)\operatorname{grad}\left(\nabla^2 - \frac{1}{a^2}\frac{\partial^2}{\partial t^2}\right)\Phi = 0.$$

For the latter equation (and therefore the elastodynamics equations) to hold, it is sufficient that the *wave equation* corresponding to the velocity of propagation of waves $P$ be satisfied:

$$\left(\nabla^2 - \frac{1}{a^2}\frac{\partial^2}{\partial t^2}\right)\Phi = 0. \tag{2.71}$$

As a simple example, we note that (2.71) has particular solutions $\Phi = f(\mathbf{k}\cdot\mathbf{x} + \mathrm{k}_0 t)$, where $a^2\mathrm{k}_j\mathrm{k}_j = \mathrm{k}_0^2$, and we arrive at a plane wave $P$,

$$\mathbf{u} = \operatorname{grad}\Phi = \mathbf{k}f'(\mathbf{k}\cdot\mathbf{x} + \mathrm{k}_0 t),$$

where the prime denotes the derivative of a function with respect to its argument.

Seeking now solutions in the form

$$\mathbf{u} = \operatorname{rot}\boldsymbol{\Psi}, \quad \boldsymbol{\Psi} = \boldsymbol{\Psi}(\mathbf{x}, t), \tag{2.72}$$

for a *vector potential* $\boldsymbol{\Psi}$ we obtain the equation

$$\operatorname{rot}(\mu\nabla^2\boldsymbol{\Psi} - \rho\ddot{\boldsymbol{\Psi}}) = \mu \operatorname{rot}\left(\nabla^2 - \frac{1}{b^2}\frac{\partial^2}{\partial t^2}\right)\boldsymbol{\Psi} = 0. \tag{2.73}$$

We get *a vector wave equation* (that is, three equations for the components of $\boldsymbol{\Psi}$)

$$\left(\nabla^2 - \frac{1}{b^2}\frac{\partial^2}{\partial t^2}\right)\boldsymbol{\Psi} = 0, \tag{2.74}$$

corresponding to the velocity of waves $S$. Noting that equation (2.74) has solutions of the form of a plane wave $\boldsymbol{\Psi} = \mathbf{m} f(\mathbf{k}\cdot\mathbf{x} + \mathrm{k}_0 t)$, where $\mathrm{k}_j\mathrm{k}_j = \mathrm{k}_0^2/b^2$, and $\mathbf{m}$ is an arbitrary constant vector, we obtain a solution

$$\mathbf{u} = \operatorname{rot}\boldsymbol{\Psi} = \mathbf{h} f'(\mathbf{k}\cdot\mathbf{x} + \mathrm{k}_0 t), \quad \mathbf{h} = \mathbf{k}\times\mathbf{m} \tag{2.75}$$

for a plane wave $S$.

The vectors representable in the form (2.70) are called *potential vectors,* whereas vectors representable in the form (2.72) are called *solenoidal vectors.* The same terminology is accepted for the time-harmonic case (see (2.76)).

### Time-harmonic case

In the time-harmonic case, the potentials are introduced by the formulas

$$\mathbf{u} = \operatorname{grad}\Phi = \operatorname{grad}\phi(\mathbf{x})e^{-i\omega t} \quad \text{and} \quad \mathbf{u} = \operatorname{rot}\boldsymbol{\Psi} = \operatorname{rot}\boldsymbol{\psi}(\mathbf{x})e^{-i\omega t}. \tag{2.76}$$

The functions $\Phi$ and $\boldsymbol{\Psi}$ satisfy, respectively, the *Helmholtz equations*

$$(\nabla^2 + \kappa^2)\phi = 0 \quad \text{and} \quad (\nabla^2 + æ^2)\boldsymbol{\psi} = 0, \tag{2.77}$$

where

$$\kappa := \frac{\omega}{a} \quad \text{and} \quad æ := \frac{\omega}{b} \tag{2.78}$$

are the *wavenumbers of waves $P$ and $S$.*

### Equations with volume forces

Potentials can be useful for solving equations with right-hand sides. Suppose we solve an equation of the form (1.62) where

$$\mathbf{F} = \operatorname{grad}\mathrm{X} + \operatorname{rot}\mathbf{Y}, \tag{2.79}$$

$\mathrm{X} = \mathrm{X}(\mathbf{x}, t)$ and $\mathbf{Y} = \mathbf{Y}(\mathbf{x}, t)$. Seeking a solution in the form

$$\mathbf{u} = \operatorname{grad}\Phi + \operatorname{rot}\boldsymbol{\Psi}, \tag{2.80}$$

we reduce the problem to those for nonhomogeneous wave equations:

$$\left(\nabla^2 - \frac{1}{a^2}\frac{\partial^2}{\partial t^2}\right)\Phi = -\frac{1}{(\lambda + 2\mu)}\mathrm{X}, \quad \left(\nabla^2 - \frac{1}{b^2}\frac{\partial^2}{\partial t^2}\right)\boldsymbol{\Psi} = -\frac{1}{\mu}\mathbf{Y}. \tag{2.81}$$

Nonhomogeneous equations with time-harmonic right-hand sides can similarly be reduced to nonhomogeneous Helmholtz equations.

## 2.4 Plane waves in unbounded anisotropic media

The theory of waves in homogeneous anisotropic media is important first of all for investigating elastic properties of crystals but is also applicable to geophysics and other fields of science and technology. We restrict our discussion to the harmonic case (1.72).

### 2.4.1 Eigenvalue problem

Substituting harmonic plane-wave ansatz (2.4) into the elastodynamics equations (2.1)–(2.2), we get

$$\mathfrak{N}\boldsymbol{h} = 0, \tag{2.82}$$

where $\mathfrak{N}$ is now the matrix with entries

$$\mathfrak{N}_{ql} = \mathfrak{N}_{ql}(\boldsymbol{k}, \omega) = c_{qjml}k_jk_m - \rho\omega^2\delta_{ql}. \tag{2.83}$$

As in the isotropic case, we obtain an eigenvalue problem,

$$\boldsymbol{\Gamma}(\boldsymbol{k})\boldsymbol{h} = \rho\omega^2\boldsymbol{h}, \tag{2.84}$$

for the matrix $\boldsymbol{\Gamma}$ with entries

$$\Gamma_{ql} = \Gamma_{ql}(\boldsymbol{k}) = c_{qjml}k_jk_m. \tag{2.85}$$

Matrix (2.85) (by the symmetry of $c_{qjlm}$ (1.18)) is symmetric and, as will be shown, positive definite.

#### Positive definiteness of $\boldsymbol{\Gamma}$

We will show that for real $\boldsymbol{k} \neq 0$ the matrix $\boldsymbol{\Gamma}$ is positive definite. Let $\mathfrak{h} = (\mathfrak{h}_1, \mathfrak{h}_2, \mathfrak{h}_3)$ be a real vector. Consider the quadratic form $(\boldsymbol{\Gamma}\mathfrak{h}, \mathfrak{h})$. It is nonnegative, because

$$(\boldsymbol{\Gamma}\mathfrak{h}, \mathfrak{h}) = \Gamma_{lm}\mathfrak{h}_l\mathfrak{h}_m = c_{ljnm}k_jk_n\mathfrak{h}_l\mathfrak{h}_m = c_{ljnm}e_{lj}e_{nm} \geqslant 0, \tag{2.86}$$

where

$$e_{pq} = \frac{1}{2}\left(\mathfrak{h}_pk_q + \mathfrak{h}_qk_p\right). \tag{2.87}$$

By the positive definiteness of the potential energy, from $(\boldsymbol{\Gamma}\mathfrak{h}, \mathfrak{h}) = 0$ it follows that $e_{pq} = 0$ for all $p$ and $q$,

$$\mathfrak{h}_pk_q + \mathfrak{h}_qk_p = 0, \quad p, q = 1, 2, 3. \tag{2.88}$$

Now we will show that this implies that $\mathfrak{h} = 0$. Multiplying both parts of relation (2.88) by $\mathfrak{h}_p$ and $k_q$ and summing over repeated indices, we get

$$\mathfrak{h}^2\boldsymbol{k}^2 + (\mathfrak{h}, \boldsymbol{k})^2 = 0, \tag{2.89}$$

from which it follows that $\mathfrak{h}^2\boldsymbol{k}^2 = 0$, and since $\boldsymbol{k} \neq 0$, we have $\mathfrak{h} = 0$. Thus, we have established the positive definiteness of the matrix $\boldsymbol{\Gamma}$.

### 2.4.2 Phase velocity

Let us return to equation (2.84). It is more convenient to consider the matrix $\frac{1}{\rho}\mathbf{\Gamma}$ rather than the matrix $\mathbf{\Gamma}$. Taking (2.5) into account, we denote its eigenvalues by $H^2(\boldsymbol{k})$ and rewrite (2.84) in the form

$$\frac{1}{\rho}\mathbf{\Gamma}(\boldsymbol{k})\boldsymbol{h} = H^2(\boldsymbol{k})\boldsymbol{h}. \tag{2.90}$$

Its three eigenvalues, which we denote by $(H^{(j)}(\boldsymbol{k}))^2$, $j = 1, 2, 3$, are positive. We consider $H^{(j)}(\boldsymbol{k})$ to be positive as well.

In the sequel, it is assumed that $(H^j)^2(\boldsymbol{k})$ is a nondegenerate eigenvalue of the matrix $\frac{1}{\rho}\mathbf{\Gamma}(\boldsymbol{k})$ (i.e., this eigenvalue corresponds to a unique, up to a scalar factor, eigenvector). Such a situation occurs, e.g., for a wave $P$ in the isotropic case. It can easily be shown that the corresponding wave has linear polarization, similarly to a wave $P$.

The second-degree homogeneity of the entries of the matrix $\frac{1}{\rho}\mathbf{\Gamma}(\boldsymbol{k})$ with respect to $\boldsymbol{k}$ results in the first-degree homogeneity of the functions $H^{(j)}(\boldsymbol{k})$ with respect to $\boldsymbol{k}$ (see (2.6)). Equation (2.5) (we will often omit the mark $^{(j)}$), where $H(\boldsymbol{k})$ is one of the functions $H^{(j)}(\boldsymbol{k})$, is called, as we know, a dispersion equation. Generally speaking, there are three of them. We said "generally speaking," because some $H^{(j)}(\boldsymbol{k})$ may coincide at some $\boldsymbol{k}$ (for example, in the isotropic case $H^{(1)} \equiv H^{(2)} = b|\boldsymbol{k}|$ for all $\boldsymbol{k}$). Since all $H^{(j)}(\boldsymbol{k})$ are first-degree homogeneous, plane waves in an anisotropic medium are nondispersive (see Section 2.2.2).

For phase velocities, relations (2.17), (2.18), and (2.23) hold. The phase velocities will be numbered as follows:

$$c^{(1)}(\mathbf{s}) \leqslant c^{(2)}(\mathbf{s}) \leqslant c^{(3)}(\mathbf{s}). \tag{2.91}$$

It can easily be seen that the squared phase velocities are eigenvalues of the matrix $\frac{1}{\rho}\mathbf{\Gamma}(\mathbf{s})$.

### 2.4.3 Group velocity

Let $\omega = H(\boldsymbol{k})$ be one of the dispersion relations $\omega = H^{(j)}(\boldsymbol{k})$. Recall that the group velocity $\boldsymbol{v}^{gr}$ is a vector with components $\frac{\partial H}{\partial k_1}, \frac{\partial H}{\partial k_2}, \frac{\partial H}{\partial k_3}$. In the isotropic case (where it coincides with the phase velocity), the group velocity is related to the propagation of energy in accordance with (2.68). In an anisotropic medium, such a relation also holds and is important. Let us proceed to its derivation.

Using (1.55), we find an expression for averaged potential energy. Obviously, $\varepsilon_{lm}(\mathbf{u}) = \frac{1}{2}i(h_l k_m + h_l k_m)e^{i\psi}$ and $\sigma_{lm}(\mathbf{u}) = i c_{lmnp} k_n h_p e^{i\psi}$, where $\psi = \boldsymbol{k} \cdot \mathbf{x} - \omega t$. The symmetry of elastic stiffnesses, together with equation (2.84), gives

$$\overline{\mathcal{W}} = \frac{1}{4}\Gamma_{lm}(\boldsymbol{k})h_l h_m^* = \frac{1}{4}(\mathbf{\Gamma}(\boldsymbol{k})\boldsymbol{h}, \boldsymbol{h}) = \frac{1}{4}\rho\omega^2|\boldsymbol{h}|^2 = \overline{\mathcal{K}}. \tag{2.92}$$

It is now obvious that the virial theorem (2.66) is true once again.

We note that relation (2.66) is a form of the dispersion relation. Differentiating[4] it with respect to $k_j$, we get

$$\frac{1}{2}\rho\omega\frac{\partial\omega}{\partial k_j}|\boldsymbol{h}|^2 = \frac{1}{4}\frac{\partial\Gamma_{lm}}{\partial k_j}h_l h_m^* = \frac{1}{2}c_{ljnm}k_n h_l h_m^*, \quad \omega = H(\boldsymbol{k}).$$

We "naively" ignore here the dependence of $\boldsymbol{h}$ on $\boldsymbol{k}$. It is justified because the derivatives of $\boldsymbol{h}$ on the left-hand and right-hand sides are cancelled. Indeed, in differentiating $\overline{\mathcal{K}}$, there arises the term $\frac{1}{4}\rho\omega^2\Big(h_s\frac{\partial h_s^*}{\partial k_j} + h_s^*\frac{\partial h_s}{\partial k_j}\Big)$, whereas in differentiating $\overline{\mathcal{W}}$, the term $\frac{1}{4}\Gamma_{lm}\Big(\frac{\partial h_l}{\partial k_j}h_m^* + h_l^*\frac{\partial h_m}{\partial k_j}\Big)$ appears, and the corresponding expressions are mutually deleted. We will meet with *"naive differentiation"* in a more complicated situation in Section 2.9.3.

On the basis of relations (2.84) and (2.92), using the obvious expression for the time-averaged energy flow

$$\overline{S_j} = \frac{\omega}{2}c_{ljnm}k_n h_l h_m^*, \tag{2.93}$$

we conclude that

$$\overline{\mathcal{E}}\frac{\partial\omega}{\partial k_j} = \overline{S}_j. \tag{2.94}$$

Thus, we have proved the group velocity theorem (2.68) for our case.

Note that the direction of propagation of the phase front $\mathbf{s} = \frac{\boldsymbol{k}}{|\boldsymbol{k}|}$ and $\boldsymbol{v}^{gr}$ form an acute angle, because

$$(\boldsymbol{v}^{gr}, \mathbf{s}) = \frac{(\overline{\boldsymbol{S}}, \mathbf{s})}{\overline{\mathcal{E}}} = \frac{\omega}{2}\frac{c_{ljnm}k_n h_l h_m^* s_j}{\frac{1}{2}\rho\omega^2|\boldsymbol{h}|^2} = \frac{|\boldsymbol{k}|(\boldsymbol{\Gamma}(\mathbf{s})\boldsymbol{h}, \boldsymbol{h})}{\rho\omega|\boldsymbol{h}|^2} > 0, \tag{2.95}$$

which follows from the positive definiteness of $\boldsymbol{\Gamma}$. As follows from the remark after formula (2.91), expression (2.95) coincides with the modulus of the phase velocity, that is,

$$(\boldsymbol{v}^{gr}, \mathbf{s}) = c. \tag{2.96}$$

Hence, the phase velocity cannot be greater than the group velocity.

### 2.4.4 Slowness, slowness surface, velocity surface

#### Slowness surface

Let us return to the dispersion equation $\omega = H^{(j)}(\boldsymbol{k})$. In consequence of (2.20) and (2.21), we have

$$H^{(j)}(\boldsymbol{p}) = 1, \quad j = 1, 2, 3. \tag{2.97}$$

[4] In the case where the corresponding eigenvalue of the matrix $\boldsymbol{\Gamma}$ is simple, the smoothness of $\omega(\boldsymbol{k})$ and $\boldsymbol{h}(\boldsymbol{k})$ can be proved with little difficulty. We assume the smoothness if it is multiple.

The sum of the surfaces $H^{(j)}(\boldsymbol{p}) = 1$ is called a *slowness surface.* It can be obtained geometrically as the union of ends of all the corresponding slowness vectors $\boldsymbol{p}$ starting at the origin of coordinates. The set of ends of these vectors forms the $j$-th *sheet*, and the slowness surface consists of three sheets. The slowness surface can be described by the equation

$$\det(\boldsymbol{\Gamma}(\boldsymbol{p}) - \rho \mathbf{I}) = \det \|c_{plmq} p_l p_m - \rho \delta_{pq}\| = 0, \tag{2.98}$$

where $\mathbf{I}$ is the identity matrix, and $\delta_{pq}$ is the Kronecker symbol.

In the case of an isotropic medium, a slowness surface is the sum of two concentric spheres $|\boldsymbol{p}| = \frac{1}{a}$ and $|\boldsymbol{p}| = \frac{1}{b}$, where the sphere $|\boldsymbol{p}| = \frac{1}{b}$ of the bigger radius is twofold, because it corresponds to a double eigenvalue.

A slowness surface is always centrally symmetric, i.e., with its point $\boldsymbol{p}$ it contains also the point $-\boldsymbol{p}$. In the general case, it is not necessarily smooth, and its sheets are not necessarily convex.

## ⋆ Legendre transformation and velocity surface

We will see that the slowness and the group velocity are in a sense dual. We emphasize that the fundamental notion of group velocity (relation (2.68) illustrates its fundamental nature) is dual to the slowness, not to the phase velocity. The slowness is no less important than the phase velocity.

Let us write the equation of a sheet of the slowness surface in the form

$$\frac{H^2(\boldsymbol{p})}{2} = \frac{1}{2}, \tag{2.99}$$

where $H(\boldsymbol{p})$ is one of $H^{(j)}(\boldsymbol{p})$. We will find the *Legendre transformation* (see, e.g., Arnold 1989 [4], Dubrovin, Fomenko, and Novikov 1984 [18], Courant and Hilbert 1962b [16]) of the function on the left-hand side of (2.99). For this purpose, we introduce new variables $(v_1, v_2, v_3) = \boldsymbol{v}$,

$$v_m := \frac{\partial}{\partial p_m}\left(\frac{H^2(\boldsymbol{p})}{2}\right), \quad m = 1, 2, 3. \tag{2.100}$$

Since on the slowness surface $H(\boldsymbol{p}) = 1$, from relation (2.100) it follows that on this surface,

$$v_m = \frac{\partial H}{\partial p_m} \tag{2.101}$$

(see (2.61)). Therefore, $\boldsymbol{v} = \frac{\partial \mathbf{x}}{\partial t}$ coincides with the group velocity. From formulas (2.101) and (2.97) it can be seen that the group velocity is directed along the normal to the corresponding sheet of the slowness surface.

We find $(p_1(\boldsymbol{v}), p_2(\boldsymbol{v}), p_3(\boldsymbol{v})) = \boldsymbol{p}(\boldsymbol{v})$ from (2.100) (assuming that this is possible) and substitute it into the function

$$\frac{\mathscr{L}^2}{2} := \frac{\mathscr{L}^2(\boldsymbol{v})}{2} = \sum_{j=1}^{3} p_j v_j - \frac{\omega^2(\boldsymbol{p})}{2}. \tag{2.102}$$

From the *Euler homogeneity relation*,[5] it follows that the functions $\frac{\mathscr{L}^2(\boldsymbol{p})}{2}$ and $\frac{1}{2}H^2(\boldsymbol{p})$ coincide. Thus,

$$\frac{\mathscr{L}^2(\boldsymbol{v})}{2} = \frac{1}{2}H^2(\boldsymbol{p}(\boldsymbol{v})). \tag{2.103}$$

When the vector $\boldsymbol{p} = (p_1, p_2, p_3)$ runs over the slowness surface, the vector $\boldsymbol{v} = (v_1, v_2, v_3)$ runs over the surface called a *velocity surface*. Let us return to formula (2.100). As is known, the Legendre transformation is involutive: after the Legendre transformation of function (2.102) we return to the function $\frac{1}{2}H^2(\boldsymbol{p})$. The slowness and velocity surfaces are dual in such a sense.

### 2.4.5 ⋆ Rayleigh principle

Let us compare properties of two media I and II characterized by elastic stiffnesses $c^{\mathrm{I}}_{lpnm}$ and $c^{\mathrm{II}}_{lpnm}$, and densities $\rho^{\mathrm{I}}$ and $\rho^{\mathrm{II}}$, respectively. We assume that for a direction $\mathbf{s}$, the matrix $\frac{1}{\rho^{\mathrm{I}}}\boldsymbol{\Gamma}^{\mathrm{I}}(\mathbf{s}) = \left\| \frac{1}{\rho^{\mathrm{I}}} c^{\mathrm{I}}_{lpnm} s_p s_n \right\|$ is greater than the matrix $\frac{1}{\rho^{\mathrm{II}}}\boldsymbol{\Gamma}^{\mathrm{II}}(\mathbf{s}) = \left\| \frac{1}{\rho^{\mathrm{II}}} c^{\mathrm{II}}_{lpnm} s_p s_n \right\|$, i.e., for any real $\mathfrak{h} \neq 0$,

$$\frac{1}{\rho^{\mathrm{I}}}(\boldsymbol{\Gamma}^{\mathrm{I}}(\mathbf{s})\mathfrak{h}, \mathfrak{h}) > \frac{1}{\rho^{\mathrm{II}}}(\boldsymbol{\Gamma}^{\mathrm{II}}(\mathbf{s})\mathfrak{h}, \mathfrak{h}) \tag{2.104}$$

or, in other words, if the matrix $\frac{1}{\rho^{\mathrm{I}}}\boldsymbol{\Gamma}^{\mathrm{I}} - \frac{1}{\rho^{\mathrm{II}}}\boldsymbol{\Gamma}^{\mathrm{II}}$ is positive definite.

As is known (see, e.g., Courant and Hilbert 1962a [15]), the eigenvalues of a bigger matrix are no less than that of a smaller matrix, so that for the eigenvalues $(H^2)^{\mathrm{I,II}}(\mathbf{s})$ of the matrices $\frac{1}{\rho^{\mathrm{I,II}}}\boldsymbol{\Gamma}^{\mathrm{I,II}}(\mathbf{s})$ the following inequalities hold:

$$(H^2)^{\mathrm{I}(j)}(\mathbf{s}) \geqslant (H^2)^{\mathrm{II}(j)}(\mathbf{s}), \quad j = 1, 2, 3.$$

Therefore, for the corresponding velocities we have

$$c^{\mathrm{I}(j)}(\mathbf{s}) \geqslant c^{\mathrm{II}(j)}(\mathbf{s}), \quad j = 1, 2, 3. \tag{2.105}$$

We call this statement *the Rayleigh principle*, because Rayleigh was first to note the relation between the parameters of a system and the corresponding eigenvalues (see, e.g., Strutt 1945 [53]).

---

## 2.5 ⋆ Local velocities and domain of influence

Let us assume that $\mathbf{u} \equiv 0$ in a domain $\Omega \subset \mathbb{R}^3$. In such a case, it is said that there is no perturbation in the domain $\Omega$. If, on the contrary, $\mathbf{u}|_{\mathbf{x}\in\Omega} \not\equiv 0$, then there is *perturbation* in $\Omega$.

[5] If for any $\Lambda > 0$ the relation $f(\Lambda y_1, \ldots, \Lambda y_n) = \Lambda^p f(y_1, \ldots, y_n)$ holds, then $y_1 \frac{\partial f(y_1,\ldots,y_n)}{\partial y_1} + \ldots + y_n \frac{\partial f(y_1,\ldots,y_n)}{\partial y_n} = p f(y_1, \ldots, y_n)$.

We will show that the velocity of propagation of perturbation is bounded above by the maximal velocity of plane waves, which is true in the case of arbitrary anisotropy. Moreover, it is true also for inhomogeneous media if the velocities of plane waves at each point are to be understood as those in a homogeneous medium with "frozen" coefficients. The presence of interfaces and internal cavities (on the boundaries of which the traction-free conditions are satisfied) has no effect on the result.

### 2.5.1 Statement of the problem

#### Local velocities of propagation

Let us introduce the fundamental notion of local phase velocities of propagation in the most general case of an inhomogeneous anisotropic medium. Let the elastic stiffnesses $c_{ijkl}$ and the volume density $\rho$ be smooth at a point $\mathbf{x} \in \mathbb{R}^3$. We determine three *local phase velocities* $c^{(1)}(\mathbf{x};\mathbf{s})$, $c^{(2)}(\mathbf{x};\mathbf{s})$, and $c^{(3)}(\mathbf{x};\mathbf{s})$ of propagation in the direction of a given unit vector $\mathbf{s}$, $\mathbf{s}^2 = 1$. These are the velocities of plane waves in a homogeneous medium corresponding to the "frozen" elastic stiffnesses and volume density. In accordance with (2.91), $0 < c^{(1)}(\mathbf{x};\mathbf{s}) \leqslant c^{(2)}(\mathbf{x};\mathbf{s}) \leqslant c^{(3)}(\mathbf{x};\mathbf{s})$.

#### Mixed problem

Let $\Omega_1, \ldots, \Omega_N$ be (finite or infinite) domains with good boundaries, and

$$\Omega := \Omega_1 + \ldots + \Omega_N.$$

We consider a *mixed problem* for the elastodynamics equations:

$$\partial_j(c_{ijkl}\partial_k \mathrm{u}_l) - \rho\ddot{\mathrm{u}}_i = -F_i, \quad t > 0, \ \ \mathbf{x} \in \mathbb{R}^3 \setminus \Omega, \tag{2.106a}$$

$$\mathbf{u}|_{t=0} = \mathbf{u}^{\circ}(\mathbf{x}), \quad \dot{\mathbf{u}}|_{t=0} = \mathbf{u}^{\circ\circ}(\mathbf{x}), \quad \mathbf{x} \in \mathbb{R}^3 \setminus \Omega, \tag{2.106b}$$

$$\mathbf{t}^{\boldsymbol{n}}(\mathbf{u})|_{\partial\Omega} = \mathbf{f}, \tag{2.106c}$$

where $\mathbf{u}^{\circ}(\mathbf{x})$ and $\mathbf{u}^{\circ\circ}(\mathbf{x})$ are given initial data, namely, the displacement and the velocity of the medium. The sources of oscillations $\mathbf{F} = \mathbf{F}(\mathbf{x}, t)$ and $\mathbf{f} = \mathbf{f}(\mathbf{x}, t)$ (that is, the densities of external forces, respectively, of volume and surface ones) are also given. Furthermore, we assume that the right-hand sides in (2.106) vanish at all time instants outside a given finite domain $\mathscr{D} \subset \mathbb{R}^3$ with a good boundary $\partial\mathscr{D}$ (see Fig. 2.2).

#### Statement of the theorem on a domain of influence

**Theorem.** *Let $\mathscr{M}(t)$, $t > 0$, be an expanding smooth surface in $\mathbb{R}^3$ such that at $t = 0$ it coincides with $\partial\mathscr{D}$, $\mathscr{M}(0) = \partial\mathscr{D}$, and then moves from $\mathscr{D}$ (see Fig. 2.3) with normal velocity $v = v(\mathbf{x}, t)$. We assume that the normal velocity of $\mathscr{M}(t)$ is no less than the maximum of local phase velocities of elastic waves,*

$$v(\mathbf{x}, t) \geqslant c^{(3)}(\mathbf{x}; \boldsymbol{n}), (\mathbf{x}, t) \in \mathscr{M}, \tag{2.107}$$

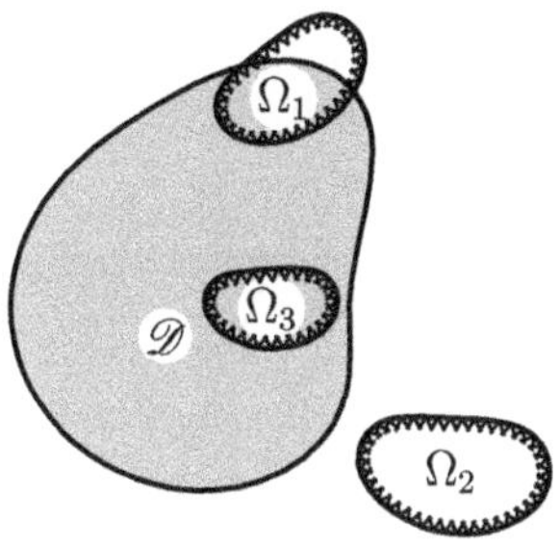


**Figure 2.2**: Domain with a good boundary

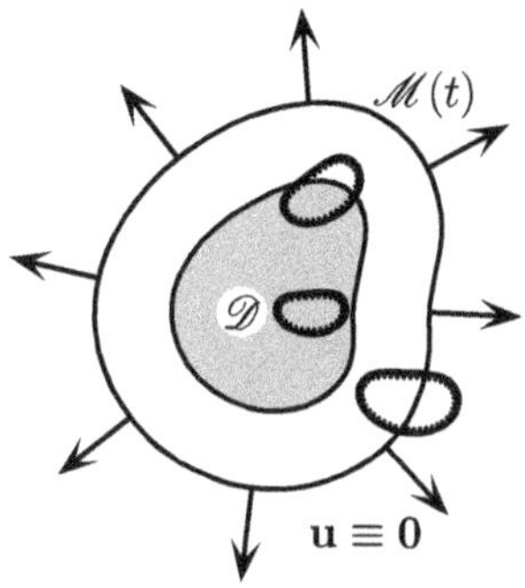


**Figure 2.3**: Domain $\mathscr{D}$ and surface $\mathscr{M}(t)$

*and ahead of $\mathscr{M}(t)$ all the right-hand sides in* (2.106) *vanish. We also assume the uniform boundedness of the local phase velocities:*

$$\max_{\mathbf{x}\in\mathbb{R}^3,\, \mathbf{s}^2=1} c^{(3)}(\mathbf{x}; \mathbf{s}) < +\infty. \tag{2.108}$$

*Then there is no perturbation (i.e., $\mathbf{u} \equiv 0$) in the medium ahead of $\mathscr{M}(t)$. In other words, if outside $\mathscr{D}$ there are no sources of oscillations and the initial data equal zero, then $\mathbf{u} \equiv 0$ ahead of $\mathscr{M}(t)$.*

Thus, initial data and sources defined inside $\mathscr{M}(t)$ do not influence the wavefield outside $\mathscr{M}(t)$. Their *domain of influence* lies inside an expanding domain, where the normal velocity of the expansion of the domain equals the maximum of the local phase velocities.

The proof of this statement is based on energy considerations, which will be presented with some details omitted.

### 2.5.2 Energy lemma

#### Energy in a collapsing domain without sources does not grow

**Energy lemma.** *Let $\mathscr{C}(t)$ be a collapsing surface that bounds a domain* $\operatorname{Int}\mathscr{C}(t)$ *without sources inside, that is,* $\mathbf{F}(\mathbf{x})|_{\mathbf{x}\in\operatorname{Int}\mathscr{C}(t)} \equiv 0$, $\mathbf{f}(\mathbf{x})|_{\mathbf{x}\in\operatorname{Int}\mathscr{C}(t)} \equiv 0$.

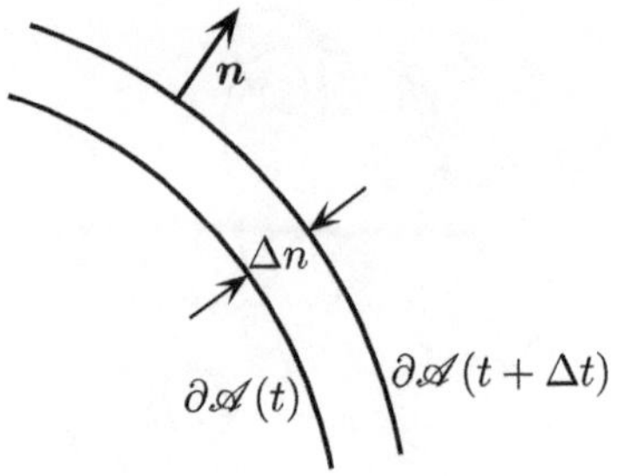


**Figure 2.4**: To the derivation of (2.111)

*If the normal velocity of* $\mathscr{C}(t)$ *at each point* $\mathbf{x}$ *is no less than* $c^{(3)}(\mathbf{x};\boldsymbol{n})$, *then the (total) field energy in* $\operatorname{Int}\mathscr{C}(t)$, $E := \int_{\operatorname{Int}\mathscr{C}(t)\setminus\Omega}\mathcal{E}d\mathbf{x}$, *does not grow:*

$$\frac{dE}{dt} = \frac{d}{dt}\int_{\operatorname{Int}\mathscr{C}(t)\setminus\Omega}\mathcal{E}d\mathbf{x} \leqslant 0. \tag{2.109}$$

Let us first understand on the basis of heuristic considerations that it would be a surprise if this statement is wrong. Indeed, no energy inside the domain $\operatorname{Int}\mathscr{C}(t)$ is produced, but it can only go away through the surface of the domain. At the same time, energy cannot come from the outside because it is carried by waves the velocity of which is no greater than $c^{(3)}(\mathbf{x};\boldsymbol{n})$. Thus, waves cannot overtake $\operatorname{Int}\mathscr{C}(t)$ and add energy in $\operatorname{Int}\mathscr{C}(t)$.

Having assured the plausibility of the lemma, we proceed to its proof.

### Differentiation of an integral with respect to a time-dependent domain

We need a general formula for differentiating an integral with respect to a time-dependent domain. Let a function $y = y(\mathbf{x},t)$ be continuously differentiable in a domain $\mathscr{A}(t)$, then

$$\frac{d}{dt}\int_{\mathscr{A}(t)} y(\mathbf{x},t)d\mathbf{x} = \int_{\mathscr{A}(t)}\frac{\partial y(\mathbf{x},t)}{\partial t}d\mathbf{x} \pm \int_{\partial\mathscr{A}(t)} v(\mathbf{x},t)y(\mathbf{x},t)dS, \tag{2.110}$$

where $v(\mathbf{x},t) \geqslant 0$ is the normal velocity of the surface $\partial\mathscr{A}(t)$ of the domain $\mathscr{A}(t)$ (see Fig. 2.4), and $dS$ is a surface element. The latter integral is to be taken with the plus sign if the velocity is directed along the external normal, and with the minus sign if the velocity is directed along the internal normal.

We prove the above formula, omitting details. Let us consider an increment of the integral on the left-hand side of (2.110) for a time interval $\Delta t$, assuming first that the function $y$ does not depend on $\mathbf{x}$:

$$\int_{\mathscr{A}(t+\Delta t)} y(\mathbf{x})d\mathbf{x} - \int_{\mathscr{A}(t)} y(\mathbf{x})d\mathbf{x} = \int_{\mathscr{A}_{\Delta t}} y(\mathbf{x})dS, \tag{2.111}$$

where $\mathscr{A}_{\Delta t}$ is a layer between $\partial\mathscr{A}(t+\Delta t)$ and $\partial\mathscr{A}(t)$ (see Fig. 2.4) and the domain $\mathscr{A}(t)$ is shown expanding.

Now let $\Delta t$ be small; then a layer volume element is defined by $\Delta n dS = v\Delta t dS$, where $\Delta n$ is the distance between $\mathscr{A}(t)$ and $\mathscr{A}(t+\Delta t)$. Hence, if $y$ does not depend on $t$,

$$\frac{d}{dt}\int_{\mathscr{A}(t)} y(\mathbf{x})d\mathbf{x} = \lim_{\Delta t\to 0}\left(\frac{1}{\Delta t}\int_{\partial\mathscr{A}(t)} y\Delta n dS\right) = \int_{\partial\mathscr{A}(t)} yvdS.$$

Admitting now the dependence of $y$ on $t$, we immediately obtain (2.110).

**Proof of the lemma**

Taking in (2.109) the density of energy in $\operatorname{Int}\mathscr{C}(t)\setminus\Omega$ as $y(\mathbf{x},t)$, we obtain (see Fig. 2.3)

$$\frac{dE}{dt} = \frac{d}{dt}\int\limits_{\operatorname{Int}\mathscr{C}(t)\setminus\Omega} \mathcal{E}(\mathbf{x},t)d\mathbf{x} = \int\limits_{\operatorname{Int}\mathscr{C}(t)\setminus\Omega} \frac{\partial\mathcal{E}(\mathbf{x},t)}{\partial t}d\mathbf{x} - \int\limits_{\mathscr{C}(t)\setminus\partial\Omega} \mathcal{E}vdS. \tag{2.112}$$

The domain $\Omega$ was introduced in the beginning of Section 2.5.1. Thanks to the balance energy equation (1.68) and with (2.107) taken into account, we can replace here $\frac{\partial\mathcal{E}(\mathbf{x},t)}{\partial t}$ by $-\operatorname{div}\mathbf{S}$. Applying the Ostrogradsky – Gauss divergence theorem, we get

$$\frac{dE}{dt} = -\int\limits_{\partial\mathscr{C}(t)/\partial\Omega} (\mathbf{S}\cdot\boldsymbol{n} + \mathcal{E}v)\, dS. \tag{2.113}$$

Let us proceed to the proof of the nonnegativity of the integrand. Note that $v \equiv 0$ on the motionless boundary $\Omega = \Omega_1 + \ldots + \Omega_N$, and it is also obvious that

$$\mathbf{S}\cdot\boldsymbol{n}|_{\partial\Omega_m} = -\sigma_{ij}\dot{\mathrm{u}}_i\cos\widehat{nx_j}|_{\partial\Omega_m} = -\mathbf{t}^{\boldsymbol{n}}(\mathbf{u})\cdot\dot{\mathbf{u}}|_{\partial\Omega_m} = 0.$$

On the moving boundary $\mathscr{C}(t)$,

$$\mathcal{E}v + \mathbf{S}\cdot\boldsymbol{n} = \frac{v}{2}\left[\rho\dot{\mathbf{u}}^2 + \sigma_{ij}(\mathbf{u})\varepsilon_{ij}(\mathbf{u})\right] - \sigma_{ij}(\mathbf{u})\dot{\mathrm{u}}_i n_j \tag{2.114a}$$

$$= \frac{1}{2v}\left[\rho v^2\dot{\mathbf{u}}^2 + v^2\sigma_{ij}(\mathbf{u})\varepsilon_{ij}(\mathbf{u}) - 2v\sigma_{ij}(\mathbf{u})\dot{\mathrm{u}}_i n_j\right] \tag{2.114b}$$

$$= \frac{1}{2v}\left[\rho v^2\dot{\mathbf{u}}^2 + c_{ijkp}(v\varepsilon_{ij} - \widetilde{\varepsilon}_{ij})(v\varepsilon_{kp} - \widetilde{\varepsilon}_{kp}) - c_{ijkp}\widetilde{\varepsilon}_{ij}\widetilde{\varepsilon}_{kp}\right], \tag{2.114c}$$

where

$$\widetilde{\varepsilon}_{sq} = \frac{1}{2}\left[\dot{\mathrm{u}}_s n_q + \dot{\mathrm{u}}_q n_s\right], \quad n_p := \cos\widehat{nx_p}.$$

We have used Hooke's law (1.30) and the symmetry of elastic stiffnesses (1.18).

The middle term on the right-hand side of (2.114c) is obviously nonnegative; we prove that the sum of the other two terms is also nonnegative, that is,

$$\rho v^2 \dot{\mathbf{u}}^2 - c_{ijkp}\widetilde{\varepsilon}_{ij}\widetilde{\varepsilon}_{kp} \geqslant 0. \tag{2.115}$$

From the symmetry of elastic stiffnesses it follows that

$$c_{ijkp}\widetilde{\varepsilon}_{ij}\widetilde{\varepsilon}_{kp} = c_{ijkp}n_j n_k \dot{u}_i \dot{u}_p. \tag{2.116}$$

Recall a simple fact from the theory of quadratic forms in a real space. If $\boldsymbol{C} = \|C_{ij}\|$ is a positive definite symmetric matrix, then its eigenvalues are positive, and the maximal eigenvalue $\Lambda$ coincides with the maximum (with respect to $\boldsymbol{\psi}$) of the *Rayleigh quotient*

$$\frac{(\boldsymbol{C}\boldsymbol{\psi}, \boldsymbol{\psi})}{(\boldsymbol{\psi}, \boldsymbol{\psi})} = \frac{C_{ij}\psi_i\psi_j}{\psi_s\psi_s}.$$

Therefore, for all $\boldsymbol{\psi}$

$$(\boldsymbol{C}\boldsymbol{\psi}, \boldsymbol{\psi}) \leqslant \Lambda(\boldsymbol{\psi}, \boldsymbol{\psi}).$$

We apply this statement to the matrix $\boldsymbol{\Gamma} = \boldsymbol{\Gamma}(\mathbf{x}; \boldsymbol{n})$, $\Gamma_{il} = c_{ijkl}n_j n_k$, where $n_s = \cos\widehat{nx_s}$ are the components of the unit external normal to $\mathscr{C}(t)$. Its maximal eigenvalue was denoted by $\rho(c^{(3)}(\mathbf{x}; \boldsymbol{n}))^2$. Using (2.107), for any real vector $\boldsymbol{\psi}$ we have

$$(\boldsymbol{\Gamma}\boldsymbol{\psi}, \boldsymbol{\psi}) = c_{ijkl}n_j n_k \psi_i \psi_l \leqslant \rho(c^{(3)}(\mathbf{x}; \boldsymbol{n}))^2(\boldsymbol{\psi}, \boldsymbol{\psi}) \leqslant \rho v^2(\boldsymbol{\psi}, \boldsymbol{\psi}).$$

Taking $\boldsymbol{\psi} = \dot{\mathbf{u}}$, we arrive at (2.115) and therefore at (2.109). The lemma is proved.

### Proof of the theorem

Let us fix an arbitrary $t_0 > 0$ and take the sphere $\mathscr{C}(t) = \{\mathbf{x} : |\mathbf{x}| = R - Vt\}$ as the collapsing surface mentioned in the lemma, where $V \geqslant \max c^{(3)}(\mathbf{x}, \boldsymbol{n})$, and $\boldsymbol{n}$ is the normal to $\mathscr{C}$ (the maximum is taken over a very large compact and over all directions). The constant $R$ defining the initial radius of the sphere is so large that for $0 \leqslant t \leqslant t_0$ $\mathscr{D}$ lies inside $\mathscr{C}(t)$.

We denote by $\mathscr{B}(t)$ the domain bounded by $\mathscr{C}(t)$ and $\mathscr{M}(t)$ (Fig. 2.5), $\partial\mathscr{B}(t) = \mathscr{C}(t) + \mathscr{M}(t)$. This domain is collapsing, and we can apply the lemma. Therefore $E(t) = \int_{\mathscr{B}(t)} \mathcal{E} d\mathbf{x}$ has a nonpositive derivative.

Let us show that $E(t) \equiv 0$. Indeed, on the one hand,

$$E(t) = E(0) + \int_0^t \frac{dE}{dt}dt = \int_0^t \frac{dE}{dt}dt \leqslant 0, \tag{2.117}$$

because $E|_{t=0} = 0$ by the zero initial conditions. On the other hand, the energy $E(t)$ is an integral of the nonnegative function $\frac{1}{2}\rho\dot{\mathbf{u}}^2 + \mathcal{W}$, whence

$$E(t) \geqslant 0. \tag{2.118}$$

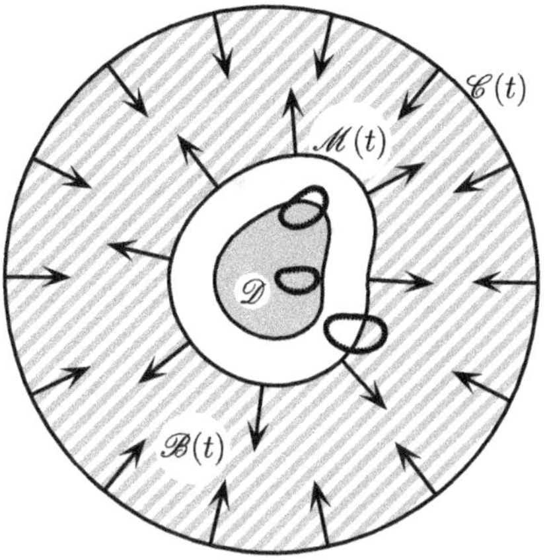


**Figure 2.5**: Collapsing domain $\mathscr{B}(t)$

From (2.117) and (2.118) it can be seen that $E(t) \equiv 0$. Therefore $\dot{\mathbf{u}} \equiv 0$. Further,

$$\mathbf{u}(\mathbf{x}, t) = \mathbf{u}(\mathbf{x}, 0) + \int_0^t \dot{\mathbf{u}}(\mathbf{x}, t')dt', \tag{2.119}$$

and since $R$ can be chosen arbitrarily large, $\mathbf{u}(\mathbf{x}, t) \equiv 0$ everywhere outside the domain bounded by the surface $\mathscr{M}(t)$. The proof of this statement was the purpose of this section.

The above results can easily be generalized to the case of inclusions of slipping contact conditions.

### 2.5.3 Uniqueness theorem

**Uniqueness theorem.** *The solution of the mixed problem* (2.106) *in the class of finite energy displacements is unique.*

Indeed, suppose that there are two solutions, then their difference $\widehat{\mathbf{u}}$ shall satisfy a homogeneous equation with zero initial data (2.106b). From the above it is clear that the energy of a solution $\widehat{\mathbf{u}}$ does not grow in time and equals zero at the initial time instant. Therefore, it equals zero identically. Thus, $\widehat{\mathbf{u}} \equiv 0$.

The above proof requires the smoothness of the solution. The uniqueness theorem can be proved also in the case where the displacement vector is nonsmooth and even is a generalized function. In this case, considerations are much more complicated. Even a strict mathematical statement of the problem in the presence of boundaries is not an easy matter. The uniqueness of the solution of the Cauchy problem in the class of generalized functions for the case of a homogeneous anisotropic space will be considered in Section 3.5.3.

## 2.6 Reflection of plane waves from a free boundary of an isotropic half-space

The problems of reflection of plane waves on plane boundaries of elastic media and at interfaces between different media are important for applications. We consider the most simple case, namely, that of reflection of plane waves from a free boundary of an isotropic half-space. Except for Section 2.6.9, we will deal with the time-harmonic case.

### 2.6.1 Upgoing and downgoing waves

Following the seismological tradition, we describe an elastic half-space by the inequality

$$z := x_3 \geqslant 0.$$

We assume that the boundary $z = 0$ of the half-space is traction-free, that is, the boundary condition

$$\mathbf{t}(\mathbf{u}) := \mathbf{t}^{\mathbf{e}_3}(\mathbf{u})\big|_{z=0} = 0 \tag{2.120}$$

holds. Here, $\mathbf{e}_3$ is the unit coordinate vector of the axis $z$. The axis $z$ is called a *vertical*, while the directions orthogonal to it are called *horizontal* or *lateral.* We use the traditional notation

$$x_1 := x, \quad x_2 := y.$$

The unit coordinate vectors of the axes $x$ and $y$ are denoted by $\mathbf{e}_1$ and $\mathbf{e}_2$.

Let us call a plane harmonic wave *upgoing* or incident on the interface from below, if the projection of its group velocity to the vector $\mathbf{e}_3$ is negative,

$$\boldsymbol{v}^{gr} \cdot \mathbf{e}_3 < 0, \tag{2.121}$$

and *downgoing* if

$$\boldsymbol{v}^{gr} \cdot \mathbf{e}_3 > 0. \tag{2.122}$$

The case where

$$\boldsymbol{v}^{gr} \cdot \mathbf{e}_3 = 0 \tag{2.123}$$

corresponds to the total internal reflection which will be considered in Section 2.6.7.

The reflection problem is stated as follows. An upgoing (harmonic or non-harmonic) *incident plane wave* $\mathbf{u}^{\mathrm{i}}$ is given, and it is required to find the *reflected wave* being the sum of downgoing plane waves such that the total *wavefield*

$$\mathbf{u} = \mathbf{u}^{\mathrm{i}} + \mathbf{u}^{\mathrm{r}} \tag{2.124}$$

satisfies the boundary condition (2.120).

### 2.6.2 Waves of polarizations $SH$ and $P - SV$

We assume that the wave vector of the incident wave lies in the plane $XOZ$, that is,

$$\boldsymbol{k} \cdot \mathbf{e}_2 = 0, \tag{2.125}$$

and therefore

$$\boldsymbol{k} = k_1\mathbf{e}_1 + k_3\mathbf{e}_3. \tag{2.126}$$

Represent the total wavefield as follows:

$$\mathbf{u} = u\mathbf{e}_1 + v\mathbf{e}_2 + w\mathbf{e}_3. \tag{2.127}$$

Further we will see that if the incident wave has the form

$$\mathbf{u}^{\mathrm{i}} = v^{\mathrm{i}}(x, z)\mathbf{e}_2, \quad u^{\mathrm{i}} \equiv w^{\mathrm{i}} \equiv 0, \tag{2.128}$$

then the reflected (and therefore the total wavefield) has the same form. If the incident wave is

$$\mathbf{u}^{\mathrm{i}} = u^{\mathrm{i}}(x, z)\mathbf{e}_1 + w^{\mathrm{i}}(x, z)\mathbf{e}_3, \quad v^{\mathrm{i}} \equiv 0, \tag{2.129}$$

the same is true for the reflected and total wavefield as well.

In mechanics, the case described by (2.129) is called the case of *plane strain*, that is, strain in the plane $XOZ$, whereas the case described by (2.128) is called *antiplane strain.* According to the terminology accepted in geophysics, the solutions satisfying (2.126) and (2.128) are called waves of *polarization* $SH$, that is, waves $S$ with horizontal polarization. The reason is that only waves $S$ can satisfy (2.126) and (2.128) simultaneously. The alternative case described by (2.126) and (2.129) is called the case of *polarization* $P - SV$, that assumes the possibility of the presence of waves $P$ and waves $S$ having vertical displacement components.

### 2.6.3 The case of polarization $SH$

Assuming that (2.128) holds for the total field, that is,

$$\mathbf{u} = v(x, y)\mathbf{e}_2, \tag{2.130}$$

we easily derive that $\operatorname{div} \mathbf{u} = 0$. The elastodynamics equations reduce to the scalar one $\mu(\partial_1^2 + \partial_3^2)v - \rho\ddot{v} = 0$, $\partial_{1,3} := \frac{\partial}{\partial x_{1,3}}$, or

$$\frac{\partial^2 v}{\partial x_1^2} + \frac{\partial^2 v}{\partial x_3^2} - \frac{1}{b^2}\frac{\partial^2 v}{\partial t^2} = 0. \tag{2.131}$$

This is the two-dimensional wave equation with velocity $b$. Traction on the free boundary (2.120) takes the form

$$\mathbf{t}^{\mathbf{e}_3}(\mathbf{u}) = \mu\partial_3 v\big|_{z=0}\, \mathbf{e}_2 = 0. \tag{2.132}$$

In accordance with (2.126) and (2.128), the incident wave can be defined as follows:

$$\mathbf{u}^{\mathrm{i}} = \mathrm{const}\, \mathbf{e}_2\, e^{i(k_1 x + k_3 z - \omega t)}, \qquad (2.133)$$

where, in consequence of (2.131), the dispersion equation (2.44) has the form

$$(k_1^{\mathrm{i}})^2 + (k_3^{\mathrm{i}})^2 = \frac{\omega^2}{b^2}. \qquad (2.134)$$

Denote the horizontal component of the wave vector by

$$k_1^{\mathrm{i}} =: \omega\xi, \qquad (2.135)$$

then its vertical component has the form

$$k_3^{\mathrm{i}} = k_3^{\mathrm{i}}(\xi) = \pm\omega q(\xi), \qquad (2.136)$$

where

$$q(\xi) := \sqrt{\frac{1}{b^2} - \xi^2}. \qquad (2.137)$$

The parameter $\xi$ characterizing the direction of propagation of the incident wave is called its *horizontal slowness*. Since the wave vector is assumed to be real, $\xi$ lies within the interval

$$-\frac{1}{b} < \xi < \frac{1}{b}, \qquad (2.138)$$

and the minus sign shall be taken in (2.136) because of (2.121). Thus, we can assume (const = 1 is chosen in (2.133)) that

$$\mathbf{u}^{\mathrm{i}} = \mathbf{e}_2 e^{i\omega(\xi x - q(\xi) z - t)}. \qquad (2.139)$$

This wave has the unit amplitude, and its group velocity is directed to the traction-free boundary.

Similarly, any downgoing wave $SH$ can be represented in the form

$$\mathbf{u}^{\mathrm{r}} = R\mathbf{e}_2 e^{i\omega(\xi' x + q(\xi') z - t)}, \qquad (2.140)$$

$q(\xi') := \sqrt{\frac{1}{b^2} - (\xi')^2}$, where $\xi'$ is its horizontal slowness,

$$-\frac{1}{b} < \xi' < \frac{1}{b},$$

and $R$ is an arbitrary number.

Seeking the total wavefield in the form (2.124), where $\mathbf{u}^{\mathrm{i}}$ and $\mathbf{u}^{\mathrm{r}}$ are given by (2.139) and (2.140), we aim at finding $\xi'$ and $R$ from the boundary condition (2.120). The number $R = R(\xi)$ determined in this way is called the *reflection coefficient (of the wave SH from a free boundary)*. The condition (2.120) in the form (2.132) immediately gives

$$q(\xi) \exp(i\omega\xi x) - Rq(\xi') \exp(i\omega\xi' x) = 0,$$

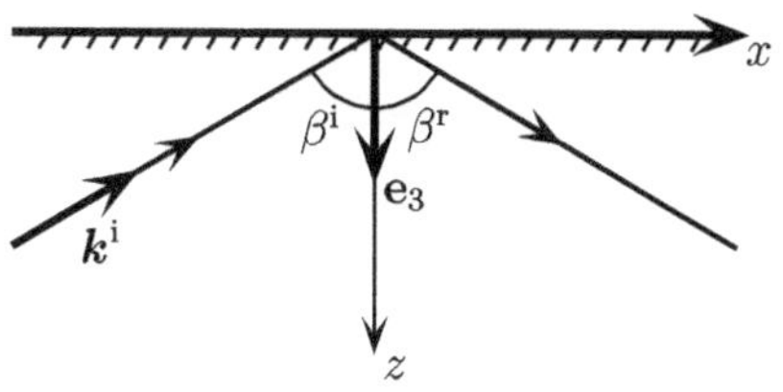


**Figure 2.6**: Reflection of a wave $SH$

or

$$\exp\left(i\omega(\xi - \xi')x\right)q(\xi) - Rq(\xi') = 0.$$

The latter condition shall hold identically with respect to $x$ for all $-\infty < x < +\infty$, and this is possible only if horizontal slownesses are equal:

$$\xi' = \xi. \tag{2.141}$$

We thus find the reflection coefficient:

$$R = R(\xi) = 1. \tag{2.142}$$

The relation (2.141) is known as the *conservation of horizontal slowness* under reflection. It has a simple geometric interpretation. Let us introduce angles $\beta^{\mathrm{i}}$ and $\beta^{\mathrm{r}}$, which the wave vectors of the incident and reflected waves form with the internal normal $\mathbf{e}_3$ to the boundary (see Fig. 2.6), where, for the sake of definiteness, it is assumed that $\xi > 0$. Obviously,

$$\xi = \frac{\sin\beta^{\mathrm{i}}}{b}, \quad \xi' = \frac{\sin\beta^{\mathrm{r}}}{b},$$

and therefore

$$\sin\beta^{\mathrm{r}} = \sin\beta^{\mathrm{i}}. \tag{2.143}$$

We arrive at a law known from the time of antiquity (to say the truth, in optics): the *reflection angle* $\beta^{\mathrm{r}}$ equals the *incidence angle* $\beta^{\mathrm{i}}$.

### 2.6.4 The case of polarization $P - SV$

In the case (2.129), one can easily derive that

$$\begin{aligned} \mathbf{t}(\mathbf{u})|_{z=0} &= \{\sigma_{13}\mathbf{e}_1 + \sigma_{33}\mathbf{e}_3\}|_{z=0} \\ &= \left\{\mu\left(\frac{\partial u}{\partial z} + \frac{\partial w}{\partial x}\right)\mathbf{e}_1 + \left(\lambda\frac{\partial u}{\partial x} + \nu\frac{\partial w}{\partial z}\right)\mathbf{e}_3\right\}\Bigg|_{z=0} = 0, \end{aligned} \tag{2.144}$$

where $\nu := \lambda + 2\mu$.

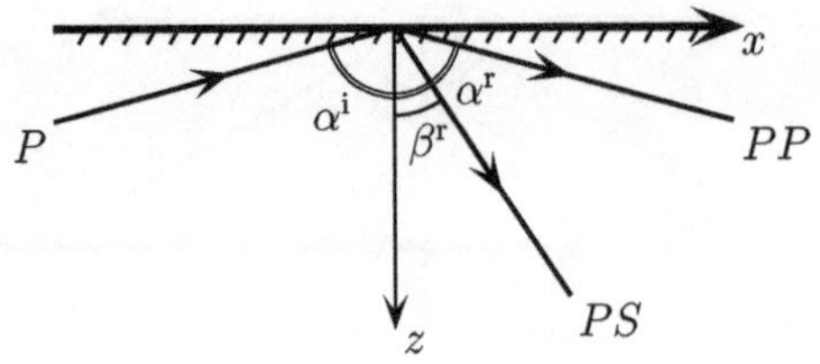


**Figure 2.7**: Reflection of an incident wave $P$

**Incident wave $P$**

The upgoing wave $P$ (see Fig. 2.7) can be represented in the form

$$\mathbf{u}^{\mathrm{i}} = \mathrm{const}\, \boldsymbol{k}^{\mathrm{i}} e^{i(k_1^{\mathrm{i}} x + k_3^{\mathrm{i}} z - \omega t)}, \quad k_3^{\mathrm{i}} < 0. \tag{2.145}$$

We parameterize the components of its wave vector by the horizontal slowness $\xi$ (see (2.135)). As follows from the dispersion equation (2.41) for the wave $P$, in the case (2.29) we have

$$k_3^{\mathrm{i}} = k_3^{\mathrm{i}}(\xi) = -\omega p(\xi), \tag{2.146}$$

where

$$p(\xi) := \sqrt{\frac{1}{a^2} - \xi^2}. \tag{2.147}$$

We assume that the inequality

$$-\frac{1}{a} < \xi < \frac{1}{a} \tag{2.148}$$

holds, which guarantees that $p(\xi)$ is real. The square root in the expression (2.147) is assumed to be positive. For simplicity of formulas, we put const $=$ $1/\omega$ in (2.145) and for the incident wave $P$ obtain

$$\mathbf{u}^{ai} = (\xi \mathbf{e}_1 - p(\xi)\mathbf{e}_3) e^{i\omega(\xi x - p(\xi) z - t)}. \tag{2.149}$$

Let us note that the modulus of the vector $\boldsymbol{h}^a = \xi \mathbf{e}_1 - p(\xi)\mathbf{e}_3$ equals $1/a$, not 1, dissimilarly to waves $SH$. We seek the total field in the form (2.124), and the reflected field as the sum of downgoing waves $P$ and $S$:

$$\mathbf{u}^r = \mathbf{u}^a + \mathbf{u}^b, \tag{2.150}$$

where

$$\mathbf{u}^a = A^a(\xi' \mathbf{e}_1 + p(\xi')\mathbf{e}_3) e^{i\omega(\xi' x + p(\xi') z - t)}, \tag{2.151}$$

$$\mathbf{u}^b = B^a(-q(\xi'')\mathbf{e}_1 + \xi'' \mathbf{e}_3) e^{i\omega(\xi'' x + q(\xi'') z - t)}, \tag{2.152}$$

their horizontal slownesses $\xi'$ and $\xi''$ lie within the intervals

$$-\frac{1}{a} < \xi' < \frac{1}{a}, \quad -\frac{1}{b} < \xi'' < \frac{1}{b}, \tag{2.153}$$

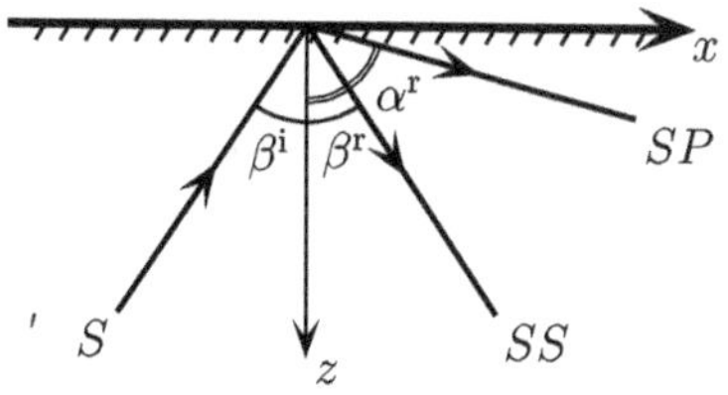


**Figure 2.8**: Reflection of an incident wave $SV$

and $A^a$ and $B^a$ are constants to be found. The upper mark $^a$ is used for the case of incidence of a wave $P$. According to the terminology accepted in seismology, the reflected waves of the same type as the incident wave are called *monotype waves*, and those of another type *converted waves*. The converted wave $S$ is called the *wave PS*, and the reflected wave $P$ the *wave PP*. The quantity $A^a$ is called the *(monotype) reflection coefficient*, and $B^a$ is called the *conversion coefficient* for the incident wave $P$. Note that the moduli of the vectors $(\xi'\mathbf{e}_1+p(\xi')\mathbf{e}_3)$ and $(-q(\xi'')\mathbf{e}_1+\xi''\mathbf{e}_3)$ equal $1/a$ and $1/b$, respectively.

Similarly to the case of waves $SH$, the boundary condition (2.120) holding identically with respect to $x$ implies that horizontal slownesses of all three waves are equal:

$$\xi' = \xi'' = \xi. \tag{2.154}$$

Substituting (2.124) and (2.150)–(2.152) into the boundary conditions (2.120), we get

$$\begin{gathered} 2p\xi A^a - \left(\frac{1}{b^2} - 2\xi^2\right) B^a = 2p\xi, \\ \left(\frac{1}{b^2} - 2\xi^2\right) A^a + 2q\xi B^a = -\left(\frac{1}{b^2} - 2\xi^2\right), \end{gathered} \tag{2.155}$$

from which

$$A^a = \frac{\widetilde{\mathcal{R}}}{\mathcal{R}}, \quad B^a = -\frac{4p\xi\left(\frac{1}{b^2} - 2\xi^2\right)}{\mathcal{R}}, \tag{2.156}$$

where

$$\mathcal{R} := \mathcal{R}(\xi) = \left(\frac{1}{b^2} - 2\xi^2\right)^2 + 4pq\xi^2 \tag{2.157}$$

is the so-called *Rayleigh denominator*, and

$$\widetilde{\mathcal{R}} = \widetilde{\mathcal{R}}(\xi) = 4pq\xi^2 - \left(\frac{1}{b^2} - 2\xi^2\right)^2. \tag{2.158}$$

Since $\mathcal{R}$ is the principal determinant of the system (2.155), and, obviously, $\mathcal{R}(\xi) > 0$ in the interval (2.148), the solution exists and is unique. $A^a$ and $B^a$ are both real. In Figures 2.7 and 2.8, for the sake of definiteness, it is assumed that $\xi > 0$.

**Incident wave $SV$**

Now consider an incident wave $S$ of the form

$$\mathbf{u}^{bi} = [q(\xi)\mathbf{e}_1 + \xi\mathbf{e}_3]e^{i\omega(\xi x - q(\xi)z - t)}, \tag{2.159}$$

where $\xi$ lies in the interval (2.138). We seek the reflected wavefield as the sum of the reflected wave $SV$ and the converted wave $P$ with the same horizontal slowness

$$\mathbf{u}^{\mathrm{r}} = B^b[-q(\xi)\mathbf{e}_1 + \xi\mathbf{e}_3]e^{i\omega(\xi x + q(\xi)z - t)} + A^b[\xi\mathbf{e}_1 + p(\xi)\mathbf{e}_3]e^{i\omega(\xi x + p(\xi)z - t)}, \tag{2.160}$$

where $B^b = B^b(\xi)$ and $A^b = A^b(\xi)$ are called the *reflection* and *conversion coefficients* of the wave $SV$. The converted wave $P$ is called the *wave* $SP$, and the reflected monotype one is called the *wave* $SS$. From the boundary condition (2.120), we obtain a system for their determination with the same matrix as in (2.155) but with a different right-hand side,

$$\begin{aligned} 2p\xi A^b - \left(\frac{1}{b^2} - 2\xi^2\right) B^b &= \frac{1}{b^2} - 2\xi^2, \\ \left(\frac{1}{b^2} - 2\xi^2\right) A^b + 2q\xi B^a &= 2q\xi. \end{aligned} \tag{2.161}$$

Solving the system, we derive that

$$A^b = \frac{4q\xi\left(\frac{1}{b^2} - 2\xi^2\right)}{\mathcal{R}}, \quad B^b = \frac{\widetilde{\mathcal{R}}}{\mathcal{R}}. \tag{2.162}$$

Both expressions in (2.162) are real if the horizontal slowness lies within the interval

$$-\frac{1}{a} \leqslant \xi \leqslant \frac{1}{a}, \tag{2.163}$$

and are complex if

$$\frac{1}{a} < |\xi| < \frac{1}{b}. \tag{2.164}$$

It can easily be seen that the function $\mathcal{R}$ vanishes neither in the case of (2.163) where it is a sum of nonnegative terms, nor in the case of (2.164) where its imaginary part is positive. The expressions (2.162) in both cases give the unique solution.

## 2.6.5 Snell's law

The fact that horizontal slownesses of all waves participating in a reflection process are equal to each other has a simple geometric interpretation. Let $\alpha^{\mathrm{i}}$, $\alpha^{\mathrm{r}}$, and $\beta^{\mathrm{r}}$ be the incidence, reflection and conversion angles for the case of incidence of a wave $P$ defined as follows:

$$\sin\alpha^{\mathrm{i}} = a\xi, \quad \sin\alpha^{\mathrm{r}} = a\xi, \quad \sin\beta^{\mathrm{r}} = b\xi, \tag{2.165}$$

(see Fig. 2.7), where $\xi > 0$. These are the angles between the wave vectors and the normal to the surface. The relation (2.154) is equivalent to the *Snell law*

$$\frac{\sin\alpha^{\mathrm{i}}}{a} = \frac{\sin\alpha^{\mathrm{r}}}{a} = \frac{\sin\beta^{\mathrm{r}}}{b}. \tag{2.166}$$

Similarly, in the case of incidence of a wave $SV$, the respective incidence, reflection and conversion angles $\beta^{\mathrm{i}}$, $\beta^{\mathrm{r}}$, and $\alpha^{\mathrm{r}}$ (see Fig. 2.8) satisfy the condition

$$\frac{\sin\beta^{\mathrm{i}}}{b} = \frac{\sin\beta^{\mathrm{r}}}{b} = \frac{\sin\alpha^{\mathrm{r}}}{a}. \tag{2.167}$$

### 2.6.6 Total internal reflection

Let us define the *critical angle* $\beta_*$ by the relation

$$\sin\beta_* := \frac{b}{a}. \tag{2.168}$$

It is important to note that if the incidence angle $\beta^{\mathrm{i}}$ of a wave $SV$ is greater than the critical one, i.e.,

$$\sin\beta^{\mathrm{i}} > \sin\beta_*, \tag{2.169}$$

then the conversion angle $\alpha^r$ defined by the relation

$$\sin\alpha^{\mathrm{r}} = \frac{a}{b}\sin\beta^{\mathrm{i}}$$

or

$$\cos\alpha^{\mathrm{r}} = \sqrt{1-\sin^2\alpha^{\mathrm{r}}} = \sqrt{1-\frac{a^2}{b^2}\sin^2\beta^{\mathrm{i}}}, \tag{2.170}$$

becomes imaginary. This case is called the *total internal reflection*, and the corresponding incidence angles $\beta^{\mathrm{i}}$ are called *supercritical*.

In the case of the total internal reflection, where a wave $SV$ falls on the boundary under an angle (2.169), which means that its horizontal slowness lies in the interval (2.164), formally, the wave vector of the converted wave $P$ is complex. According to the generally accepted (though somewhat strange) terminology, such plane waves, that is, the solutions of the form (2.4) with complex $\boldsymbol{k}$ are called *inhomogeneous plane waves* (whereas the solutions with real $\boldsymbol{k}$ are called *homogeneous*).

To ensure the decay of a wave $P$ with the distance from the boundary, we choose a branch of the square root in (2.147) as follows:

$$p(\xi) = i\sqrt{\xi^2-\frac{1}{a^2}} = i|p(\xi)| \quad \text{for} \quad \xi > \frac{1}{a} \quad \text{or} \quad \xi < -\frac{1}{a}. \tag{2.171}$$

Henceforth we accept this choice of branch of the square root when considering the inhomogeneous waves $P$.

Note that an inhomogeneous wave $P$, in contrast to a homogeneous one, has elliptic polarization.

### 2.6.7 Energy flow in an inhomogeneous wave $P$

Calculate the averaged density of the energy flow in an inhomogeneous wave $P$. Let its wave vector $\boldsymbol{k}$ in (2.4) have the form

$$\boldsymbol{k} = k_1\mathbf{e}_1 + ik_3'\mathbf{e}_3 \tag{2.172}$$

where $k_1 = \omega\xi$ and $k_3' = \omega\sqrt{\xi^2 - 1/a^2} > 0$ are real. Using (1.91), we derive that (see (2.40))

$$\begin{gathered}\overline{S_1} = -\frac{\omega}{2}\left[(\lambda+2\mu)k_1^2 + (2\mu-\lambda)k_3'^2\right]|\phi^2|^2 e^{-2k_3'z}k_1,\\ \overline{S_2} = 0, \quad \overline{S_3} = 0.\end{gathered} \tag{2.173}$$

Therefore, the energy flow vector in the inhomogeneous plane wave (2.4), (2.172) is directed along the boundary. Such a wave, in contrast to the homogeneous waves, "does not draw energy" from the boundary.

Now we extend the definition of downgoing waves to inhomogeneous waves. In the case of (2.123), a plane wave is said to be *downgoing* if

$$|\boldsymbol{u}| \to 0 \quad \text{as} \quad z \to \infty. \tag{2.174}$$

### 2.6.8 ⋆ Energy flow under reflection from a boundary and the unitarity of the reflection matrix

Here, we consider a relation between the reflection coefficients $A^a$, $A^b$, $B^a$, and $B^b$. This relation possesses great generality; it is a consequence of energy conservation under reflection from a traction-free boundary.

#### Statement of the problem

Consider the incidence on a traction-free boundary of an arbitrary linear combination of waves $P$ and $SV$ (see (2.149) and (2.159))

$$\mathbf{u}^{\mathrm{i}} = A^{\mathrm{i}}\mathbf{u}^{a\mathrm{i}} + B^{\mathrm{i}}\mathbf{u}^{b\mathrm{i}}. \tag{2.175}$$

The reflected wavefield is given by the expression

$$\mathbf{u}^{\mathrm{r}} = A^{\mathrm{r}}\mathbf{u}^{a\mathrm{r}} + B^{\mathrm{r}}\mathbf{u}^{b\mathrm{r}}, \tag{2.176}$$

where

$$\mathbf{u}^{a\mathrm{r}} = (\xi\mathbf{e}_1 + p\mathbf{e}_3)e^{i\omega(\xi x + pz - t)}, \quad \mathbf{u}^{b\mathrm{r}} = (-q\mathbf{e}_1 + \xi\mathbf{e}_3)e^{i\omega(\xi x + qz - t)} \tag{2.177}$$

with $p = p(\xi)$ and $q = q(\xi)$. Here, $A^{\mathrm{r}}$ and $B^{\mathrm{r}}$ are constants (dependent on the horizontal slowness $\xi$), which can be expressed via $A^{\mathrm{i}}$ and $B^{\mathrm{i}}$ and the reflection coefficients.

Consider a linear transformation that maps the coefficients $A^{\mathrm{i}}$ and $B^{\mathrm{i}}$ characterizing the incoming wave to the coefficients $A^{\mathrm{r}}$ and $B^{\mathrm{r}}$ defining the outgoing wave:

$$\boldsymbol{V}^{\mathrm{r}} = \boldsymbol{C}\,\boldsymbol{V}^{\mathrm{i}}, \qquad \boldsymbol{V}^{\mathrm{i}} := \begin{pmatrix} A^{\mathrm{i}} \\ B^{\mathrm{i}} \end{pmatrix}, \qquad \boldsymbol{V}^{\mathrm{r}} := \begin{pmatrix} A^{\mathrm{r}} \\ B^{\mathrm{r}} \end{pmatrix}. \tag{2.178}$$

As follows from the explicit formulas (see (2.156) and (2.162)), the entries of the *reflection matrix* $\boldsymbol{C}$ are:

$$C_{11} = A^a = \frac{\widetilde{\mathcal{R}}}{\mathcal{R}}, \quad C_{12} = A^b = \frac{4p\xi\left(\frac{1}{b^2} - 2\xi^2\right)}{\mathcal{R}},$$

$$C_{21} = B^a = -\frac{4q\xi\left(\frac{1}{b^2} - 2\xi^2\right)}{\mathcal{R}}, \quad C_{22} = B^b = A^a = C_{11}.$$

One can observe that the matrix $\boldsymbol{C}$ has some specific properties, namely, its diagonal entries are equal, and

$$\det(\boldsymbol{C}) = 1. \tag{2.179}$$

These specific relations are not accidental (as a rule, accidents are rare in mathematical physics). The physical cause is that the energy flow, averaged over a period, of the waves coming to the boundary coincides (with the opposite sign) with the value of the flow of energy of the outgoing waves. This "makes" the coefficients satisfy a relation which will be derived below.

### Averaging the energy flow of the total wavefield

Now we assume that $|\xi| < \frac{1}{b}$, i.e., no total internal reflection occurs. In the case of waves $P - SV$, from the boundary condition (2.120) it follows that $\sigma_{3j}|_{z=0} = 0$, $j = 1, 2, 3$, whence (see (1.91))

$$\overline{S_3}|_{z=0} = -\frac{\omega}{2}\,\mathrm{Im}\left[\sigma_{3j}\mathrm{u}_j^*\right]\Big|_{z=0} = 0. \tag{2.180}$$

We apply the energy conservation law in the differential form $\dot{\mathcal{E}} + \operatorname{div}\boldsymbol{S} = 0$ to the total wavefield $\mathbf{u} = \mathbf{u}^{\mathrm{i}} + \mathbf{u}^{\mathrm{r}}$ and average it over the time period $T = \frac{2\pi}{\omega}$. By the periodicity of $\mathcal{E}$ in time, we have $\overline{\dot{\mathcal{E}}} = 0$, whence $\operatorname{div}\overline{\boldsymbol{S}} = 0$. Further, $\mathbf{u}^{\mathrm{i}}$ and $\mathbf{u}^{\mathrm{r}}$ do not depend on $x$ and $t$ separately but only on the difference $t - \xi x$, where $\xi$ is a real number. Therefore, $\overline{\boldsymbol{S}}$ does not depend on $x$, and we obtain

$$\frac{\partial}{\partial z}\overline{\boldsymbol{S}} = 0. \tag{2.181}$$

Integrating (2.181) from 0 to $z$ and taking (2.180) into account, we derive that

$$\overline{S_3} = 0 \tag{2.182}$$

for any depth $z$.

From the formula for averaging over time (1.91), it follows that $\overline{S_3}$ is a quadratic form with respect to the variables $A^{\mathrm{i}}e^{-ipz}$, $B^{\mathrm{i}}e^{-iqz}$, $A^{\mathrm{r}}e^{ipz}$, and $B^{\mathrm{r}}e^{iqz}$, where $A^{\mathrm{i}}$, $B^{\mathrm{i}}$, $A^{\mathrm{r}}$, and $B^{\mathrm{r}}$ are the constants introduced by (2.175) and (2.176). Let us average this form over depth by integrating from 0 to $Z > 0$, dividing the result by $Z$ and taking the limit as $Z$ tends to infinity. We denote the limit by

$$\langle\overline{S_3(\mathbf{u})}\rangle := \lim_{Z\to\infty} \frac{1}{Z}\int_0^Z \overline{S_3(\mathbf{u})}dz = 0. \tag{2.183}$$

The expression $\overline{S_3(\mathbf{u})} = \overline{S_3(\mathbf{u}^{\mathrm{i}} + \mathbf{u}^{\mathrm{r}})}$ includes terms having factors in the form of oscillating exponents $e^{\pm i\omega(q-p)z}$, $e^{\pm 2i\omega pz}$, and $e^{\pm 2i\omega qz}$. Averaging (2.183) over $z$ cancels all these terms. In particular, it cancels the terms containing the product of the components $\mathbf{u}^{\mathrm{i}}$ and $\mathbf{u}^{\mathrm{r}}$, and we can write

$$\langle\overline{S_3(\mathbf{u})}\rangle = \langle\overline{S_3(\mathbf{u}^{\mathrm{i}})}\rangle + \langle\overline{S_3(\mathbf{u}^{\mathrm{r}})}\rangle = 0. \tag{2.184}$$

We temporarily denote the component of the density of energy flow along the axis $z$ by $S_z := S_3 = (\boldsymbol{S}, \mathbf{e}_3)$, and its component in the opposite direction by $S_{-z}$, $S_{-z} := (\boldsymbol{S}, -\mathbf{e}_3)$. The following simple relation

$$S_z = -S_{-z} \tag{2.185}$$

holds, which enables us to write the expression (2.184) in the form

$$\langle\overline{S_z(\mathbf{u}^{\mathrm{i}})}\rangle = \langle\overline{S_{-z}(\mathbf{u}^{\mathrm{r}})}\rangle. \tag{2.186}$$

The relation (2.186) means that the averaged density of the energy flow of the waves coming to the boundary is equal to the averaged density of the energy flow of the outgoing waves.

Calculation (which is not complicated, because only the terms that include $|A^{\mathrm{i}}|^2$, $|B^{\mathrm{i}}|^2$, $|A^{\mathrm{r}}|^2$, and $|B^{\mathrm{r}}|^2$ "survive" after averaging) shows that

$$\begin{aligned}\langle\overline{S_3(\mathbf{u}^{\mathrm{i}})}\rangle &= \frac{1}{2}|A^{\mathrm{i}}|^2(\lambda+2\mu)p(p^2+\xi^2) + \frac{1}{2}|B^{\mathrm{i}}|^2\mu q(q^2+\xi^2)\\ &= \frac{1}{2}\rho p|A^{\mathrm{i}}|^2 + \frac{1}{2}\rho q|B^{\mathrm{i}}|^2,\end{aligned} \tag{2.187}$$

and

$$\langle\overline{S_3(\mathbf{u}^{\mathrm{r}})}\rangle = \frac{1}{2}\rho p|A^{\mathrm{r}}|^2 + \frac{1}{2}\rho q|B^{\mathrm{r}}|^2. \tag{2.188}$$

### Properties of the quadratic form (2.187)

Denote the doubled matrix of the quadratic form (2.187) by $\boldsymbol{D}$:

$$\boldsymbol{D} = \begin{pmatrix}\rho p & 0\\ 0 & \rho q\end{pmatrix}. \tag{2.189}$$

We write the quadratic form (2.187) as $\left(\boldsymbol{D}\boldsymbol{V}^{\mathrm{i}}, \boldsymbol{V}^{\mathrm{i}}\right)$ (see (2.178)). It remains unchanged after the linear transformation defined by the matrix $\boldsymbol{C}$ (2.178) (see (2.188) and (2.186)), i.e., $\left(\boldsymbol{D}\boldsymbol{V}^{\mathrm{i}}, \boldsymbol{V}^{\mathrm{i}}\right) = \left(\boldsymbol{D}\boldsymbol{V}^{\mathrm{r}}, \boldsymbol{V}^{\mathrm{r}}\right)$. Using the reality of the reflection matrix $\boldsymbol{C}$, we obtain

$$\left(\boldsymbol{D}\boldsymbol{V}^{\mathrm{i}}, \boldsymbol{V}^{\mathrm{i}}\right) = \left(\boldsymbol{D}\boldsymbol{V}^{\mathrm{r}}, \boldsymbol{V}^{\mathrm{r}}\right) = \left(\boldsymbol{D}\boldsymbol{C}\boldsymbol{V}^{\mathrm{i}}, \boldsymbol{C}\boldsymbol{V}^{\mathrm{i}}\right) = \left(\boldsymbol{C}^{T}\boldsymbol{D}\boldsymbol{C}\boldsymbol{V}^{\mathrm{i}}, \boldsymbol{V}^{\mathrm{i}}\right),$$

where the symbol $^{T}$ stands for transposition. In view of the arbitrariness of the vector $\boldsymbol{V}^{\mathrm{i}}$, we get the relation

$$\boldsymbol{C}^{T}\boldsymbol{D}\boldsymbol{C} = \boldsymbol{D} \tag{2.190}$$

or

$$\boldsymbol{D}^{-\frac{1}{2}}\boldsymbol{C}^{T}\boldsymbol{D}^{\frac{1}{2}}\boldsymbol{D}^{\frac{1}{2}}\boldsymbol{C}\boldsymbol{D}^{-\frac{1}{2}} = \mathbf{I}, \tag{2.191}$$

where $\mathbf{I}$ is the identity matrix, and

$$\boldsymbol{D}^{\pm\frac{1}{2}} = \begin{pmatrix} (\rho p)^{\pm\frac{1}{2}} & 0 \\ 0 & (\rho q)^{\pm\frac{1}{2}} \end{pmatrix}.$$

The relation (2.190) can be written in the form

$$\left(\boldsymbol{D}^{\frac{1}{2}}\boldsymbol{C}\boldsymbol{D}^{-\frac{1}{2}}\right)^{T}\boldsymbol{D}^{\frac{1}{2}}\boldsymbol{C}\boldsymbol{D}^{-\frac{1}{2}} = \mathbf{I}, \tag{2.192}$$

and the matrix $\boldsymbol{D}^{\frac{1}{2}}\boldsymbol{C}\boldsymbol{D}^{-\frac{1}{2}}$ is thus orthogonal.

Let us return to the quadratic form (2.187), (2.188). If we replace $A^{\mathrm{i}}$, $B^{\mathrm{i}}$, $A^{\mathrm{r}}$, and $B^{\mathrm{r}}$ by $\hat{A}^{\mathrm{i}} = \sqrt{\rho p}A^{\mathrm{i}}$, $\hat{B}^{\mathrm{i}} = \sqrt{\rho q}B^{\mathrm{i}}$, $\hat{A}^{\mathrm{r}} = \sqrt{\rho p}A^{\mathrm{r}}$, and $\hat{B}^{\mathrm{r}} = \sqrt{\rho q}B^{\mathrm{r}}$, then (2.187) and (2.188) transform into $\frac{1}{2}(|A^{\mathrm{i}}|^2 + |B^{\mathrm{i}}|^2)$ and $\frac{1}{2}(|A^{\mathrm{r}}|^2 + |B^{\mathrm{r}}|^2)$, respectively. The matrix of transformation of $\begin{pmatrix} \hat{A}^{\mathrm{i}} \\ \hat{B}^{\mathrm{i}} \end{pmatrix}$ into $\begin{pmatrix} \hat{A}^{\mathrm{r}} \\ \hat{B}^{\mathrm{r}} \end{pmatrix}$ under reflection is the orthogonal matrix $\boldsymbol{D}^{\frac{1}{2}}\boldsymbol{C}\boldsymbol{D}^{-\frac{1}{2}}$. An orthogonal matrix is a real case of a unitary matrix.

### The case of the total internal reflection

In the case of the total internal reflection where (2.164) holds, considerations analogous to the above again lead to the relation (2.186), in which only the averaged flows of the incident and reflected waves $S$ are present on the left-hand and right-hand sides. The flow of the averaged (as defined by the relation (2.183)) wave $P$ vanishes. The relation (2.185) means that the modulus of the (complex) reflection coefficient of the wave $S$ is equal to one. This, of course, agrees with the explicit formulas (2.157) and (2.158).

## 2.6.9 Reflection of non-time-harmonic waves

### Statement of the problem

Consider the reflection of non-time-harmonic waves of the form (2.3) from a traction-free boundary. We restrict the discussion to the case of waves $P-SV$, where we choose a less trivial case of an incident wave $S$. The incident wave is defined by the formulas (2.124) and (2.150), where

$$\mathbf{u}^{\mathrm{i}} = [q\mathbf{e}_1 + \xi\mathbf{e}_3]f(\xi x - qz - t), \quad q = q(\xi), \tag{2.193}$$

and $f$ is a prescribed function of one variable. We seek the reflected wave in the form

$$\mathbf{u}^{\mathrm{r}} = \mathbf{u}^a + \mathbf{u}^b = A^b[\xi\mathbf{e}_1 + p\mathbf{e}_3]f^a(\xi x + pz - t) + B^b[-q\mathbf{e}_1 + \xi\mathbf{e}_3]f^b(\xi x + qz - t), \tag{2.194}$$

where the functions $f^a$ and $f^b$ are to be determined, and $p = p(\xi)$, $q = q(\xi)$. We are interested in a real solution of the problem.

The cases with and without the total internal reflection differ in the intricacy of analysis, as well as in the intricacy of the result.

## The case of (2.163) without the total internal reflection

Let the waveform of an incident wave be an arbitrary real function $f(s)$ of a real variable $s$ on the whole axis. We seek a real reflected wave. Substituting the total wavefield $\mathbf{u}^i + \mathbf{u}^r$ into the boundary condition (2.120), we find out that the problem is solved by taking

$$f^a(s) = f^b(s) = f(s), \quad s \in \mathbb{R}. \tag{2.195}$$

Since the reflection and conversion coefficients are real, the desired real solution is found.

## The case of (2.164) with the total internal reflection

In this case, the formulas (2.193) and (2.194) are not applicable. Now the reflection coefficients are complex, and the question arises as to what should be meant by the waveform of a wave $P$, which proves to be a function of a complex variable ranging over the upper half-plane when $x, z$ and $t$ take real values.

Consider an auxiliary complex problem of the form (2.124) and (2.150)

$$\mathbf{u} = \mathbf{U}^{\mathrm{i}} + \mathbf{U}^{\mathrm{r}}, \quad \mathbf{U}^{\mathrm{r}} = \mathbf{U}^a + \mathbf{U}^b \tag{2.196}$$

where

$$\mathbf{U}^{\mathrm{i}} = (q\mathbf{e}_1 + \xi\mathbf{e}_3)\mathfrak{f}(\xi x - q(\xi)z - t), \tag{2.197}$$

and $\mathfrak{f}$ is a function of a complex variable, which is analytic in the upper half-plane and has "not too bad" limiting values on the real axis (for example, they can define a generalized function).

It can easily be verified that the reflected field can be found in the form (2.196) with

$$\mathbf{U}^a = A^b[\xi\mathbf{e}_1 + p\mathbf{e}_3]\mathfrak{f}(\xi x + pz - t), \quad \mathbf{U}^b = B^b[-q\mathbf{e}_1 + \xi\mathbf{e}_3]\mathfrak{f}(\xi x + qz - t).$$

The function $\mathbf{u} = \mathbf{U}^i + \mathbf{U}^r$ is now defined for $x \in \mathbb{R}$, $0 \leqslant z < \infty$ and $t \in \mathbb{R}$. By simple direct calculation, one can show that the traction-free condition at $z = 0$ is now satisfied. We have presented a complex solution of the reflection problem.

Finding a real solution is now an easy matter. Since the traction-free condition is real and the elastodynamics equations are also real, the real and imaginary parts of the found complex solution separately satisfy both. Thus, $\mathrm{Re}(\mathbf{U}^{\mathrm{i}} + \mathbf{U}^{\mathrm{r}})$ can be taken as a solution of the problem. Then, for the incident wave we obtain the formula (2.193) with $f(s) = \mathrm{Re}\,\mathfrak{f}(s)$, and for the reflected wave we have

$$\mathbf{u}^{\mathrm{r}} = \mathrm{Re}\,\mathbf{U}^{\mathrm{r}} = \mathrm{Re}\,\mathbf{U}^{a} + \mathrm{Re}\,\mathbf{U}^{b}. \tag{2.198}$$

Denote

$$f(s) = \mathrm{Re}\,\mathfrak{f}(s), \quad g(s) = \mathrm{Im}\,\mathfrak{f}(s). \tag{2.199}$$

The functions $f(s)$ and $g(s)$, $s \in \mathbb{R}$, are called harmonic conjugate. The field on the boundary $z = 0$ is their linear combination.

The expression for the reflected wave $S$ can be written in the form

$$\mathbf{u}^{b} = \mathrm{Re}\,\mathbf{U}^{b} = (-q\mathbf{e}_1 + \xi\mathbf{e}_3)\big\{\mathrm{Re}(B^b)f(\xi x + qz - t) - \mathrm{Im}(B^b)g(\xi x + qz - t)\big\}. \tag{2.200}$$

We omit a similar expression for a wave $P$. In this case, the waveforms in the reflected field are linear combinations of the waveform of the incident wave $f$ and its harmonic conjugate function $g$.

If it is not required that $\mathfrak{f}$ decay with the distance from the real axis, then the converted wave does not decay with depth. The issues related to causality are not discussed here.

Considerations of particular functions $f$ in problems with the total internal reflection and some references to the corresponding literature can be found in Brekhovskykh and Godin 1991 [12].

---

## 2.7 Classical plane surface waves in isotropic media

We will consider a vertically layered or *vertically stratified medium*, that is, a medium with elastic stiffnesses and density dependent solely on one variable $z$ (an important particular case is a homogeneous half-space). A surface wave in such a medium is a solution of the elastodynamics equations that satisfies a boundary condition on some surface $z = \text{const}$ and rapidly decays with the distance from the boundary.

In this section, we consider *plane surface waves* having, in respect to the horizontal variables, the same form as the volume plane waves considered above,

$$\mathbf{u}(x, z, t) = e^{i(k_1 x + k_2 y - \omega t)}\boldsymbol{v},$$

where $\boldsymbol{v} = \boldsymbol{v}(z) = u(z, \boldsymbol{k})\mathbf{e}_1 + v(z, \boldsymbol{k})\mathbf{e}_2 + w(z, \boldsymbol{k})\mathbf{e}_3$. The real vector lying in the horizontal plane

$$\boldsymbol{k} = k_1\mathbf{e}_1 + k_2\mathbf{e}_2 \tag{2.201}$$

is called the *wave vector of a surface wave.* The issue whether it is expedient to regard $\boldsymbol{k}$ as a function of $\omega$ or vice versa will be discussed later.

Important classical examples of the Rayleigh and Love waves relate to the case of a half-space described here by the inequality $z \geqslant 0$. We require that a solution decay with depth, i.e.,

$$|\boldsymbol{v}| \to 0 \quad \text{as} \quad z \to \infty. \tag{2.202}$$

For the sake of definiteness, it is convenient to consider the propagation of a surface wave along the axis $x$, assuming that $k_2 = 0$, $k_1 =: k$,

$$\mathbf{u}(x, z, t) = e^{ikx - i\omega t}\boldsymbol{v}, \tag{2.203}$$

where $\boldsymbol{v} = \boldsymbol{v}(z, k) = u(z)\mathbf{e}_x + v(z)\mathbf{e}_y + w(z)\mathbf{e}_z$.

Accordingly, the *wavefront* is a moving line $kx - \omega t = \text{const}$ on the surface $z = 0$.

We assume that the boundary $z = 0$ is traction-free, $\mathbf{t}^{\mathbf{e}_3}\mathbf{u}|_{z=0} = 0$, i.e.,

$$[\sigma_{13}(\mathbf{u})]_{z=0} = [\sigma_{23}(\mathbf{u})]_{z=0} = [\sigma_{33}(\mathbf{u})]_{z=0} = 0. \tag{2.204}$$

First we consider two simple classical solutions for a homogeneous half-space and for a homogeneous layer in a homogeneous half-space.

## 2.7.1 Classical Rayleigh wave

The best-known surface wave is, of course, the Rayleigh wave running along the free boundary of an isotropic homogeneous half-space, which was discovered by Rayleigh in the 1880s (see Lord Rayleigh 1885 [38]). Initially, the interest in this wave was motivated by the desire to understand the nature of movements of the Earth's surface during earthquakes. A half-space is rather a rough model of the Earth's surface. Shallow seismic sources prove to excite intensive Rayleigh waves responsible for most of the damage caused by earthquakes. Later the Rayleigh wave found various technical applications, e.g., in acoustics and radio electronics.

### Rayleigh ansatz

Rayleigh sought a solution in an elastic half-space as a linear combination of two inhomogeneous plane waves $P-SV$ with the same horizontal slownesses $\xi$. Following Rayleigh (see Lord Rayleigh 1885 [38]), we deal with potentials and therefore the representation of plane waves here differs a bit from that in the theory of reflection from the boundary given in Section 2.6,

$$\begin{gathered}\mathbf{u} = \operatorname{grad}\varphi + \operatorname{rot}(\psi\mathbf{e}_2),\\ \varphi = Ae^{i(kx-\omega t)-\omega \mathrm{p}(\xi)z}, \quad \psi = Be^{i(kx-\omega t)-\omega \mathrm{q}(\xi)z},\end{gathered} \tag{2.205}$$

where $\xi = \frac{k}{\omega}$ lies in the interval

$$\frac{1}{b} < \xi < \infty, \tag{2.206}$$

$\mathrm{p} = \mathrm{p}(\xi)$ and $\mathrm{q} = \mathrm{q}(\xi)$ are defined by the relations

$$\mathrm{p} = \sqrt{\xi^2 - 1/a^2} > 0, \quad \mathrm{q} = \sqrt{\xi^2 - 1/b^2} > 0, \tag{2.207}$$

and $A = A(\xi)$ and $B = B(\xi)$ are unknown constants. The inequality (2.206) enables the positiveness of p and q, which guarantees the surface nature of a solution, that is, its decay with depth $z$. Obviously,

$$\mathrm{p} > \mathrm{q}.$$

**Rayleigh dispersion equation**

The traction-free boundary condition (2.120) after simple calculations leads to the system of two equations:

$$\begin{aligned} \sigma_{13}(\mathbf{u})|_{z=0} &= -\omega^2 e^{i\omega(\xi x - t)} \mu[2i\mathrm{p}\xi A + (\mathrm{q}^2 + \xi^2)B] = 0, \\ \sigma_{33}(\mathbf{u})|_{z=0} &= \omega^2 e^{i\omega(\xi x - t)} \{[(\lambda + 2\mu)\mathrm{p}^2 - \lambda\xi^2]A - 2i\mu\mathrm{q}\xi B\} = 0 \end{aligned} \tag{2.208}$$

(the condition $\sigma_{23}(\mathbf{u})|_{z=0} = 0$ is satisfied automatically). The consistency condition for this system with respect to $A$ and $B$ is that its principal determinant equals zero, which immediately yields the *Rayleigh dispersion equation*

$$\mathcal{R}(\xi) = 0. \tag{2.209}$$

The function $\mathcal{R}(\xi) = \left(2\xi^2 - \frac{1}{b^2}\right)^2 - 4\xi^2\mathrm{pq}$ has already appeared in Section 2.6. It can be obtained by replacing in (2.157) $p$ by $ip$, and $q$ by $iq$.

**Existence of the Rayleigh wave**

To ascertain the existence of a real solution of the equation (2.209) in the interval (2.206), we note that at the left end of the interval

$$\mathcal{R}|_{\xi=\frac{1}{b}} = \frac{1}{b^4} > 0,$$

and when approaching its right end

$$\begin{aligned} \mathcal{R}|_{\xi\to\infty} &= 4\xi^4 \left\{ \left(1 - \frac{1}{2b^2\xi^2}\right)^2 - \left(1 - \frac{1}{a^2\xi^2}\right)^{\frac{1}{2}} \left(1 - \frac{1}{b^2\xi^2}\right)^{\frac{1}{2}} \right\} \\ &\approx 4\xi^4 \left\{ 1 - \frac{1}{b^2\xi^2} - \left(1 - \frac{1}{2a^2\xi^2} - \frac{1}{2b^2\xi^2}\right) \right\} = 2\xi^2 \left(\frac{1}{a^2} - \frac{1}{b^2}\right) < 0. \end{aligned}$$

Therefore, the equation (2.209) has at least one solution within the interval (2.206).

**Uniqueness of the Rayleigh wave**

Note that

$$\begin{aligned}\mathcal{R}'_\xi &= 8p\left(2\xi^2 - \frac{1}{b^2}\right) - 8\xi \mathrm{pq} - 4\xi^3\left(\frac{\mathrm{p}}{\mathrm{q}} + \frac{\mathrm{q}}{\mathrm{p}}\right)\\ &= \frac{4\xi}{\mathrm{pq}}\left[2\left(2\xi^2 - \frac{1}{b^2}\right)\mathrm{pq} - 2\mathrm{p}^2\mathrm{q}^2 - \xi^2(\mathrm{p}^2+\mathrm{q}^2)\right]\\ &= \frac{4\xi}{\mathrm{pq}}\left[2\left(\xi^2 - \frac{1}{b^2}\right)\mathrm{pq} - 2\mathrm{p}^2\mathrm{q}^2 - \xi^2(\mathrm{p}^2+\mathrm{q}^2-2\mathrm{pq})\right]\\ &= -\frac{4\xi}{\mathrm{pq}}\left[\xi^2(\mathrm{p}-\mathrm{q})^2 + 2\mathrm{pq}^2(\mathrm{p}-\mathrm{q})\right] < 0.\end{aligned}$$

Thus, the function $\mathcal{R}(\xi)$ is monotonic in the interval (2.206). Therefore the positive root of the Rayleigh equation is unique.

**Some properties of the Rayleigh wave**

So, for any $\lambda$ and $\mu$ the equation (2.209) has only one positive solution

$$\xi =: \frac{1}{c}. \tag{2.210}$$

Here, $c$ is called the velocity of the Rayleigh wave (it coincides with the phase velocity and the group velocity as well, according to the immediately verifiable formula $\omega = ck$). The inequality

$$c < b. \tag{2.211}$$

always holds. In applications, $c$ and $b$ may be close to each other. In seismic exploration, it is often accepted, in rough approximation, that $\lambda = \mu$, i.e., $a = \sqrt{3}b$ (exactly for this case Rayleigh established the presence of a root), then $c \approx 0.92b$.

It can easily be shown that for the boundary condition (1.39) describing the contact with an absolutely rigid body, no counterpart for the surface Rayleigh wave exists.

**Expressions for displacements**

Let us introduce the notation

$$\mathbf{u} = e^{i\omega\xi x - i\omega t}\boldsymbol{V}, \tag{2.212}$$

$V_y \equiv V_2 = 0$. Since the determinant of the system (2.208) vanishes, it is sufficient to use its first equation, which gives the expressions for $V_x \equiv V_1$ and $V_z \equiv V_3$,

$$\begin{aligned}V_x &= -\frac{2}{\xi}\left[\left(2\xi^2 - \frac{1}{b^2}\right)e^{-\omega \mathrm{p}z} - 2\mathrm{pq}e^{-\omega \mathrm{q}z}\right],\\ V_z &= \frac{2i\mathrm{p}}{\xi^2}\left[-\left(2\xi^2 - \frac{1}{b^2}\right)e^{-\omega \mathrm{p}z} + 2\xi^2 e^{-\omega \mathrm{q}z}\right].\end{aligned} \tag{2.213}$$

Note that in this case (and, in general, for any plane surface wave of polarization $P - SV$), in the normalization in which the vertical component is imaginary, the horizontal one proves to be real.

**Averaged density of energy**

In the sequel, we will need an expression for the density of energy $\overline{\mathcal{E}}$, averaged over a period and integrated by $z$, of the Rayleigh wave. As follows from the virial theorem stated in Section 2.9.2, this quantity equals the doubled averaged density of the kinetic energy $\overline{\mathcal{K}}$. Therefore

$$\int_0^\infty \overline{\mathcal{E}(\mathbf{u})}dz = 2\int_0^\infty \overline{\mathcal{K}(\mathbf{u})}dz = 2\omega^2\frac{\rho}{4}\int_0^\infty (V_xV_x^* + V_zV_z^*)dz.$$

By direct, though tiresome, calculations with the use of formulas (1.90) and (2.213), we obtain

$$\int_0^\infty \overline{\mathcal{E}(\boldsymbol{u})}dz = \frac{2\rho c^2}{\omega\sqrt{\frac{1}{c^2}-\frac{1}{b^2}}}\left[\frac{6}{c^2}\left(\frac{1}{b^2}-\frac{1}{a^2}\right)+\frac{4}{a^2b^2}-\frac{6}{b^4}+\frac{c^2}{b^6}\right]\frac{\omega^2}{2}. \tag{2.214}$$

We note that the positiveness of the right-hand side of the latter formula, which is not so obvious from (2.214), follows from the positiveness of the energy of a nonzero wavefield. The group velocity theorem for the classical Rayleigh wave can easily be derived on the basis of the virial theorem given in Section 2.9.2.

### 2.7.2 Classical Love wave

**Boundary-value problem**

Consider a homogeneous layer welded to the underlying homogeneous half-space, that is, a medium with the piecewise constant parameters

$$\lambda(z), \mu(z), \rho(z) = \begin{cases} \lambda_1, \mu_1, \rho_1 & \text{as } 0 < z < h, \\ \lambda_2, \mu_2, \rho_2 & \text{as } h < z < \infty, \end{cases} \tag{2.215}$$

$h > 0$. Following Love's idea (see Love 1911 [40]), we seek a wave $SH$, that is, a solution of the form $\mathbf{u} = e^{i(kx-\omega t)}v(z)\mathbf{e}_2$. The equations in the layer and in the underlying half-space take the form $\mu_j\big(\frac{d^2}{dz^2} - k^2\big)v + \omega^2\rho_j v = 0$, or

$$D^2v + \left(\frac{\omega^2}{b_1^2} - k^2\right)v = 0, \quad 0 < z < h, \tag{2.216}$$

$$D^2v + \left(\frac{\omega^2}{b_2^2} - k^2\right)v = 0, \quad h < z < \infty, \tag{2.217}$$

where $D := \frac{d}{dz}$, and $b_j = \sqrt{\mu_j/\rho_j}$ are the velocities of waves $S$. The boundary conditions on the free surface take the form

$$Dv|_{z=0} = 0. \tag{2.218}$$

On the *interface* we have

$$[v]_{z=h} = 0, \quad [\mu D v]_{z=h} = 0, \tag{2.219}$$

and $v \to 0$ as $z \to \infty$.

**Derivation of the dispersion equation**

The equation (2.217) has a solution decreasing at infinity if

$$k > \frac{\omega}{b_2}. \tag{2.220}$$

If $z > h$, then

$$v(z) = m_2 e^{-Q_2 z}, \quad Q_2 = \sqrt{k^2 - \frac{\omega^2}{b_2^2}} \tag{2.221}$$

with some constant $m_2$.

The general solution of equation (2.216) has the form $v = m_1 \cos(Q_1 z) + \widetilde{m_1} \sin(Q_1 z)$, where

$$Q_1 = \sqrt{\frac{\omega^2}{b_1^2} - k^2}. \tag{2.222}$$

The boundary condition on a free surface (2.218) implies putting $\widetilde{m_1} = 0$. Within the layer $0 < z < h$, we have

$$v(z) = m_1 \cos(Q_1 z), \quad 0 < z < h. \tag{2.223}$$

It remains to satisfy the boundary conditions for a welded interface (2.219). The first condition, obviously, has the form

$$m_1 \cos(Q_1 h) - m_2 e^{-Q_2 h} = 0, \tag{2.224}$$

whereas the second one is

$$m_1 \mu_1 Q_1 \sin(Q_1 h) - m_2 \mu_2 Q_2 e^{-Q_2 h} = 0. \tag{2.225}$$

The consistency condition for this linear homogeneous system with respect to $m_1$ and $m_2$ requires that its determinant vanish, which gives the *Love dispersion equation*

$$L = 0, \tag{2.226}$$

$$L := \operatorname{tg}(Q_1 h) - \frac{\mu_2}{\mu_1} \frac{Q_2}{Q_1} = \operatorname{tg}\left( h \sqrt{\frac{\omega^2}{b_1^2} - k^2} \right) - \frac{\mu_2}{\mu_1} \frac{\sqrt{k^2 - \frac{\omega^2}{b_2^2}}}{\sqrt{\frac{\omega^2}{b_1^2} - k^2}}.$$

This equation binds $k$ and $\omega$.

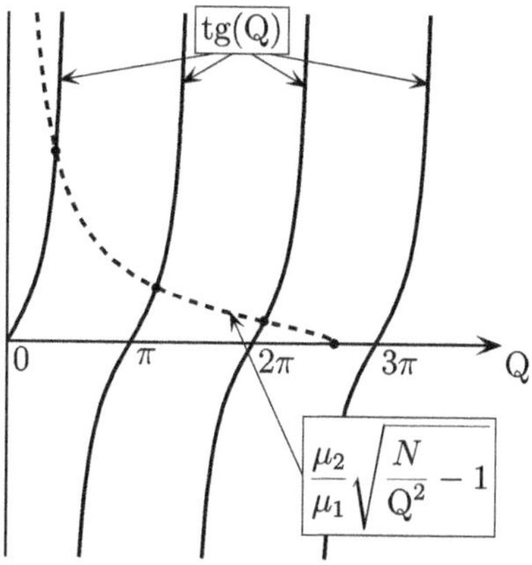


**Figure 2.9**: Graphic solution of equation (2.226)

### Analysis of the Love dispersion equation

In the case where

$$b_1 < b_2 \tag{2.227}$$

(i.e., we consider a *low-velocity layer* welded to a half-space), a wave propagating along this layer, like along a waveguide, always exists. We will show that such a wave may be nonunique.

Fix $k > 0$ and seek roots $\omega$ of the equation (2.226) in the interval

$$b_1 k < \omega < b_2 k. \tag{2.228}$$

In this interval, both square roots $Q_1$ and $Q_2$ are real and strictly positive. We rewrite (2.226) in the form

$$\mathrm{tg}(\mathrm{Q}) = \frac{\mu_2}{\mu_1}\sqrt{\frac{\mathrm{N}}{\mathrm{Q}^2} - 1}, \tag{2.229}$$

$$\mathrm{Q} = Q_1 h = h\sqrt{\frac{\omega^2}{b_1^2} - k^2}, \quad \mathrm{N} = \omega^2 h^2 \left(\frac{1}{b_1^2} - \frac{1}{b_2^2}\right).$$

If condition (2.227) holds, N is positive. The number of solutions of the equation (2.229) (and therefore of the equivalent equation (2.226)) equals the number of zeros of the function tg(Q) in the interval $0 \leqslant \mathrm{Q} < \sqrt{\mathrm{N}}$, i.e., it equals

$$\mathrm{E}\left(\frac{\sqrt{\mathrm{N}}}{\pi}\right) = \mathrm{E}\left(\frac{\omega^2 h}{\pi}\sqrt{\frac{1}{b_1^2} - \frac{1}{b_2^2}}\right).$$

Here, $\mathrm{E}(n)$, $n \geqslant 0$, is the integer part of $n$ (see Fig. 2.9).

Generally speaking, we have obtained several *dispersion curves* $\omega = H_l(k)$. Here, $H_l(k)$ is not a homogeneous function of $k$, which easily follows from the dispersion equation (2.226); thus, any Love wave is dispersive.

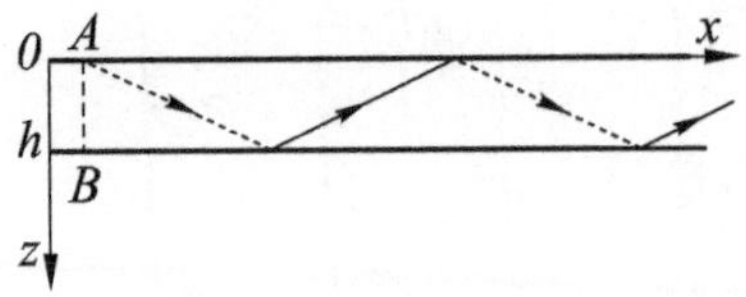


**Figure 2.10**: Rays in a layer

### 2.7.3 ⋆ The total internal reflection and constructive interference

Under the same assumptions, the theory of the classical Love wave can be exposed in a less formal, but more illustrative, manner based on ray considerations.

#### Outline of further calculation

Let a downgoing plane wave $SH$ propagate in the layer $0 \leqslant z \leqslant h$ and fall on the interface $z = h$ under a supercritical angle (see Fig. 2.10). The rays of the incident wave, after reflection from the interface, turn into the rays of the reflected wave shown with a dashed line, which after reflection from the boundary $z = 0$ again turn into the rays of a downgoing wave. However, this "secondary" downgoing wave will not, generally speaking, be identical to the initial one: propagating along the rays and reflecting from the interface and the boundary, it acquires an additional phase $\Phi$ which is the sum of the geometrical phase increment along the ray path and the phases of the reflection coefficients,[6] i.e., it acquires the phase factor $e^{i\Phi}$. If

$$e^{i\Phi} = 1, \tag{2.230}$$

i.e., $\Phi = 2\pi m$, $m = 0, \pm 1, \pm 2, \ldots$, then after two reflections the initial wave turns into itself. We arrive at a wave being a superposition of two plane waves and decaying as $z \to \infty$. The related calculations will show that this wave is identical to the Love wave considered above.

The calculation of phase can be split into several steps.

#### Coefficient of reflection from an interface

Consider a downgoing wave incident on the interface $z = h$ in the form

$$\mathbf{u}^{\downarrow} = \mathbf{e}_2 \mathrm{A}^{\downarrow} e^{-i\omega t + ikx + iQ_1(z-h)}, \quad \mathrm{A}^{\downarrow} = \text{const} \neq 0. \tag{2.231}$$

Then the upgoing reflected wave and the twice reflected downgoing wave equal

$$\mathbf{u}^{\uparrow} = \mathbf{e}_2 \mathrm{A}^{\uparrow} e^{-i\omega t + ikx - iQ_1(z-h)} \text{ and } \mathbf{u}^{\downarrow\downarrow} = \mathbf{e}_2 \mathrm{A}^{\downarrow\downarrow} e^{-i\omega t + ikx - Q_2(z-h)}, \tag{2.232}$$

[6]The modulus of the reflection coefficient under the total internal reflection equals 1.

respectively, where $Q_1$ and $Q_2$ are defined in (2.222) and (2.221), and $\mathrm{A}^{\uparrow}$ and $\mathrm{A}^{\downarrow\downarrow}$ are yet unknown constants. Formulas (2.231) and (2.232) related to a free boundary at $z = h$ are closely connected with those of Section 2.6, where a boundary at $z = 0$ is considered, and can be obtained from them via replacing $z$ by $z - h$.

The continuity conditions for displacement and traction (2.219) at $z = h$ imply the system of equations (cf. (2.224) and (2.225))

$$\begin{aligned} \mathrm{A}^{\downarrow} + \mathrm{A}^{\uparrow} &= \mathrm{A}^{\downarrow\downarrow}, \\ i\mu_1 Q_1(\mathrm{A}^{\downarrow} - \mathrm{A}^{\uparrow}) &= -\mu_2 Q_2 A^{\downarrow\downarrow}. \end{aligned} \tag{2.233}$$

This gives

$$\mathrm{A}^{\uparrow} = \mathrm{R}\mathrm{A}^{\downarrow}, \tag{2.234}$$

where

$$\mathrm{R} = \frac{i\mu_1 Q_1 + \mu_2 Q_2}{i\mu_1 Q_1 - \mu_2 Q_2} \tag{2.235}$$

is called the *coefficient of reflection (of a wave SH) from an interface.*

Obviously, $|\mathrm{R}| = 1$ for real $Q_1$ and $Q_2$, similarly to the case of the total internal reflection of waves $P - SV$. We rewrite the expression for R in the form

$$\mathrm{R} = -\frac{\zeta}{\zeta^*} = \frac{\zeta}{\zeta^*}e^{-i\pi}, \quad \zeta := \mu_2 Q_2 + i\mu_1 Q_1.$$

One can easily verify the formula

$$\mathrm{R} = e^{i\Psi - i\pi}, \quad \text{where} \quad \Psi = 2\,\mathrm{arctg}\,\frac{\mu_1 Q_1}{\mu_2 Q_2}. \tag{2.236}$$

**Phase increments of downgoing and upgoing waves**

Let us calculate the phase increment of a downgoing wave along the segment $AB$, $A = (x^{\circ}, 0)$, $B = (x^{\circ}, h)$ (see Fig. 2.10). From the formula (2.231), it is clear that it equals $Q_1 h$. The increment in the backward propagation is the same, so that the total phase increment in propagation up and down is

$$\psi = 2Q_1 h. \tag{2.237}$$

Calculate, on the basis of (2.237) and (2.232), the phase increment of the aforementioned wave resulting from its passage to the interface $z = h$, reflection from it, and passage backward to the free boundary. It equals

$$Q_1 h + (\Psi - \pi) + Q_1 h = 2Q_1 h + 2\,\mathrm{arctg}\,\frac{\mu_1 Q_1}{\mu_2 Q_2} - \pi. \tag{2.238}$$

Thus, running twice through the layer with two reflections, this wave acquires the factor

$$e^{2i\psi} R = e^{i\Phi}, \quad \Phi = 2\,Q_1 h + \Psi. \tag{2.239}$$

### Dispersion equation (2.226) as a quantization condition

In order that the wave running up and down with reflection from the lower boundary "turn into itself," it is necessary and sufficient that the expression (2.238) change by an integer multiple of $2\pi$. We arrive at the equality

$$\Phi = 2Q_1 h + \Psi = 2\pi m, \quad m = 0, \pm 1, \pm 2, \ldots, \tag{2.240}$$

which is called, according to the terminology of quantum mechanics, a *quantization condition*. It is convenient to rewrite the condition (2.240) in the form

$$2Q_1 h = 2 \operatorname{arctg} \frac{\mu_2 Q_2}{\mu_1 Q_1} + 2\pi m. \tag{2.241}$$

Here, we used the relation $\operatorname{arctg} x + \operatorname{arctg} \frac{1}{x} = \frac{\pi}{2}$, $x > 0$, which can be verified by proving the constancy of its left-hand side by differentiating it and finding its value at $x = 1$.

The formula (2.241) is equivalent to the relation (2.226). This shows the identity of the classical Love wave introduced in Section 2.7.2 to the surface wave represented in Section 2.7.3 by the sum of two *constructively interfering* plane waves.

### Concluding remarks

Considerations similar to the above can be applied to various waves in layered media. They lead either to exact (as in our case) or to high-frequency approximate dispersion equations. Such a derivation is often referred to as the *method of constructive interference*.

---

## 2.8 Plane surface waves in isotropic layered media

We will consider a *vertically layered* or vertically stratified half-space characterized by piecewise-smooth functions $\lambda = \lambda(z)$, $\mu = \mu(z)$, and $\rho = \rho(z)$. As before, the boundary is traction-free (see (2.204)). A finite number of jumps of the coefficients is allowed; the jumps, if they occur, are located at depth $z = z_1, \ldots, z_N$. The contacts are welded,

$$[\mathbf{u}]_{z=z_m} = 0, \tag{2.242}$$

$$[\sigma_{13}(\mathbf{u})]_{z=z_m} = [\sigma_{23}(\mathbf{u})]_{z=z_m} = [\sigma_{33}(\mathbf{u})]_{z=z_m} = 0, \tag{2.243}$$

$m = 1, \ldots, N$, where $[\ ]_{z=z_m}$ denotes the jump at $z = z_m$.

### Waves localized near interfaces and waveguide waves

Waves in a layered elastic space $-\infty < z < +\infty$ can be considered in a similar way while omitting the traction-free boundary condition (2.204) and replacing (2.202) by the requirement that $\mathbf{u}$ decay as $z \to \pm\infty$. Of particular interest for applications is a solution of this kind known as the *Stoneley wave* (Stoneley 1924 [52]), that may run along the plane interface of two welded homogeneous elastic half-spaces exponentially decaying on both sides from the interface. Also widely known is the *Schölte – Gogoladze wave* at the interface between elastic and liquid half-spaces.

The presence of sharp boundaries is not necessary for existence of waves that do not decay in lateral directions; the waveguide propagation in a low-velocity layer is also possible in the case of smooth coefficients.

The problem on waves in a layer of finite thickness (which is sometimes called a plate) $0 \leqslant z \leqslant h$, with traction-free boundaries, splits into those for the Rayleigh waves $P - SV$ (or plane deformations) or for the Love waves $SH$ (or antiplane deformations).

### 2.8.1 Waves $P - SV$

In the case of waves $P - SV$ where $v \equiv 0$, for $z > 0$ we have the equations

$$\begin{aligned} (D\mu D + \rho\omega^2 - \nu k^2)u + ik(\lambda D + D\mu)w = 0, \\ (D\nu D + \rho\omega^2 - \mu k^2)w + ik(\mu D + D\lambda)u = 0, \end{aligned} \tag{2.244}$$

where $D := \frac{d}{dz}$, and $\nu = \nu(z) := \lambda(z) + 2\mu(z)$. The conditions on the traction-free boundary $\sigma_{13}|_{z=0} = 0$ and $\sigma_{33}|_{z=0} = 0$ give

$$\mu(Du + ikw)|_{z=0} = 0, \quad (\nu Dw + ik\lambda u)|_{z=0} = 0, \tag{2.245}$$

and the condition $\sigma_{23}|_{z=0} = 0$ holds automatically. The condition of decaying at infinity (2.202) takes the form

$$|u| + |w| \to 0, \quad z \to \infty. \tag{2.246}$$

In the presence of *interfaces* (that is, planes where $\lambda(z)$, $\mu(z)$, and $\rho(z)$ have jumps with welded contacts), (2.242) and (2.243) lead to the conditions

$$[u]_{z=z_m} = [w]_{z=z_m} = 0, \quad [\mu(Du + ikw)]_{z=z_m} = [\nu Dw + ik\lambda u]_{z=z_m} = 0. \tag{2.247}$$

An example of such a solution is the classical Rayleigh wave. Here, $\omega$ can be regarded as given, and $k$ thus plays the role of a spectral parameter. Alternatively, $k$ can be regarded as given, and $\omega = \omega(k)$.

### 2.8.2 Waves $SH$

In the case of waves $SH$ ($u \equiv w \equiv 0$, but $v \not\equiv 0$), for $z > 0$ we obtain the scalar equation

$$(D\mu D + \rho\omega^2 - \mu k^2)v = 0. \tag{2.248}$$

The traction-free condition gives

$$Dv|_{z=0} = 0. \tag{2.249}$$

We add the condition at infinity:

$$v \to 0, \quad z \to \infty. \tag{2.250}$$

In the presence of welded contacts, from (2.242) and (2.243) we get

$$[v]_{z=z_m} = 0, \quad [\mu D v]_{z=z_m} = 0. \tag{2.251}$$

An example of such a solution is the classical Love wave.

---

## 2.9 Plane waves in arbitrarily layered media

Waves in layered media are important in applications of the theory of elastic waves to seismics, acoustics, etc. Here, we consider the case of layered media.

A medium is called *layered* (or *stratified*) if its elastic parameters and volume density depend solely on one Cartesian coordinate; in our case, it is depth $x_3 = z$:

$$c_{ijkl} = c_{ijkl}(z), \quad \rho = \rho(z). \tag{2.252}$$

The presence of the preferred direction in layered media allows specific waves to propagate in the horizontal direction.

### 2.9.1 Eigenvalue problem

It is assumed that a medium (which is sometimes called a plate) is bounded by the planes $z = z_< = \text{const}$ and $z = z_> = \text{const}$,

$$-\infty < x_1 = x < -\infty, \quad -\infty < x_2 = y < -\infty, \quad z_< < z < z_>. \tag{2.253}$$

Let the boundaries $z = z_<$ and $z = z_>$ be traction-free,

$$\mathbf{t}^{\mathbf{e}_3}(\mathbf{u})\big|_{z=z_<} = \mathbf{t}^{\mathbf{e}_3}(\mathbf{u})\big|_{z=z_>} = 0. \tag{2.254}$$

The medium can be separated with plane interfaces $z = z_1$, $z = z_2$, $\ldots$, $z = z_N$, $z_< < z_1 < z_2 < \ldots < z_N < z_>$, at which the matching conditions hold, that is, the conditions of welded (1.44) or slipping contact (1.46). Not excluded are the cases of half-infinite and infinite media, where one or both of the quantities $z_{\lessgtr}$ become infinite; in such a case, the solution is assumed to tend to zero there; for $z_< = -\infty$ and/or $z_> = +\infty$, we assume that $\mathbf{u} \to 0$ as $z_< \to -\infty$ and/or $\mathbf{u} \to 0$ as $z_> \to +\infty$.

Here, we consider the solution of the form

$$\mathbf{u} = e^{-i\omega t + i(k_1 x_1 + k_2 x_2)} \boldsymbol{v}(z,\omega) = e^{-i\omega t + i k_\alpha x_\alpha} \boldsymbol{v}(z,\omega), \quad \alpha = 1,2. \tag{2.255}$$

The summation from 1 to 2 with respect to repeated lower Greek subscripts is implied. The vector with horizontal components $\boldsymbol{k} = (k_1, k_2, 0)$ is called the *wave vector* (of the surface wave under consideration). Dealing with waves which neither grow nor attenuate in a horizontal direction (along $x_1$ and $x_2$), we assume that the components $\boldsymbol{k}$ are real. Any solution of the form (2.255) is called a *(homogeneous) plane wave propagating horizontally* in the direction of the vector $\boldsymbol{k}$. It is important to note that the energy flow in these waves is directed along the boundary. As before, we take $\omega > 0$.

In the case of a half-space, the solutions of the form (2.255) are called *plane surface waves.* The Stoneley and Schölte – Gogoladze waves are suitably termed *"interfacial waves."* Unfortunately, a nice term "channel waves" introduced by Levshin 1973 [35] has not become accustomed.

Substituting the ansatz (2.255) in the homogeneous elastodynamics equations (2.1), (1.35) and using (2.252), we obtain the system of ordinary differential equations of the form

$$\gamma \boldsymbol{v} = \rho \omega^2 \boldsymbol{v}, \tag{2.256}$$

where

$$(\gamma \boldsymbol{v})_i = -\left\{ D c_{i33j} D + i k_\beta (D c_{i\beta 3j} + c_{i3\beta j} D) - k_\alpha k_\beta c_{i\alpha\beta j} \right\} v_j, \tag{2.257}$$
$$D := \frac{d}{dz}, \quad j = 1,2,3.$$

Repeated Greek subscripts imply summation from 1 to 2, while Latin subscripts imply summation from 1 to 3. The conditions on free boundaries take the form

$$\mathcal{T} \boldsymbol{v}\big|_{z = z_{\lessgtr}} = 0, \tag{2.258}$$

where

$$(\mathcal{T} \boldsymbol{v})_i = (i k_\beta c_{i3\beta j} + c_{i33j} D) v_j, \quad i = 1,2,3, \tag{2.259}$$

and the *matching conditions*, that is, the conditions of welded contacts at interfaces, are as follows:

$$[\boldsymbol{v}]\big|_{z=z_m} = 0, \quad [\mathcal{T} \boldsymbol{v}]\big|_{z=z_m} = 0, \tag{2.260}$$

where $[\psi]|_{z=h}$ denotes the jump of the function $\psi(z)$ at $z = h$. In the cases of one or two infinite $z_{<,>}$, we assume one or two conditions of attenuation at infinity.

As it already occurred several times, we arrive at an eigenvalue problem, in which $\omega^2$ plays the role of an eigenvalue. It is convenient to regard $\omega$ as a function of the components $k_1$ and $k_2$ of the vector $\boldsymbol{k}$,

$$\omega = H(k_1, k_2). \tag{2.261}$$

Similarly to Section 2.1, the equation (2.261) is called the *dispersion equation*, and a surface wave is called *nondispersive* if the function $H(k_1, k_2)$ is first-degree homogeneous, similarly to the Rayleigh waves in a homogeneous half-space and to the Stoneley wave at the interface between homogeneous half-spaces. In any other case, a surface wave is dispersive.

In the general case, if the frequency $\omega$ is fixed, we can say nothing on the existence of real wave vectors and their number. For $\boldsymbol{k}^2$, we can obtain, generally speaking, both positive and negative values. In the case where $\boldsymbol{k}^2 < 0$, no waves propagate in a horizontal direction. In the case of a finite interval, an infinite set of eigenvalues and respective eigenfunctions exists (the proof can be found, e.g., in a book by Naimark 1968 [45]).

In the case of the classical Rayleigh wave, an eigenvector exists and, by the relation $\omega = c|\boldsymbol{k}|$, $\omega$ determines its length uniquely. The Stoneley wave localized near the interface of two homogeneous half-spaces exists only if a certain inequality containing medium parameters holds. The existence of the Rayleigh wave in a homogeneous half-space for arbitrary anisotropy will be discussed in Section 2.10. In the sequel, the eigenvalue $H^2$ (see (2.261)) is assumed to exist and be fixed.

We will address the problem (2.256)–(2.260) using an approach based on variational calculus.

### 2.9.2 Virial theorem

Let

$$\mathbf{u} = \mathrm{Re}[e^{-i\omega t + ik_\alpha x_\alpha}\boldsymbol{v}(z,\omega)], \tag{2.262}$$

and let the relations (2.256)–(2.260) hold. Therefore, $\mathbf{u}$ obeys the Hamilton principle, i.e.,

$$\delta \int_{t_1}^{t_2} dt \int_\Omega \mathcal{L}(\mathbf{u}) d\mathbf{x} = \int_{t_1}^{t_2} dt \int_\Omega \left\{ \frac{\partial \mathcal{L}}{\partial \dot{\mathrm{u}}_m} \delta \dot{\mathrm{u}}_m + \frac{\partial \mathcal{L}}{\partial \varepsilon_{ij}} \varepsilon_{ij}(\delta \mathbf{u}) \right\} d\mathbf{x} = 0, \tag{2.263}$$

$\mathcal{L} = \mathcal{K} - \mathcal{W}$, for any smooth vector $\delta\mathbf{u}$ that satisfies boundary conditions and matching conditions (2.260) and equals zero at $t = t_{1,2}$ and on the boundary $\partial\Omega$ of the domain $\Omega \subset \mathbb{R}^3$. If the latter two conditions do not hold, then, generally speaking, (2.263) is not satisfied.

Now we describe two other ways of choosing $\delta\mathbf{u}$. Let us examine the situation more carefully. Integration by parts shows that

$$\delta \int_{t_1}^{t_2} dt \int_\Omega \mathcal{L}(\mathbf{u}) d\mathbf{x} = \int_{t_1}^{t_2} dt \int_\Omega \left( \frac{\partial \sigma_{ij}}{\partial x_i} - \rho \frac{\partial^2 \mathrm{u}_j}{\partial t^2} \right) \delta \mathrm{u}_j d\mathbf{x} + \int_\Omega \frac{\partial \mathcal{L}}{\partial \dot{\mathrm{u}}_j} \delta \mathrm{u}_j d\mathbf{x} \Big|_{t=t_1}^{t=t_2}$$
$$- \int_{t_1}^{t_2} dt \int_{\partial\Omega} \mathrm{t}_j^{\boldsymbol{n}} \delta \mathrm{u}_j dS = 0. \tag{2.264}$$

In the last term, $\boldsymbol{n}$ is the outward normal to the surface $\partial\Omega$, over which we integrate. Since $\mathbf{u}$ satisfies the elastodynamics equations (2.3), (1.35), for equation (2.263) to hold it is sufficient that the integrated terms vanish. In particular, this occurs when the domain of integration is taken as

$$0 \leqslant t \leqslant T, \quad z_< < z < z_>, \quad \alpha = 1, 2, \quad 0 \leqslant x_\alpha \leqslant T_\alpha, \tag{2.265}$$

where

$$T = \frac{2\pi}{\omega}, \quad T_\alpha = \frac{2\pi}{|k_\alpha|}, \tag{2.266}$$

and the variation $\delta\mathbf{u}$ is taken as

$$\delta\mathbf{u} = \mathbf{u} = \mathrm{Re}[e^{-i\omega t + ik_\beta x_\beta}\boldsymbol{v}(z,\omega)]. \tag{2.267}$$

Recall that the variation is not necessarily small.

Indeed, the integrated terms in (2.264), which are the integrals over $t =$ const and $x_\alpha =$ const, cancel because of the periodicity of the integrand over $t$ and $x_\alpha$; the integrals over the planes $z = z_k$ cancel by the matching conditions; the integrals over the horizontal boundaries $z_\lessgtr$ cancel by virtue of the traction-free conditions (2.254). Where one or both of the quantities $z_\lessgtr$ are infinite, the integrals vanish, because $\boldsymbol{v}$ tends to zero at infinity.

Also note that the integrated terms vanish and the equation (2.263) holds again under the assumption that

$$\delta\mathbf{u} = \mathrm{Re}\left[e^{-i\omega t + ik_\beta x_\beta}\frac{\partial \boldsymbol{v}}{\partial k_\alpha}\right]. \tag{2.268}$$

It is important that the function $\delta\mathbf{u}$ vanishes at infinity. These observations will be used in proving the group velocity theorem.

Denote the average of a function $\psi(x_1, x_2, z, t)$ over two lateral variables by the symbol $\langle\!\langle\ \rangle\!\rangle$:

$$\langle\!\langle\psi\rangle\!\rangle(z,t) := \frac{1}{T_1 T_2}\int_0^{T_1} dx_1 \int_0^{T_2} dx_2\, \psi(x_1, x_2, z, t). \tag{2.269}$$

Recall also the notation $\overline{\phantom{x}}$ introduced in Chapter 1 for averaging over a time period.

Thus, the equation (2.263) is satisfied as the expression (2.267) is substituted instead of $\delta\mathbf{u}$, and integration over time and over the lateral variables is performed over the periods $T_0$, $T_1$, and $T_2$, whereas integration over $z$ is performed from $z_<$ to $z_>$. In this case, we arrive at the relation

$$\int_{z_<}^{z_>}\left\{\left\langle\!\!\left\langle\overline{\frac{\partial\mathcal{L}}{\partial\dot{u}_j}\dot{u}_j}\right\rangle\!\!\right\rangle + \left\langle\!\!\left\langle\overline{\frac{\partial\mathcal{L}}{\partial\varepsilon_{ij}(\mathbf{u})}\varepsilon_{ij}(\mathbf{u})}\right\rangle\!\!\right\rangle\right\} dz = 0. \tag{2.270}$$

Recall that $\mathcal{L}$ is a second-degree homogeneous function with respect to $\dot{\mathbf{u}}$ and $\varepsilon_{ij}$. The Euler homogeneity relation (see Section 2.4.4) implies that the integrand in the formula (2.270) equals $2\langle\!\langle\overline{\mathcal{L}}\rangle\!\rangle$. This expression can be further simplified thanks to the fact that the function $\mathbf{u}$ is $2\pi$-periodic with respect to the expression $-\omega t + k_\alpha x_\alpha$, which is linear in $t$, $x_1$ and $x_2$. Averaging over $t$ or $x_1$, or $x_2$, gives the same result, therefore (2.270) can be written as

$$\int_{z_<}^{z_>} \overline{\mathcal{L}} dz = 0. \tag{2.271}$$

Since the density of the Lagrangian equals $\mathcal{L} = \mathcal{W} - \mathcal{K}$, the relation (2.271) implies that

$$\int_{z_<}^{z_>} \overline{\mathcal{W}} dz = \int_{z_<}^{z_>} \overline{\mathcal{K}} dz. \tag{2.272}$$

The statement that the potential and kinetic energy densities averaged over time and integrated over depth are equal is called *the virial theorem for surface waves*.

### 2.9.3 Group velocity theorem

In this section, we aim at proving the fundamental formula

$$\boldsymbol{v}^{gr} \int_{z_<}^{z_>} \overline{\mathcal{E}} dz = \int_{z_<}^{z_>} \overline{\mathbf{S}} dz, \tag{2.273}$$

where, as before, $\mathbf{S}$ is the density of the energy flow, $\mathcal{E}$ is the density of energy, and $\boldsymbol{v}^{gr}$ is the *group velocity of a surface wave* defined as follows:

$$\boldsymbol{v}^{gr} := \frac{\partial \omega}{\partial k_\alpha} \mathbf{e}_\alpha = \frac{\partial \omega}{\partial k_1} \mathbf{e}_1 + \frac{\partial \omega}{\partial k_2} \mathbf{e}_2. \tag{2.274}$$

The group velocity vector of a surface wave lies in a plane $z = \text{const}$ and can be naturally regarded as a two-dimensional vector. Instead, we will consider it to be a three-dimensional vector with zeroth third component.

The proof of the formula (2.273) comprises differentiation $\int_{z_<}^{z_>} \langle\!\langle \overline{\mathcal{L}} \rangle\!\rangle dz$ with respect to $k_\alpha$. We employ the auxiliary relations

$$\omega \frac{\partial}{\partial \omega} \int_{z_<}^{z_>} \overline{\mathcal{L}} dz = 2 \int_{z_<}^{z_>} \overline{\mathcal{K}} dz = \int_{z_<}^{z_>} \langle\!\langle \overline{\mathcal{E}} \rangle\!\rangle dz \tag{2.275}$$

and

$$\omega \frac{\partial}{\partial k_\alpha} \int_{z_<}^{z_>} \overline{\mathcal{L}} dz = - \int_{z_<}^{z_>} \overline{\mathcal{S}_\alpha} dz, \quad \alpha = 1, 2. \tag{2.276}$$

Here, we use "naive differentiation", i.e., we do not take into account the dependence of $\boldsymbol{v}$ on $k_\alpha$ (cf. Section 2.4.3). The first relation (2.275) follows from (2.272), the relations $\overline{\mathcal{L}} = \overline{\mathcal{K}} - \overline{\mathcal{W}}$ and $\overline{\mathcal{E}} = \overline{\mathcal{K}} + \overline{\mathcal{W}}$, and the fact easily seen from (1.91) that $\overline{\mathcal{K}}$ is proportional to $\omega^2$. The second relation (2.276) can be obtained by direct calculation.

Differentiate the relation (2.271) with respect to $k_\alpha$, noting that its left-hand side depends on $k_\alpha$, first, directly, second, via $\omega$ and, third, via the amplitudes $v_j$. In consequence of (2.275) and (2.276), the first and the second contributions give the expression

$$\frac{1}{\omega}\left(\frac{\partial\omega}{\partial k_\alpha}\int_{z_<}^{z_>}\overline{\mathcal{E}}\,dz - \int_{z_<}^{z_>}\overline{\mathcal{S}}_\alpha dz\right). \tag{2.277}$$

The third contribution equals zero, because it can be regarded as a variation (see (2.268)).

Thus, we have proved the group velocity theorem (2.273).

The consideration in this section is applicable to the classical Rayleigh wave, as well as to the Rayleigh wave in an anisotropic homogeneous half-space. Their existence will be established in the subsequent section.

---

## 2.10 ⋆ Existence of the Rayleigh wave in an anisotropic homogeneous half-space

A plane surface wave in an anisotropic half-space with a traction-free boundary (also known as a *subsonic surface wave*) generalizes the classical Rayleigh wave discussed in Section 2.7.1. For constant elastic stiffnesses and density (only this case is considered in the present section), it is a linear combination of, generally speaking, three inhomogeneous plane waves propagating in a given direction along the boundary. Its phase velocity is obviously less than that of each of the homogeneous plane waves running in the same direction (otherwise, it would not decay with depth). In the case of general anisotropy, a simple analysis of the corresponding dispersion equation similar to that of Section 2.7.1 is hardly possible, which makes investigation of the existence and uniqueness of the Rayleigh wave a tricky task. Essential for the existence is the *condition of the general position* (or the *nondegeneracy condition*) due to Barnett and Lothe (Lothe and Barnett 1976, Barnett and Lothe 1985 [7, 39]), that is, the slowest homogeneous plane wave in a given direction does not satisfy the traction-free condition.

We will outline a proof of the existence based on techniques of the spectral theory of operators and variational methods. In this approach, the main subject is an ordinary differential operator with respect to a variable corresponding to depth. The existence of the Rayleigh wave is equivalent to the existence of an eigenvalue of this operator in some interval (more precisely, an eigenvalue of the corresponding self-adjoint operator $\widehat{\gamma}$ in a Hilbert space). The existence of the eigenvalue is established by variational methods. The operator $\widehat{\gamma}$ also has a continuous spectrum (associated with homogeneous plane waves), which fact implies a certain technical complication. Of crucial importance is the observation that the traction-free boundary condition belongs to a class of conditions that are called (in the calculus of variations) natural conditions.

The reader is assumed to be familiar with the basics of the theory of self-adjoint operators in a Hilbert space.

### 2.10.1 One-dimensional problem and the corresponding energy quadratic form

Let

$$\boldsymbol{k} = (k_1, 0, 0), \tag{2.278}$$

where $k_1 > 0$. Then $\boldsymbol{u}(\mathbf{x}) = e^{ik_1x_1}\boldsymbol{v}(z)$, where $\boldsymbol{v}(z)$ satisfies the relation (2.256) for $z > 0$ and the boundary condition

$$\mathcal{T}\boldsymbol{v}\big|_{z=0} = 0. \tag{2.279}$$

Represent the differential expressions for $\gamma$ and $\mathcal{T}$ as follows:

$$\gamma\boldsymbol{v} = -\mathbf{C}\boldsymbol{v}'' - ik_1\check{\mathbf{C}}\boldsymbol{v}' + k_1^2\widetilde{\mathbf{C}}\boldsymbol{v}, \quad {}' := D = \frac{d}{dz}, \tag{2.280}$$

$$\mathcal{T}\boldsymbol{v}\big|_{z=0} = \{\mathbf{C}\boldsymbol{v}' + ik_1\mathbf{c}\boldsymbol{v}\}\,|_{z=0} = 0, \tag{2.281}$$

$$\begin{gathered} \mathbf{C} = \|\mathrm{C}_{jm}\| := \|c_{j33m}\|, \ \check{\mathbf{C}} = \|\check{\mathrm{C}}_{jm}\| := \|c_{j13m} + c_{j31m}\|, \\ \widetilde{\mathbf{C}} = \|\,\widetilde{\mathrm{C}}_{jm}\| := \|c_{j11m}\|, \quad \mathbf{c} = \|\mathrm{c}_{jm}\| := \|c_{j13m}\|. \end{gathered} \tag{2.282}$$

The solution is required to decay as $z \to \infty$, and the decay is exponential, because $\boldsymbol{v}$ satisfies an ordinary differential equation with constant coefficients. The above eigenvalue problem is naturally related to a positive self-adjoint operator in the space $L_2(0,\infty)$ of vector functions with the scalar product

$$(\mathbf{f}, \mathbf{g})_{L_2(0,\infty)} = \int_0^\infty (\mathbf{f}(z), \mathbf{g}(z))dz = \int_0^\infty f_j(z)g_j^*(z)dz. \tag{2.283}$$

We will need an analytical expression for a one-dimensional quadratic form $\mathscr{E}$ (dependent on $k_1$) that corresponds to the problem (2.256), (2.279). First multiply the first term on the right-hand side of (2.280) by $\boldsymbol{v}^*$ and integrate the obtained relation by parts:

$$-\int_0^\infty (\mathbf{C}\boldsymbol{v}'', \boldsymbol{v})dz = \int_0^\infty (\mathbf{C}\boldsymbol{v}', \boldsymbol{v}')dz - (\mathbf{C}\boldsymbol{v}', \boldsymbol{v}')|_{z=0}. \tag{2.284}$$

Next, represent the result of multiplying the second term in (2.280) by $\boldsymbol{v}^*$ as the sum of two equal terms, in one of which integrate by parts:

$$\begin{aligned} -ik_1\int_0^\infty (\check{\mathbf{C}}\boldsymbol{v}', \boldsymbol{v})dz &= -\frac{ik_1}{2}\int_0^\infty \{(\check{\mathbf{C}}\boldsymbol{v}', \boldsymbol{v}) - (\check{\mathbf{C}}\boldsymbol{v}, \boldsymbol{v}')\}\,dz - \frac{ik_1}{2}\,(\check{\mathbf{C}}\boldsymbol{v}, \boldsymbol{v})\big|_{z=0} \\ &= k_1\,\mathrm{Im}\int_0^\infty (\check{\mathbf{C}}\boldsymbol{v}', \boldsymbol{v})dz - \frac{ik_1}{2}\,(\check{\mathbf{C}}\boldsymbol{v}, \boldsymbol{v})\big|_{z=0}. \end{aligned} \tag{2.285}$$

We transform the sum of the integrated terms in (2.284) and (2.285), taking into account the fact that $\frac{1}{2}(\check{\mathbf{C}}\boldsymbol{v},\boldsymbol{v}) = (\mathbf{c}\boldsymbol{v},\boldsymbol{v})$. Thus, we arrive at the relation

$$(\gamma \boldsymbol{v}, \boldsymbol{v})_{L_2(0,\infty)} = \mathscr{E}(\boldsymbol{v}, \boldsymbol{v}) + (\mathcal{T}\boldsymbol{v}, \boldsymbol{v})|_{z=0}, \tag{2.286}$$

where the form $\mathscr{E} = \mathscr{E}(\boldsymbol{v}, \boldsymbol{v}) = \mathscr{E}(k_1, \boldsymbol{v}, \boldsymbol{v})$ is defined by

$$\mathscr{E}(k_1, \boldsymbol{v}, \boldsymbol{v}) = \int_0^\infty \mathcal{E}(k_1, \boldsymbol{v}, \boldsymbol{v})dz \tag{2.287}$$

with the density

$$\begin{aligned}\mathcal{E}(k_1, \boldsymbol{v}, \boldsymbol{v}) &= (\mathbf{C}\boldsymbol{v}', \boldsymbol{v}') + k_1 \operatorname{Im}(\check{\mathbf{C}}\boldsymbol{v}', \boldsymbol{v}) + k_1^2(\widetilde{\mathbf{C}}\boldsymbol{v}, \boldsymbol{v}) \\ &= c_{j33m}v_j'v_m'^* + k_1 \operatorname{Im}[(c_{j13m} + c_{j31m})v_j'v_m^*] + k_1^2 c_{j11m}v_j v_m^*. \end{aligned} \tag{2.288}$$

The quantity (2.288) can be obtained from the time-averaged density of potential energy (1.90) $c_{jklm}\partial_j u_k \partial_l u_m^*$ multiplied by 4 via replacing $\partial_1$ by $ik_1$. The form $\mathscr{E}$ is positive definite (which follows from the inequality (1.20)) and can therefore be extended in a standard way onto the space $\mathrm{H}^1(0,\infty)$ of vector functions square integrable with their first derivatives (see, e.g., Birman and Solomjak 1987 [9]). The self-adjoint operator corresponding to this form is denoted by $\widehat{\gamma}$.

The Rayleigh wave exists if and only if $\widehat{\gamma}$ has an eigenvalue. We need some information on the continuous spectrum of $\widehat{\gamma}$.

## 2.10.2 Inhomogeneous plane waves and the continuous spectrum of the operator $\widehat{\gamma}$

### Plane waves and the minimal frequency along the axis $x_1$

Consider a plane wave

$$\boldsymbol{U} = \boldsymbol{h}e^{i(k_1x_1 + k_3z)} \tag{2.289}$$

with the same $k_1$. As we know from Section 2.4, the vector $\boldsymbol{h} = \boldsymbol{h}(\boldsymbol{k})$, $\boldsymbol{k} = (k_1, 0, k_3)$, is an eigenvector of the positive definite matrix $\boldsymbol{\Gamma}$

$$\boldsymbol{\Gamma}(\boldsymbol{k})\boldsymbol{h} = \Lambda\boldsymbol{h}, \quad \Lambda = \rho\omega^2. \tag{2.290}$$

If $\omega$ is fixed, $k_3$ can be regarded as a function of $k_1$. For sufficiently large values of the horizontal slowness $\xi = k_1/\omega$ of the plane wave (2.290), no homogeneous plane wave exists (which follows, e.g., from the boundedness of the slowness surface). If $\xi^\star = k_1^\star/\omega$ is the maximal value of the horizontal slowness for homogeneous plane waves,[7] then, obviously, plane waves with $k_1 > k_1^*$ are inhomogeneous, i.e., they decay or grow as $z \to \infty$. We are not interested

[7] In what follows, it does not matter whether such a plane wave is unique or not.

here in growing solutions. Now we regard the system (2.290) as a problem of determining eigenvalues $\Lambda = \rho\omega^2$, in which $k_1$ and $k_3$ are parameters. Let us fix a real $k_1$ and restrict the discussion to real $k_3$. We denote the eigenvalues of the matrix $\boldsymbol{\Gamma}(\boldsymbol{k})$ by $\Lambda^{(1)}(k_3)$, $\Lambda^{(2)}(k_3)$, and $\Lambda^{(3)}(k_3)$, numbering them so that $\Lambda^{(1)}(k_3) \leqslant \Lambda^{(2)}(k_3) \leqslant \Lambda^{(3)}(k_3)$. Obviously, there exists

$$\Lambda_\star = \min_{\operatorname{Im} k_3 = 0} \{\Lambda^{(1)}(k_3)\}, \quad \Lambda_\star > 0. \tag{2.291}$$

Let $k_3^\star$ be the value of $k_3$ (possibly, not unique) at which this minimum is achieved, and $\boldsymbol{h}^\star$ be the corresponding eigenvector of the matrix $\boldsymbol{\Gamma}$. We choose $\boldsymbol{h}^\star$ to be real. We put

$$\omega_\star = \sqrt{\Lambda_\star/\rho}. \tag{2.292}$$

Homogeneous plane waves exist for $\omega > \omega_\star$ (and fixed $k_3$), and do not for $\omega < \omega_\star$.

**Minimal frequency along the axis $x_1$ and the continuous spectrum of $\widehat{\gamma}$**

It is obvious that any $\rho\omega^2$, where $\omega \geqslant \omega_\star$, is a point of the continuous spectrum of the operator $\widehat{\gamma}$, because the corresponding function of the continuous spectrum is a solution of the problem of incidence of the homogeneous plane wave $\boldsymbol{h}e^{i(k_1x_1 - k_3z)}$ on the traction-free boundary $z = 0$. One can naturally expect that the quantity $\rho\omega_\star^2$ introduced by the relations (2.291) and (2.292) in discussing plane waves is the lower boundary of the continuous spectrum. In fact, true is

**Theorem.** *The interval $(-\infty, \rho\omega_\star^2)$ includes no points of the continuous spectrum of the operator $\widehat{\gamma}$.*

A rather technical proof of this fact is presented by Kamotskii and Kiselev 2009 [28].

### 2.10.3 Variational principle

The lower boundary $\underline{\sigma(\widehat{\gamma})}$ of the spectrum of the positive operator $\widehat{\gamma}$ (in our case, its minimal eigenvalue) can be found from the variational principle traceable to Rayleigh

$$\underline{\sigma(\widehat{\gamma})} = \inf \left\{ \frac{(\widehat{\gamma}\boldsymbol{u}, \boldsymbol{u})_{L_2(0,\infty)}}{(\boldsymbol{u}, \boldsymbol{u})_{L_2(0,\infty)}} \right\}, \quad \boldsymbol{\mathcal{T}u}\big|_{z=0} = 0; \quad \boldsymbol{u} \in \mathrm{H}^1(0,\infty), \tag{2.293}$$

where $\boldsymbol{u} \neq 0$ belongs to the domain of definition of $\mathscr{E}$. The quantity, the infimum of which is taken in (2.293), is called the *Rayleigh quotient*. For the boundary conditions called in variational calculus natural (the free-traction condition (2.281) is of this type), we can use the following form of the variational principle (see a textbook by Birman and Solomjak 1987 [9]):

$$\underline{\sigma(\widehat{\gamma})} = \inf \left\{ \frac{\mathscr{E}(\boldsymbol{u}, \boldsymbol{u})}{(\boldsymbol{u}, \boldsymbol{u})_{L_2(0,\infty)}} \right\}, \quad \boldsymbol{u} \in \mathrm{H}^1(0,\infty). \tag{2.294}$$

Here, in contrast to (2.293), it is not required that the boundary conditions (2.279) be satisfied. The fact that we can use in (2.294) the solutions of the equation (2.256) without any boundary conditions enables us to prove the existence of the sought-for eigenvalue.

### 2.10.4 Discrete spectrum of $\widehat{\gamma}$

Let $\boldsymbol{h}^\star$ be the eigenvector of the matrix $\boldsymbol{\Gamma}$ that corresponds to the eigenvalue $\rho\omega_\star^2$. We introduce the vector functions

$$\boldsymbol{U}^\star(\mathbf{x}) = \boldsymbol{h}^\star \exp(ik_1^\star x_1 + ik_3^\star z), \quad \boldsymbol{u}^\star(z) = \boldsymbol{h}^\star \exp(ik_3^\star z). \tag{2.295}$$

They satisfy the equations (2.1)–(2.2) and (2.256), respectively. In the case of a general position, they do not satisfy the boundary conditions at $z = 0$. The function $\boldsymbol{U}^\star$ is a homogeneous plane wave that propagates in the direction $\boldsymbol{k} = (k_1, 0, k_3)$ with the minimal possible frequency for a fixed value of $k_3$. The vector $\boldsymbol{h}^\star$ is taken real.

Obviously, the relations

$$\mathbf{t}^{\mathbf{e}_3}\boldsymbol{U}^\star|_{z=0} = 0 \quad \text{and} \quad \boldsymbol{\mathcal{T}}\boldsymbol{u}^\star|_{z=0} = 0 \tag{2.296}$$

are equivalent.

**The Barnett and Lothe theorem.**[8] *If*

$$\mathbf{t}^{\mathbf{e}_3}\boldsymbol{U}^\star|_{z=0} \neq 0, \tag{2.297}$$

*then the interval* $(0, \rho\omega_\star^2)$ *includes at least one eigenvalue of the operator* $\widehat{\gamma}$.

We will establish that (2.294) takes values not only on the half-line $(\rho\omega_\star^2, \infty)$, but also in the interval $(0, \rho\omega_\star^2)$. Therefore, there is a spectrum to the left of $\rho\omega_\star^2$. In view of the statement at the end of Section 2.10.2, the spectrum consists of eigenvalues. It is sufficient to establish that the quadratic form

$$\mathcal{B}(\boldsymbol{u}, \boldsymbol{u}) := \mathscr{E}(\boldsymbol{u}, \boldsymbol{u}) - \rho\omega_\star^2(\boldsymbol{u}, \boldsymbol{u}) \tag{2.298}$$

takes negative values. For this purpose, consider its values on the sequence of vector functions

$$\boldsymbol{u}_n^\star(z) = e^{-\frac{z}{n}}\boldsymbol{u}^\star(z) = e^{-\frac{z}{n}+ik_3^\star z}\boldsymbol{h}^\star, \qquad n = 1, 2, \ldots \tag{2.299}$$

**Lemma.** *The following inequality holds:*

$$|\mathcal{B}(\boldsymbol{u}_n^\star, \boldsymbol{u}_n^\star)| \leqslant \frac{\mathrm{const}|\boldsymbol{h}^\star|^2}{n}, \quad \mathrm{const} > 0. \tag{2.300}$$

Let us prove it. As is clear from (2.287) and (2.298),

$$\mathcal{B}(\boldsymbol{u}_n^\star, \boldsymbol{u}_n^\star) = \int_0^\infty \left[\mathcal{E}(k_1^\star, \boldsymbol{u}_n^\star, \boldsymbol{u}_n^\star) - \rho\omega_\star^2|\boldsymbol{u}_n^\star|^2\right] dz. \tag{2.301}$$

[8] This statement is also called the Lothe and Barnett theorem.

First consider the middle term on the right-hand side of (2.288) for $v = u_n^\star$. From (2.287) and (2.299), we have

$$k_1^\star \operatorname{Im}(\check{\mathbf{C}} u_n^{\star\prime}, u_n^\star) = k_1^\star e^{-2\frac{z}{n}}(\check{\mathbf{C}} h^\star, h^\star)\operatorname{Im}\Big(-\frac{1}{n} + ik_3^\star\Big) = e^{-2\frac{z}{n}} k_1^\star k_3^\star (\check{\mathbf{C}} h^\star, h^\star)$$
$$= e^{-2\frac{z}{n}}\left[k_1^\star k_3^\star (c_{j13m} + c_{j31m}) h_j^\star h_m^\star\right]. \tag{2.302}$$

The definition (2.282) implies that the terms of order $O(\frac{1}{n})$ vanish in the sum of the outside left and right terms in (2.288), and the expression (2.302) equals

$$e^{-2\frac{z}{n}}\left\{\left[c_{j11m}k_1^{\star 2} + c_{j33m}k_3^{\star 2}\right] h_j^\star h_m^\star + n^{-2} c_{j33m} h_j^\star h_m^\star\right\}. \tag{2.303}$$

Thus,

$$\mathcal{E}(k_1^\star, u_n^\star, u_n^\star) = e^{-2\frac{z}{n}}\left\{\left[c_{j11m}k_1^{\star 2} + c_{j33m}k_3^{\star 2}\right.\right.$$
$$\left.\left. + (c_{j13m} + c_{j31m})k_1^\star k_3^\star\right] h_j^\star h_m^\star n^{-2} c_{j33m} h_j^\star h_m^\star\right\}. \tag{2.304}$$

The sum of the terms in braces on the right-hand side of (2.304) that do not comprise the factor $n^{-2}$ equals $(\mathbf{\Gamma}(\boldsymbol{k}^\star)\boldsymbol{h}^\star \cdot \boldsymbol{h}^\star) = \rho\omega_\star^2 |\boldsymbol{h}^\star|^2$, $\boldsymbol{k}^\star := (k_1^\star, 0, k_3^\star)$ and does not depend on $z$.[9] After cancellation in (2.301), only the product of the exponent and $O(n^{-2})$ is integrated. Thus,

$$\mathcal{B}(u_n^\star, u_n^\star) = \int_0^\infty e^{-2\frac{z}{n}} \frac{c_{j33m} h_j^\star h_m^\star}{n^2} dz = \frac{(\mathbf{C} h^\star \cdot h^\star)}{2n}, \tag{2.305}$$

where $\mathbf{C}$ is the symmetric positive definite matrix introduced in (2.282). We have thus proved the lemma.

Let us proceed with the proof of the theorem. By the lemma, the assumption that all values of the form (2.298) are nonnegative will be reduced to a contradiction. Assume that the form $\mathcal{B}(\cdot,\cdot)$ is nonnegative (that is, $\mathcal{B}(u, u) \geqslant 0$, but $\mathcal{B}(u, u) = 0$ does not necessarily imply that $u = 0$). Then for two arbitrary vector functions $u = u(z)$ and $\mathrm{v} = \mathrm{v}(z)$ from $\mathrm{H}^1(0, \infty)$, the Cauchy–Bunyakowsky inequality holds:

$$|\mathcal{B}(u, \mathrm{v})|^2 \leqslant \mathcal{B}(u, u)\mathcal{B}(\mathrm{v}, \mathrm{v}).$$

The left-hand side of the inequality includes the corresponding bilinear form. The substitution $u = u_n^\star$ and the use of the inequality (2.300) show that

$$|\mathcal{B}(u_n^\star, \mathrm{v})|^2 \leqslant \mathrm{const}/n, \qquad \mathrm{const} > 0. \tag{2.306}$$

At the same time, integration by parts in the relation (2.298) gives

$$(\mathcal{B} u_n^\star, \mathrm{v}) = (\mathcal{T} u_n^\star, \mathrm{v})\big|_{z=0} + \int_0^{+\infty} \left((\gamma u_n^\star - \rho\omega_\star^2 u_n^\star), \mathrm{v}\right) dz. \tag{2.307}$$

[9] Cancellation of the terms independent of $n$ can be observed in applying a version of the virial theorem to the plane wave $U^\star$ (see (2.295)). However, we give a preference to direct calculation.

As follows from the definition of $\boldsymbol{u}_n^\star$ (see (2.299)), the integral on the right-hand side tends to zero as $n \to \infty$. The arbitrariness of $\mathbf{v}$ in the relation (2.307), together with the inequality (2.306), gives

$$\mathcal{T}\boldsymbol{u}_n^\star|_{z=0} \longrightarrow 0, \quad n \to +\infty.$$

The relations (2.299) and (2.256) imply that

$$\mathcal{T}\boldsymbol{u}_n^\star|_{z=0} = \mathcal{T}\boldsymbol{u}^\star|_{z=0} + \frac{\mathbf{C}\boldsymbol{u}^\star}{n} \longrightarrow \mathcal{T}\boldsymbol{u}^\star\Big|_{z=0}, \quad n \to +\infty.$$

Therefore, $\mathcal{T}\boldsymbol{u}^\star|_{z=0} = 0$. By the equivalence of the relations (2.296), the latter means that $\mathbf{t}^{\mathbf{e}_3}\, \boldsymbol{U}^\star|_{z=0} = 0$, which is incompatible with the condition of the theorem. Since the assumption on the positiveness of the form $\mathcal{B}(\cdot,\cdot)$ is reduced to a contradiction, the form takes negative values.

Thus, the functional $\mathscr{E}(\boldsymbol{u},\boldsymbol{u})/(\boldsymbol{u},\boldsymbol{u})_{L_2(0,\infty)}$ (2.294) takes values in the interval $(0, \rho\omega_\star^2)$. Accordingly (see Birman and Solomjak 1987 [9]), the operator $\widehat{\gamma}$ has a spectrum in the interval $(0, \rho\omega_\star^2)$; the spectrum is discrete because of the remark at the end of Section 2.10.2.

The theorem is proved.

We arrive at the following statement. Let $\xi^\star = k_1^\star/\omega > 0$ be the largest horizontal slowness of homogeneous plane waves with the wavenumber $\boldsymbol{k} = (k_1, 0, k_3)$. If the condition of the general position holds, i.e., the plane wave $\mathbf{U}^\star$ with the minimal phase velocity along the axis $x_1$ does not satisfy the traction-free boundary condition (2.279), the Rayleigh wave exists with horizontal slowness $1/c = \xi$, where $\xi > \xi^\star$.

The condition that the phase fronts propagate along the axis $x_1$ doesn't matter, and the wave vector $\boldsymbol{k} = (k_1, 0, k_3)$ can be replaced by $\boldsymbol{k} = (k_1, k_2, k_3)$.

---

## 2.11 ⋆ Comments to Chapter 2

The theory of plane waves is discussed in virtually any book which regards linear waves in elastic media. A considerable amount of material on the theory and applications of plane waves can be found in a classical treatise by Brekhovskikh 1960 [11] (see also Brekhovskikh and Godin 1991 [12]).

The history of the relation (2.68) is traceable to the research due to Umov, Lord Rayleigh, Reynolds, Leontovich, Biot, Lighthill (see Umov 1874, 1950 [54, 55], Gulo 1977 [24], Lord Rayleigh 1885 [38], Biot 1957 [8], and Lighthill 1965 [37]) and several others, who established similar particular facts with various degrees of lucidity. The fundamental notions of slowness and group velocity are traceable to Hamilton (see Arnold 1989 [4] and Levin 1978 [34], respectively). The group velocity is important in studying waves of various nature (see, e.g., Whitham 1974 [56]).

The boundedness of the velocity of propagation of perturbation is a characterizing feature of hyperbolic equations (the nonstationary elastodynamics equations are of the hyperbolic type). A standard technique for proving this fact is based on energy considerations (for the scalar case, see, e.g., Ladyzhenskaya 1985 [32]). Section 2.5 is based on lectures by V.M. Babich.

A cumbersome problem on reflection – transmission of plane waves on a welded interface, which is important for applications, is brilliantly analyzed by Aki and Richards 1980 [1]. In particular, they discussed the unitarity of the corresponding reflection – transmission matrix. The problem of the total internal reflection of an arbitrary nonstationary impulse was first solved by Sobolev 1937 [50], and later, independently, by Friedlander 1948 [22].

A detailed analysis of waves in anisotropic media was given by Musgrave 1970 [44], who depicted slowness surfaces of many crystals. Musgrave (and many others following him, e.g., Every 1986 [21] and Norris 2007 [46]) paid attention to a relation between umbilic points of the velocity surface and conical points of the slowness surface.

The classical Rayleigh wave in an isotropic homogeneous half-space was discovered by Lord Rayleigh 1885 [38], who established its existence by numerically solving an equation equivalent to (2.209) in the case where $\lambda = \mu$, that is, $a = \sqrt{3}b$. The proof of the existence and uniqueness of the positive solution of this equation for arbitrary constants $a$ and $b$, $a > b > 0$, given in Section 2.7.1, follows Sobolev 1937 [50].

The construction presented in Section 2.7.3 is closely related to deriving the quantization conditions, which can be found, e.g., in books by Maslov [41], Babich and Buldyrev [5], and in earlier works by Keller and Rubinow 1960 [30], and Einstein 1917, 1966 [19, 20].

The *Stoneley wave* on a welded interface between half-spaces was noted by Stoneley 1924 [52]. For its existence, it is necessary and sufficient that some inequality involving parameters of half-spaces hold (see, e.g., Chadwick and Borejko 1994 [13]). The *Schölte – Gogoladze wave* running along the interface between an isotropic elastic and a liquid half-spaces for any values of parameters was found independently in Schölte 1947 [47] and Gogoladze 1948 [23]. Gogoladze 1948 [23], using the techniques due to Sobolev 1937 [50], has established the existence and uniqueness of this wave. We also mention the *Krauklis wave* running in a liquid layer between elastic half-spaces (see Krauklis 1962 [31]).

Literature relating to surface and waveguide waves in layered media is enormous. We mention books by Aki and Richards 1980 [1], and Levshin, Yanovskaya, Lander et al. 1973 [36], dedicated to seismology, books by Dieulesaint and Royer 1980 [17], Biryukov, Gulyaev, Krylov, and Plessky 1995 [10], and Musgrave 1970 [44], dedicated to acoustics, monographs of a general character by Brekhovskikh 1960 [11], Brekhovskikh and Godin 1991 [12], and Kaplunov and Prikazchikov 2017 [29], and also articles by Alenitsyn 1963 [2], Shuvalov 2008 [48], Shuvalov, Poncelet, and Kiselev 2008 [49], Balogun and Achenbach 2012 [6], among others.

The investigation of the Rayleigh wave in a homogeneous anisotropic half-space has a long history. On the basis of numerical simulation, the existence and uniqueness (as well as the nonexistence and nonuniqueness) of the Rayleigh wave were reported for different directions in media with special types of anisotropy (see, e.g., a monograph by Musgrave 1970 [44]). The first general result was given by Lothe and Barnett 1976 [39], who established, in the case of general anisotropy, the existence of the unique (up to a constant factor) Rayleigh wave for any direction of its propagation along the surface. The important *nondegeneracy condition* had first appeared there.

The proof first outlined by Lothe and Barnett 1976 [39] and presented in detail by Barnett and Lothe 1985 [7], Chadwick and Smith 1977 [14], Alshits, Darinskii, and Shuvalov 1992 [3] and others was based on the *Stroh formalism* — a specific technique of studying a naturally arising system of linear algebraic equations. The underlying algebraic ideas of this formalism were somehow clarified by Hansen 2012, 2014 [25], [26]. Alternatives to the Stroh formalism are now being sought (see, e.g., Mielke and Fu 2004 [42], where a novel proof of the uniqueness of the Rayleigh wave was given).

The approach presented in Section 2.10 is based on variational methods fundamental in mathematical physics and possesses more generality than the Stroh formalism. Variational methods seem to be applicable to inhomogeneous media (which can, in particular, be periodic with respect to the longitudinal variable and depth). Such methods provide a rigorous proof of the existence of a Rayleigh-type wave running along the edge of an elastic wedge (see Kamotskii 2009 [27], Zavorokhin and Nazarov 2011 [57]).

---

## References to Chapter 2

[1] Aki, K. and Richards, P. G. 1980. *Quantitative seismology: Theory and methods.* San Francisco: W. H. Freeman and Company. *Аки К., Ричардс П.* Количественная сейсмология. Т. 1, 2. М.: Мир, 1983.

[2] Alenitsyn, A. G. 1963. Rayleigh waves in a nonhomogeneous elastic half-space. *J. Appl. Math. Mech.* 27: 816–22. *Аленицын А. Г.* Волны Рэлея в неоднородном упругом полупространстве волноводного типа. Прикл. матем. мех., 1967. Т. 31(2). С. 222–229.

[3] Alshits, V. I., Darinskii, A. N. and Shuvalov, A. L. 1992. Elastic waves in infinite and semiinfinite anisotropic media. *Physica Scripta.* 44:85–93.

[4] Arnold, V. I. 1989. *Mathematical methods of classical mechanics.* New York: Springer-Verlag. *Арнольд В. И.* Математические методы классической механики. М.: Едиториал УРСС, 2003.

[5] Babich, V. M. and Buldyrev, V. S. 2007. *Asymptotic methods in short-wavelength diffraction theory.* Oxford: Alpha Science. *Бабич В. М., Булдырев В. С.* Асимптотические методы в задачах дифракции коротких волн. Метод эталонных задач. М.: Наука, 1972.

[6] Balogun, O. and Achenbach, J. D. 2012. Surface waves on a half space with depth-dependent properties. *J. Acoust. Soc. Am.* 132:1336–45.

[7] Barnett, D. M. and Lothe, J. 1985. Free surface (Rayleigh) waves in anisotropic elastic half-spaces: the surface impedance method. *Proc. Roy. Soc. London. Ser. A.* 402(1822):135–52.

[8] Biot, M. A. 1957. General theorems on the equivalence of group velocity and energy transport. *Phys. Rev.* 105:1129–37.

[9] Birman, M. S. and Solomjak, M. Z. 1987. *Spectral theory of self-adjoint operators in Hilbert space.* Dordrecht: Reidel. *Бирман М. Ш., Соломяк М. З.* Спектральная теория самосопряженных операторов в гильбертовом пространстве. СПб–Краснодар: Лань, 2010.

[10] Biryukov, S. V., Gulyaev, Y. V., Krylov, V. and Plessky, V. 1995. Surface acoustic waves in inhomogeneous media. Berlin: Springer-Verlag. *Бирюков С. В., Гуляев Ю. В., Крылов В. В., Плесский В. П.* Поверхностные акустические волны в неоднородных средах. М.: Наука, 1991.

[11] Brekhovskikh, L. M. 1960. *Waves in layered media.* New York: Academic Press. *Бреховских Л. М.* Волны в слоистых средах. М.: Изд-во АН СССР, 1957.

[12] Brekhovskikh, L. M. and Godin, O. A. 1991. *Acoustics of layered media. Vol. 1: Plane and quasi-plane waves.* Heidelberg: Springer-Verlag. *Бреховских Л. М., Годин О. А.* Акустика слоистых сред. М.: Наука, 1989.

[13] Chadwick, P. and Borejko, P. 1994. Existence and uniqueness of Stoneley waves. *Geophys. J. Intern.* 118:279–84.

[14] Chadwick, P. and Smith, G. D. 1977. Foundations of the theory of surface waves in anisotropic elastic materials. *Adv. Appl. Mech.* 17:303–76.

[15] Courant, R. and Hilbert, D. 1962a. *Methods of mathematical physics.* Vol. 1. New York: Interscience. *Курант Р., Гильберт Д.* Методы математической физики. Т. 1. М.–Л.: ГТТИ, 1934.

[16] Courant, R. and Hilbert, D. 1962b. *Methods of mathematical physics. Vol. 2.* New York: Interscience. *Курант Р.* Уравнения с частными производными. М.: Наука, 1964.

[17] Dieulesaint, E. and Royer, D. 1980. *Elastic waves in solids: applications to signal processing.* New York: John Wiley & Sons. *Дьелесан Э., Руайе Д.* Упругие волны в твердых телах. Применение для обработки сигналов. М.: Наука, 1982.

[18] Dubrovin, B. A, Fomenko, A. T. and Novikov, S. P. 1984. *Modern geometry — methods and applications. Part I. The geometry of surfaces, transformation groups, and fields.* New York: Springer-Verlag. *Дубровин Б. А., Новиков С. П., Фоменко А. Т.* Современная геометрия: Методы и приложения. Т. 1. М.: Наука, 1998.

[19] Einstein, A. 1917. Zum Quantensatz von Sommerfeld und Epstein. *Verhandl. Dtsch. Phys. Ges.* 19:82–92.

[20] Einstein, A. 1966. On the quantum condition of Sommerfeld and Epstein. *Collected papers. Vol. 3.* Moscow: Nauka. *Эйнштейн А.* К квантовому условию Зоммерфельда и Эпштейна. Собр. трудов. Т. 3. М.: Наука.

[21] Every, A. G. 1986. Formation of phonon-focusing caustics in crystals. *Phys. Rev. B.* 34:2852–62.

[22] Friedlander F. G. 1948. On the total reflection of plane waves. *Quart. J. Mech. Appl. Math.* 1:376–84.

[23] Gogoladze, V. G. 1948. Rayleigh waves on the interface between a compressible fluid medium and a solid elastic half-space. *Trudy Seismol. Inst. Acad. Nauk. USSR.* 127:27–32 [in Russian]. *Гоголадзе В. Г.* Волны Рэлея на границе сжимаемой жидкости и твердого упругого полупространства. Труды Сейсмол. ин-та АН СССР, 1948. Т. 127. С. 27–32.

[24] Gulo, D. D. 1977. *N. A. Umov.* Moscow: Prosveschenie [in Russian]. *Гуло Д. Д.* Н. А. Умов. М.: Просвещение, 1977.

[25] Hansen, S. 2012. The surface impedance tensor and Rayleigh waves. *Proc. Int. Conf. Days on Diffraction 2012*, St. Petersburg, 2012. P. 115–9.

[26] Hansen, S. 2014. Subsonic free surface waves in linear elasticity. *SIAM J. Math. Anal.* 46:2501–24.

[27] Kamotskii, I. V. 2009. Surface wave running along the edge of an elastic wedge. *St. Petersburg Math. J.* 20:59–63. *Камоцкий И. В.* О поверхностной волне, бегущей вдоль ребра упругого клина. Алгебра и анализ, 2008. Т. 20(1). С. 86–92.

[28] Kamotskii, I. V. and Kiselev, A. P. 2009. An energy approach to the proof of the existence of Rayleigh waves in an anisotropic elastic half-space. *J. Appl. Math. Mech.* 73:464–70. *Камоцкий И. В., Киселев А. П.* Энергетический подход к доказательства существования волн Рэлея в анизотропном упругом полупространстве. Прикл. матем. мех., 2009. Т. 74(4). С. 643–654.

[29] Kaplunov, J. and Prikazchikov, D. A. 2017. *Asymptotic theory for Rayleigh and Rayleigh-type waves. Advances in Applied Mechanics, Volume 50.* Amsterdam: Elsevier.

[30] Keller, J. B. and Rubinow, S. I. 1960. Asymptotic solution of eigenvalue problems. *Ann. Phys.* 9:24–75.

[31] Krauklis, P. V. 1962. On some low-frequency vibrations of a liquid layer in an elastic medium. *J. Appl. Math. Mech.* 26:1685–92. *Крауклис П. В.* О некоторых низкочастотных колебаниях жидкого слоя в упругой среде. Прикл. матем. мех., 1962. Т. 26(6). С. 1111–1115.

[32] Ladyzhenskaya, O. A. 1985. *The boundary value problems of mathematical physics.* New York: Springer. *Ладыженская О. А.* Краевые задачи математической физики. М.: Физматгиз, 1973.

[33] Landau, L. D. and Lifshitz, E. M. 2007. *Mechanics.* Amsterdam: Elsevier. *Ландау Л. Д., Лифшиц Е. М.* Механика. М.: Физматлит, 2012.

[34] Levin, M. L. 1978. How light conquers darkness (W. R. Hamilton and the concept of group velocity). *Sov. Phys. Uspekhy.* 21:639–40. *Левин М. Л.* Как свет побеждает тьму (У. Р. Гамильтон и понятие групповой скорости). Успехи физ. наук, 1978. Т. 125(7), С. 565–567.

[35] Levshin, A. L. 1973. *Surface and channel waves.* Moscow: Nauka [in Russian]. *Левшин А. Л.* Поверхностные и каналовые сейсмические волны. М.: Наука, 1973.

[36] Levshin, A. L., Yanovskaya, T. B., Lander, A. V., Bukchin, B. G. et al. 1989. *Seismic surface waves in a laterally inhomogeneous Earth.* Dordrecht: Kluwer. *Левшин А. Л., Яновская Т. Б., Ландер А. В. и др.* Поверхностные сейсмические волны в горизонтально-неоднородной Земле. М.: Наука, 1987.

[37] Lighthill, M. J. 1965. Group velocity. *IMA J. Appl. Math.* 1:1–28.

[38] Lord Rayleigh. 1885. On waves propagated along the plane surface of an elastic solid. *Proc. London Math. Soc.* 17(253):4–11.

[39] Lothe, J. and Barnett, D. M. 1976. On the existence of surface wave solutions for anisotropic elastic half-space with free surface. *J. Appl. Phys.* 47:428–33.

[40] Love, A. E. H. 1911. *Some problems in geodynamics.* Cambridge: Cambridge University Press.

[41] Maslov, V. P. 1972. *Théorie des perturbations et méthodes asymptotiques.* Paris: Dunod. *Маслов В. П.* Теория возмущений и асимптотические методы. М.: Изд-во МГУ, 1965.

[42] Mielke, A. and Fu, Y. B. 2004. Uniqueness of the surface wave speed: a proof that is independent of the Stroh formalism. *Math. Mech. Solids.* 9:5–16.

[43] Mikhlin, S. G. 1964. *Variational methods in mathematical physics.* New York: Pergamon. *Михлин С. Г.* Вариационные методы в математической физике. М.: Наука, 1970.

[44] Musgrave, M. J. P. L. 1970. *Crystal acoustics.* San Francisco: Holden-Day.

[45] Naimark, M. A. 1968. Linear differential operators, Vols. 1, 2. New York: Ungar. *Наймарк М. А.* Линейные дифференциальные операторы. М.: Наука, 1969.

[46] Norris A. N. 2007. Wavefront singularities associated with the conical point in elastic solids with cubic symmetry. *Wave Motion.* 44:513–27.

[47] Schölte J. G. 1947. The range of existence of Rayleigh and Stoneley waves. *Geophys. J. Intern.* 5(S5):120–6.

[48] Shuvalov, A. L. 2008. The high-frequency dispersion coefficient for the Rayleigh velocity in a vertically inhomogeneous anisotropic half-space. *J. Acoust. Soc. Am.* 123:2484–7.

[49] Shuvalov, A. L., Poncelet, O. and Kiselev, A. P. 2008. Shear horizontal waves in transversely inhomogeneous plates. *Wave Motion.* 45:605–15.

[50] Sobolev, S. L. 1937. Some problems in wave propagation. In Frank, P. and von Mises, R. (Eds.), *Differential and integral equations of mathematical physics.* [Russian translation.] Moscow–Leningrad: ONTI, 468–617. *Соболев С. Л.* Некоторые вопросы теории распространения колебаний. В кн.: *Франк Ф., Мизес Р.* Дифференциальные и интегральные уравнения математической физики. Л.–М.: ОНТИ, 1937.

[51] Sommerfeld, A. 1950. *Mechanics of deformable bodies.* New York: Academic Press. *Зоммерфельд А.* Механика деформируемых сред. М.: Изд-во иностр. лит., 1954.

[52] Stoneley, R. 1924. Elastic waves at the surface of separation of two solids. *Proc. Roy. Soc. London. Ser. A.* 106:416–28.

[53] Strutt, J. W. (Baron Rayleigh). 1945. *The theory of sound.* New York: Dover. *Стретт Дж. В. (лорд Рэлей).* Теория звука. М.: ГИТТЛ, 1955.

[54] Umov, N. A. 1874. *Equations of the movement of energy in bodies.* Odessa: Ulrich & Schulze Printing House [in Russian]. *Умовъ Н. А.* Уравненія движенія энергіи въ тѣлахъ. Одесса: Въ типографіи Ульриха и Шульце, 1874.

[55] Umov, N. A. 1950. *Selected papers.* Moscow: GTTI [in Russian]. *Умов Н. А.* Избранные сочинения. М.: ГТТИ, 1950.

[56] Whitham, G. B. 1974. *Linear and nonlinear waves.* New York: Wiley. *Уизем Дж.* Линейные и нелинейные волны. М.: Мир, 1977.

[57] Zavorokhin, G. L. and Nazarov, A. I. 2011. On elastic waves in a wedge. *J. Math. Sci.* 175:646–50. *Заворохин Г. Л., Назаров А. И.* Об упругих волнах в клине. Зап. научн. сем. ПОМИ, 2010. Т. 380. С. 45–52.

# Chapter 3

## Point Sources and Spherical Waves in Homogeneous Isotropic Media

In this chapter, we address wavefields generated by specialized sources of volume forces. These are sources concentrated at a single point of the space. Such right-hand sides of the elastodynamics equations (1.34) are called *point sources (of oscillation)*, while the point at which they are localized is called the *source point.*

On the one hand, point sources serve as good approximations in cases where outer forces act in a small area. On the other hand, point-source problems allow simple explicit solutions, and arbitrary sources can be conveniently represented in terms of point sources. Uniqueness for problems with point sources is considered in Section 3.5.

## 3.1 Delta functions

The mathematical description of point sources (of oscillation) requires an adequate mathematical tool, which allows dealing with functions localized at a single point. We will consider several types of such point sources, which are important in applications. Nowadays, one can hardly imagine any systematic treatment of the theory of waves generated by point sources that would not be built upon techniques of delta functions. Such functions are beyond the scope of classical analysis, they constitute a subclass of the vast class of *generalized functions.* We are going to give a short outline of the formal aspect of the theory of delta functions. As teachers of our teachers said, "we have no time for Gaussian rigor, gentleman" (Klein 1937 [18]).

### The one dimension case

First consider a one-dimensional delta function $\delta(x - x^{\circ})$. We fix a point $x^{\circ} \in \mathbb{R}$. Let $\delta_{\varepsilon}(x-x^{\circ})$ be a family of nonnegative integrable functions of $x \in \mathbb{R}$ (for example, step functions) identically vanishing as $|x - x^{\circ}| > \varepsilon$, where $\varepsilon$ is

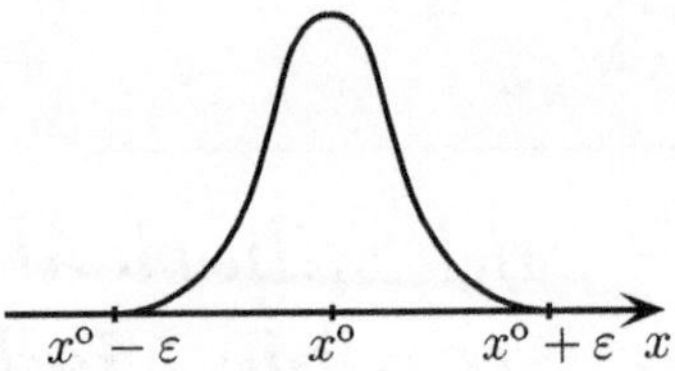


**Figure 3.1**: The function $\delta_\varepsilon(x - x^o)$

an arbitrary (fixed) positive number,

$$\begin{aligned} \delta_\varepsilon(x - x^o) &\geqslant 0, \quad |x - x^o| < \varepsilon, \\ \delta_\varepsilon(x - x^o) &= 0, \quad |x - x^o| > \varepsilon, \end{aligned} \tag{3.1}$$

see Fig. 3.1, and let us require that

$$\int_{-\infty}^{+\infty} \delta_\varepsilon(x - x^o)dx = 1. \tag{3.2}$$

A *delta function* can be defined as a limit[1]

$$\delta(x - x^o) = \lim_{\varepsilon \to 0} \delta_\varepsilon(x - x^o). \tag{3.3}$$

The formal passage to a limit in (3.3) yields

$$\delta(x - x^o) = 0, \quad x \neq x^o \tag{3.4}$$

and

$$\int_{-\infty}^{+\infty} \delta(x - x^o)dx = 1. \tag{3.5}$$

If $f$ is a continuous function, it can easily be shown that

$$\lim_{\varepsilon \to 0} \int_{-\infty}^{+\infty} f(x)\delta_\varepsilon(x - x^o)dx = f(x^o).$$

Together with (3.3), this gives the relation

$$\int_{-\infty}^{+\infty} f(x)\delta(x - x^o)dx = f(x^o). \tag{3.6}$$

As immediately follows from (3.5), for any constant $\mathrm{C} > 0$

$$\delta(\mathrm{C}x) = \frac{1}{\mathrm{C}}\delta(x), \tag{3.7}$$

that is, a one-dimensional delta function is homogeneous of degree $-1$.

[1]This limit should be understood in some special *weak sense*, as mathematicians say. A delta function is not a function in the common classical sense. We do not go into mathematical details, which can be found in many good books (see, e.g., Gelfand and Shilov 1964 [8], Jones 1982 [14], Kolmogorov and Fomin 1970 [19], Schwartz 1950 [23], Vladimirov 1984 [29]).

Let the function $f$ be $l$ times continuously differentiable. After differentiating (3.6) with respect to $x^{\circ}$, we obtain

$$\int_{-\infty}^{+\infty} f(x)\delta^{(p)}(x-x^{\circ})dx = (-1)^p \left(\frac{d}{dx^{\circ}}\right)^p f(x^{\circ}), \quad p \leqslant l, \tag{3.8}$$

where

$$\delta^{(p)}(y) := \frac{d^p \delta(y)}{dy^p} \tag{3.9}$$

is the $p$-th derivative of $\delta(y)$.

## Heaviside function

The Heaviside function is traditionally defined as

$$H(x) = \begin{cases} 0 & \text{for } x \leqslant 0, \\ 1 & \text{for } x > 0, \end{cases} \tag{3.10}$$

(see., e.g., Kolmogorov and Fomin 1970 [19]). In the theory of generalized functions (see, e.g., Gelfand and Shilov 1964 [8], Schwartz 1950 [23], Vladimirov 1984 [29]), it is proved that

$$\frac{dH(x)}{dx} = \delta(x). \tag{3.11}$$

The formula (3.11) agrees with the fact that for $x \neq 0$ the classically understood derivative $\frac{dH(x)}{dx}$ equals zero, as well as $\delta(x)$; see (3.4).

Let $h = \text{const} > 0$. Equation (3.11), together with the (formally understood) fundamental theorem of calculus (the Newton–Leibnitz formula), yields that for $x > 0$,

$$\int_{-h}^{x} \frac{dH(\xi)}{d\xi}\, d\xi = H(x) - H(-h) = 1,$$

which agrees with the formulas (3.4) and (3.5) for the case $x^{\circ} = 0$.

## The $m$ dimensions case

A *delta function* $\delta(\mathbf{x}-\mathbf{x}^{\circ})$ *of* $m$ *variables* $\mathbf{x}-\mathbf{x}^{\circ} = (x_1 - x_1^{\circ}, \dots, x_m - x_m^{\circ})$ is defined similarly. Let $\delta_\varepsilon$ be nonnegative functions such that

$$\delta_\varepsilon(\mathbf{x}-\mathbf{x}^{\circ}) = 0, \quad |\mathbf{x}-\mathbf{x}^{\circ}| > \varepsilon, \tag{3.12}$$

where $|\mathbf{x}-\mathbf{x}^{\circ}| = \sqrt{\sum_{p=1}^{p=m}(x_p - x_p^{\circ})^2}$, and

$$\int_{-\infty}^{+\infty} \cdots \int_{-\infty}^{+\infty} \delta_\varepsilon(\mathbf{x}-\mathbf{x}^{\circ})dx_1 \dots dx_m = \int_{\mathbb{R}^m} \delta_\varepsilon(\mathbf{x}-\mathbf{x}^{\circ})d\mathbf{x} = 1 \tag{3.13}$$

($d\mathbf{x} = dx_1 \dots dx_m$ is a volume element in $m$ dimensions). The formal definition of a multidimensional delta function is as follows:

$$\delta(\mathbf{x} - \mathbf{x}^{\circ}) = \lim_{\varepsilon \to 0} \delta_{\varepsilon}(\mathbf{x} - \mathbf{x}^{\circ}). \tag{3.14}$$

Obviously,

$$\delta(\mathbf{x} - \mathbf{x}^{\circ}) = \delta(x_1 - x_1^{\circ})\delta(x_1 - x_1^{\circ}) \dots \delta(x_m - x_m^{\circ}). \tag{3.15}$$

A multi-dimensional delta function can be differentiated with respect to each of $x_k$ any number of times. For example,

$$\frac{\partial}{\partial x_1}\delta(\mathbf{x} - \mathbf{x}^{\circ}) = \delta'(x_1 - x_1^{\circ})\delta(x_2 - x_2^{\circ}) \dots \delta(x_m - x_m^{\circ}), \tag{3.16}$$

$$\delta'(s) := \frac{d\delta(s)}{ds}.$$

We take notice of an $m$-dimensional analog of the equality (3.6),

$$\int_{\mathbb{R}^m} f(\mathbf{x})\delta(\mathbf{x} - \mathbf{x}^{\circ})\, d\mathbf{x} = f(\mathbf{x}^{\circ}). \tag{3.17}$$

which is valid for any continuous $f$. We emphasize the important case of the above identity

$$\int_{\mathbb{R}^m} f(\mathbf{x})(x_j - x_j^{\circ})\delta(\mathbf{x} - \mathbf{x}^{\circ})d\mathbf{x} = 0. \tag{3.18}$$

This can be written as

$$(x_j - x_j^{\circ})\delta(\mathbf{x} - \mathbf{x}^{\circ}) = 0, \quad j = 1, 2, \dots, m. \tag{3.19}$$

A delta function is spherically symmetric, i.e., for any orthogonal matrix $\boldsymbol{P}$,

$$\delta(\boldsymbol{P}\mathbf{x}) = \delta(\mathbf{x}). \tag{3.20}$$

In the case where $\boldsymbol{Q}$ is an arbitrary nondegenerate matrix,

$$\delta(\boldsymbol{Q}\mathbf{x}) = \frac{1}{|\det \boldsymbol{Q}|}\delta(\mathbf{x}). \tag{3.21}$$

Equations (3.7) and (3.15) (or, alternatively, (3.20)) imply that an $m$-dimensional delta function is homogeneous of degree $-m$:

$$\delta(\mathrm{C}\mathbf{x}) = \frac{1}{\mathrm{C}^m}\delta(\mathbf{x}) \text{ for any constant } \mathrm{C} > 0. \tag{3.22}$$

**Fundamental solution of the Laplace equation**

Solutions of a nonhomogeneous linear differential equation with the right-hand side that is (up to a constant factor) a delta function are called *fundamental solutions* (of the corresponding equations). For the case of three dimensions, $m = 3$, we are coming to derivation of the classical formula for the solution of the Laplace equation:

$$\nabla^2 \frac{1}{r} = -4\pi\delta(\mathbf{x} - \mathbf{x}^\circ). \tag{3.23}$$

Here, $r = |\mathbf{x} - \mathbf{x}^\circ| = \sqrt{(x_1 - x_1^\circ)^2 + (x_2 - x_2^\circ)^2 + (x_3 - x_3^\circ)^2}$, and $\nabla^2 := \frac{\partial^2}{\partial x_1^2} + \frac{\partial^2}{\partial x_2^2} + \frac{\partial^2}{\partial x_3^2}$ is the *Laplace operator*, or the *Laplacian.*

For $r > 0$, that is, $\mathbf{x} \neq \mathbf{x}^\circ$, the formula (3.23) can be checked by a routine calculation, which we omit. Notice that both sides of (3.23) are spherically symmetric homogeneous functions of $\mathbf{x} - \mathbf{x}^\circ$ of the same degree $-3$ and vanish for $\mathbf{x} \neq \mathbf{x}^\circ$. Thus they differ only by a constant multiplier. We have just to prove that it equals 1. For this purpose, we integrate both sides of (3.23) over a ball $r < \mathcal{R}$ of an arbitrary radius $\mathcal{R} > 0$ centered at $\mathbf{x}^\circ$ and apply the Ostrogradsky – Gauss divergence theorem. The left-hand side becomes

$$\int_{r \leqslant \mathcal{R}} \nabla^2 \left(\frac{1}{r}\right) d\mathbf{x} = \int_{r = \mathcal{R}} \frac{\partial}{\partial n}\left(\frac{1}{r}\right) dS(\mathbf{x}) = \int_{|\mathbf{x}-\mathbf{x}^\circ| = \mathcal{R}} \frac{\partial}{\partial n}\left(\frac{1}{r}\right) dS(\mathbf{x}),$$

where $\partial/\partial n = \partial/\partial r$ stands for the derivative by the outer normal to the sphere. We note that $(\partial/\partial r)r^{-1} = -r^{-2}$, and $\int_{r=\mathcal{R}} r^{-2} dS = \mathcal{R}^{-2} \int_{r=\mathcal{R}} dS = 4\pi$, which yields

$$\int_{r \leqslant \mathcal{R}} \nabla^2 \left(\frac{1}{r}\right) d\mathbf{x} = -4\pi. \tag{3.24}$$

Regarding the integral on the right-hand side of (3.23), it also equals $-4\pi$ by virtue of (3.13). The formula (3.23) is proved.

The fundamental solution $\frac{1}{r}$, which we have just found, is unique in the class of generalized functions that tend to zero at infinity (see, e.g., monographs by Gelfand and Shilov 1964 [8], Schwartz 1950 [23], and Vladimirov 1984 [29]).

---

## 3.2 Scalar point source problems

Before proceeding to generation of waves by point sources in elastic media, we are going to consider simpler scalar problems. We will address the time-harmonic and nonstationary cases independently, starting with the harmonic case.

### 3.2.1 Time-harmonic source

Consider a wave process described by the *wave equation* under the assumption of the *harmonic time dependence* of the right-hand side:

$$\nabla^2 \mathrm{u} - \frac{1}{c^2}\mathrm{u}_{tt} = -\mathscr{F}(\mathbf{x})e^{-i\omega t}, \quad \mathbf{x} \in \mathbb{R}^3. \tag{3.25}$$

Here, $c = \text{const} > 0$ stands for the velocity of wave propagation. We will deal with solutions of the equation (3.25) also having harmonic time dependence:

$$\mathrm{u}(\mathbf{x}, t) = e^{-i\omega t}u(\mathbf{x}). \tag{3.26}$$

Inserting (3.26) into (3.25), we arrive at the nonhomogeneous Helmholtz equation

$$\nabla^2 u + k^2 u = -\mathscr{F}, \quad k^2 = \frac{\omega^2}{c^2}, \tag{3.27}$$

with the *wave number* $k = \text{const} > 0$. If $\mathscr{F} \equiv 0$, the equation (3.27) becomes the classical *homogeneous Helmholtz equation*

$$\nabla^2 u + k^2 u = 0. \tag{3.28}$$

Consider the source $\mathscr{F}$ described by

$$\mathscr{F} = 4\pi\delta(\mathbf{x} - \mathbf{x}^{\mathrm{o}}). \tag{3.29}$$

The factor $4\pi$ is introduced here to simplify the resulting expression. We assume that the waves go from a source to infinity. Later we will clarify what is meant by this.

Taking into account the spherical symmetry of a delta function, it is natural to seek a solution of the equation

$$\nabla^2 u + k^2 u = -4\pi\delta(\mathbf{x} - \mathbf{x}^{\mathrm{o}}), \tag{3.30}$$

dependent solely on $r = |\mathbf{x} - \mathbf{x}_0|$. For $r > 0$, the equation (3.3.4) reduces to the homogeneous Helmholtz equation $(\nabla^2 + k^2)u = 0$. Putting $u = u(r)$, for $r > 0$ we get

$$u_{rr} + \frac{2}{r}u_r + k^2 u = 0$$

or

$$\frac{\partial^2(ru)}{\partial r^2} + k^2 ru = 0.$$

Therefore,

$$ru = C_+ e^{ikr} + C_- e^{-ikr}, \quad C_\pm = \text{const},$$

that is,

$$u = C_+ \frac{e^{ikr}}{r} + C_- \frac{e^{-ikr}}{r}. \tag{3.31}$$

To find $C_\pm$, we have two conditions, which uniquely determine $u$. The first one is that $u$ describes waves outgoing to infinity, and the second is nonhomogeneous equation (3.29). The uniqueness in the class of functions that depend only on $r$ follows from further consideration. More general uniqueness theorems can be found in Section 3.5.

We start with the first condition. Let

$$\mathrm{u} = ue^{-i\omega t} = \frac{C_+}{r}e^{-i\omega(t-\frac{r}{c})} + \frac{C_-}{r}e^{-i\omega(t+\frac{r}{c})}, \tag{3.32}$$

$k = \frac{\omega}{c}$. The phase of the first term, $-i\omega(t - \frac{r}{c})$, obviously remains constant on spheres $r = ct + \mathrm{const}$ the radii of which grow with time with the constant speed $c$. The phase of the second term is constant on spheres $r = -ct + \mathrm{const}$, which focus to the source point $\mathbf{x} = \mathbf{x}_0$. The first term thus corresponds to an outgoing wave, whence we must put $C_- = 0$. Therefore,

$$u = C_+\frac{e^{ikr}}{r} \equiv \psi(r). \tag{3.33}$$

To find $C_+$, we turn back to the nonhomogeneous equation (3.30). Compare the leading-order singularity of the expression $(\nabla^2 + k^2)\psi(r)$ at the source point $\mathbf{x}^o$ (i.e., at $r \to 0$) with that of the right-hand side of (3.30). As $r \to 0$, we have

$$u = \psi(r) = C_+\frac{1 + ikr - \frac{1}{2}(kr)^2 + O(r^3)}{r} = C_+\left(\frac{1}{r} + ik - \frac{1}{2}k^2 r + O(r^2)\right).$$

Applying the Helmholtz operator $(\nabla^2 + k^2)$ to both sides and taking into account (3.23) and (3.30), we obtain

$$-4\pi C_+\delta(\mathbf{x} - \mathbf{x}^o) + O\left(\frac{1}{r}\right) = -4\pi\delta(\mathbf{x} - \mathbf{x}^o),$$

whence $C_+ = 1$ and

$$u = \frac{e^{ikr}}{r}. \tag{3.34}$$

It is worthy of note that solutions of the equation (3.30) are not reduced to the expression (3.32). An example of another solution is

$$u = \frac{e^{ikr}}{r} + Ce^{ik\mathbf{s}\cdot\mathbf{x}}, \tag{3.35}$$

where $C$ is an arbitrary constant, and $\mathbf{s}$ is an arbitrary unit vector, $\mathbf{s}^2 = 1$. This solution is not outgoing.

### 3.2.2 Determining a unique solution. Key idea of the limiting absorption principle

*The limiting absorption principle* is an extensively used method for determining solutions, interesting from a physical point of view, of time-harmonic wave problems in infinite domains, including the whole space $\mathbb{R}^3$. Its basic idea is as follows. We consider the problems in which no waves come from infinity. Assume the presence of some attenuation in the medium; then the amplitudes of outgoing waves decrease with distance and the amplitudes of incoming waves grow. Requiring that a solution vanish at infinity allows undesired incoming waves to be cut off. Then attenuation should be turned to zero.

It is customary in mathematical physics to model the attenuation of waves by adding a small positive constant to the wave number,[2] that is, by putting in (3.30)

$$k = \operatorname{Re} k + i \operatorname{Im} k, \quad \operatorname{Re} k > 0, \quad \operatorname{Im} k > 0. \tag{3.36}$$

Since $k^2 = \frac{\omega^2}{c^2}$, it is equivalent to considering $\omega$ with a small positive imaginary part. Dealing with the attenuation of solutions of the elastodynamics equations, we take

$$\omega = \operatorname{Re} \omega + i \operatorname{Im} \omega, \quad \operatorname{Re} \omega > 0, \quad \operatorname{Im} \omega > 0. \tag{3.37}$$

We return to the equation (3.30), assuming now that $\operatorname{Im} k > 0$. The first of the two terms in the formula (3.31) tends to zero as $r \to \infty$. Discarding the growing second term and turning $\operatorname{Im} k$ to 0, we get the expression (3.32). The second term in the solution (3.35) also grows at infinity and should be discarded.

Unique solutions in time-harmonic point-source problems in elastodynamics, which we deal with in this chapter and in Chapter 6, can be described in complete analogy with the scalar case. By virtue of the analytical simplicity of these solutions (they are obviously continuous and moreover analytic in $\omega$), the passage to the limit as

$$\operatorname{Im} \omega \to 0$$

is elementary, and we do not discuss it.

A detailed discussion of the limiting absorption principle and adjacent questions for scalar and vector cases is given in Section 3.5. Also, an alternative approach to uniqueness theorems is described there, which is based on the Sommerfeld radiation condition (in the scalar case) and on the *Jones radiation condition* (in isotropic elastodynamics), instead of introduction of a fictitious attenuation.

### 3.2.3 Nonstationary source

Now we address a solution of the problem for the wave equation with the *velocity of propagation* $c$ for the source localized at a *source point* $\mathbf{x}^\circ$,

$$\nabla^2 \mathrm{u} - \frac{1}{c^2} \ddot{\mathrm{u}} = -4\pi \mathcal{F}(t) \delta(\mathbf{x} - \mathbf{x}^\circ). \tag{3.38}$$

[2]We have in mind a mathematical trick rather than an adequate modeling of attenuation of acoustic or elastic waves.

In this setting, the solution is obviously not unique: one can add to it, e.g., any plane wave $\mathrm{u} = \mathrm{f}(\mathbf{p}\cdot\mathbf{x}+p_0 t)$, where $\mathbf{p}^2 = \frac{1}{c^2}p_0^2$ and f is an arbitrary function.

From a physical point of view, of much interest is the case of a *causal time dependence* of the source, that is,

$$\mathcal{F}(t) \equiv 0 \quad \text{for} \quad t < 0. \tag{3.39}$$

In this case, it appears reasonable to require that the wavefield be identical zero before the source starts acting, i.e.,

$$\mathrm{u}\big|_{t<0} \equiv 0. \tag{3.40}$$

In the sequel, the conditions (3.40) and (3.39) will be assumed to hold.

We start by constructing a solution for $\mathbf{x} \neq \mathbf{x}^{\text{o}}$. The spherical symmetry of the problem prompts to seek it in the form $\mathrm{u} = \mathrm{u}(r,t)$, $r = |\mathbf{x} - \mathbf{x}^{\text{o}}|$, which leads to the equation

$$\mathrm{u}_{rr} + \frac{2}{r}\mathrm{u}_r - \frac{1}{c^2}\mathrm{u}_{tt} = 0, \quad r \neq 0. \tag{3.41}$$

The substitution $\mathrm{u} = \frac{\mathrm{v}}{r}$ reduces it to the one-dimensional wave equation (or the *equation of transverse motion of a string*)

$$\mathrm{v}_{rr} - \frac{1}{c^2}\mathrm{v}_{tt} = 0. \tag{3.42}$$

A general solution of the equation (3.42) is known to be

$$\mathrm{v} = \chi^+\left(t + \frac{r}{c}\right) + \chi^-\left(t - \frac{r}{c}\right), \tag{3.43}$$

where $\chi^+$ and $\chi^-$ are arbitrary functions of one variable (see, e.g., Smirnov 1964a [24], Tikhonov and Samarsky 1965 [27]). Therefore

$$\mathrm{u} = \frac{1}{r}\chi^+\left(t + \frac{r}{c}\right) + \frac{1}{r}\chi^-\left(t - \frac{r}{c}\right). \tag{3.44}$$

To satisfy the condition (3.40), we discard the first term describing the wave incoming from infinity,[3] whence

$$\mathrm{u} = \frac{1}{r}\chi^-\left(t - \frac{r}{c}\right), \tag{3.45}$$

with $\chi^-$ yet undetermined.

---

[3] Develop this argument in a formal manner. Indeed, applying the differential operator $\frac{\partial}{\partial r} - \frac{1}{c}\frac{\partial}{\partial t}$ to both sides of (3.43), we establish that ${\chi^+}'(t + \frac{r}{c}) = 0$ as $t < 0$. Since any real number can be represented as a difference of two positive ones, we get that for any $\xi$, $\xi \in \mathbb{R}$, ${\chi^+}'(\xi) = 0$, that is, $\chi^+ = \text{const}$. Attributing a constant to $\chi^-$, we arrive at (3.45).

Turn $\mathbf{x}$ to $\mathbf{x}^\circ$, i.e., $r \to 0$. Taking notice that the expression (3.45) has at $\mathbf{x} = \mathbf{x}^\circ$ a singularity with respect to $r$, which is the same (up to a factor dependent on $t$) as $1/r$, with the use of (3.23), we have

$$\left(\nabla^2 - \frac{1}{c^2}\frac{\partial^2}{\partial t^2}\right)\frac{1}{r}\chi^-\left(t - \frac{r}{c}\right) \to -4\pi\chi^-(t)\delta(\mathbf{x} - \mathbf{x}^\circ).$$

Comparison with (3.38) at $r = 0$ provides $\chi^-(s) = \mathcal{F}(s)$. The condition (3.40) is satisfied because of (3.39).

To summarize, we showed that

$$\mathrm{u} = \frac{1}{r}\mathcal{F}\left(t - \frac{r}{c}\right) \tag{3.46}$$

is a solution of the problem (3.38)–(3.40). After a proper refinement, the uniqueness of the solution of this problem can be established, see Section 3.5.

---

## 3.3 Point sources in a homogeneous, isotropic, elastic medium. Time-harmonic case

Consider the generation of time harmonic waves by point sources localized at a *source point* $\mathbf{x}^\circ$. We are solving equations of the form

$$\boldsymbol{l}\boldsymbol{u} = -\boldsymbol{F}, \tag{3.47}$$

$$\boldsymbol{l}\boldsymbol{u} := \nu \operatorname{grad}\operatorname{div}\boldsymbol{u} - \mu \operatorname{rot}\operatorname{rot}\boldsymbol{u} + \rho\omega^2\boldsymbol{u}, \qquad \nu := \lambda + 2\mu. \tag{3.48}$$

It is assumed that the vector $\boldsymbol{F}$ is a (finite) linear combination of $\delta(\mathbf{x} - \mathbf{x}^\circ)$ and its derivatives with vector coefficients.[4] We seek solutions that satisfy the *limiting absorption principle* (see Sections 3.2.2 and 3.5).

We begin with two simple but physically meaningful problems.

### Center of expansion

First, we consider the spherically symmetric point source of the form

$$\boldsymbol{F}(\mathbf{x} - \mathbf{x}^\circ) = 4\pi \operatorname{grad}\delta(\mathbf{x} - \mathbf{x}^\circ), \tag{3.49}$$

known as the *center of expansion*. Seeking the solution as

$$\boldsymbol{u} = \operatorname{grad}\phi, \tag{3.50}$$

we immediately find

$$(\nu\nabla^2 + \rho\omega^2)\phi = \nu(\nabla^2 + \kappa^2)\phi = -4\pi\delta(\mathbf{x} - \mathbf{x}^\circ), \tag{3.51}$$

[4]It can be shown that no other generalized vector functions localized at a point $\mathbf{x}^\circ$ exist (this easily follows from *L. Schwartz's theorem* (see, e.g., Vladimirov 1984 [29]).

where $\kappa$ is the wave number of $P$-waves (see (2.78)). As follows from the results of Section 3.2.1, the solution of the equation (3.51), describing waves outgoing to infinity, is $\frac{1}{\nu}A$ with

$$A = \frac{e^{i\kappa r}}{r}, \tag{3.52}$$

whence

$$\boldsymbol{u} = \operatorname{grad}\phi = \frac{1}{\lambda + 2\mu}\operatorname{grad}A = \frac{1}{\rho a^2}\operatorname{grad}A. \tag{3.53}$$

**Center of rotation**

Second, we consider the point source described by

$$\boldsymbol{F}(\mathbf{x} - \mathbf{x}^{\circ}) = 4\pi\operatorname{rot}(\boldsymbol{m}\delta(\mathbf{x} - \mathbf{x}^{\circ})), \tag{3.54}$$

where $\boldsymbol{m} \neq 0$ is an arbitrary constant vector, e.g., a unit one. This axisymmetric source is known as the *centre of rotation (around $\boldsymbol{m}$)*. Seeking the solution as

$$\boldsymbol{u} = \operatorname{rot}\boldsymbol{\psi}, \tag{3.55}$$

we find

$$(\mu\nabla^2 + \rho\omega^2)\boldsymbol{\psi} = \mu(\nabla^2 + æ^2)\boldsymbol{\psi} = -4\pi\boldsymbol{m}\delta(\mathbf{x} - \mathbf{x}^{\circ}), \tag{3.56}$$

where æ is the *wave number of $S$-waves* (see (2.78)). Therefore, $\boldsymbol{\psi} = \frac{1}{\mu}\boldsymbol{m}B$ with

$$B = \frac{e^{iær}}{r}, \tag{3.57}$$

whence

$$\boldsymbol{u} = \operatorname{rot}\boldsymbol{\psi} = \frac{1}{\mu}\operatorname{rot}(\boldsymbol{m}B) = \frac{1}{\rho b^2}\operatorname{rot}(\boldsymbol{m}B). \tag{3.58}$$

The vector function (3.58) is the required solution of the equation (3.47), (3.54), satisfying the limiting absorption principle. The same is true for (3.53) and for all other time-harmonic solutions discussed in this chapter.

### 3.3.1 ⋆ Center of expansion and center of rotation as limit problems for spherical emitters

This section is aimed at clarifying the physical meaning of the point center of expansion (3.49) and the center of rotation (3.54). They arise here in the course of solving two boundary value problems with specialized tractions acting on the boundary of a spherical cavity $r = \mathfrak{D}$, $\mathfrak{D} > 0$, within a homogeneous isotropic medium. The traction, which we call *load*, in this context is spherically symmetric in one case and is special axially symmetric in the other. Both boundary value problems are named (in the Russian literature) those of *spherical emitters*, which are of some interest for seismic exploration. They allow explicit solutions (see, e.g., Blake Jr. 1952 [3] and Gurvich 1968 [10]), which can be easily simplified when the radius of the cavity $\mathfrak{D}$ is small (as compared to the wave length). Assuming that the radius of the cavity is small and the amplitude of the load is concordantly large, in the limit as $\mathfrak{D} \to 0$ we arrive at two problems with point sources, which prove to be a center of expansion (3.49) and a center of rotation (3.54), respectively.

**Spherically symmetric load**

Let the load on the surface of the cavity be

$$\mathbf{t}^{\mathbf{s}}\boldsymbol{u}|_{r=\mathfrak{D}} = -\mathfrak{p}\mathbf{s}, \tag{3.59}$$

where $\mathbf{s} = \operatorname{grad} r$ is the unit normal to the sphere, $\mathfrak{p} \neq 0$ is a constant amplitude factor, and for $r > \mathfrak{D}$ let the *homogeneous elastodynamics equations* be satisfied

$$\boldsymbol{l}\boldsymbol{u} = 0. \tag{3.60}$$

The uniqueness is guaranteed by the requirement that the limiting absorption principle hold. The boundary condition (3.59) describes harmonic in time pressure uniform over the surface of the cavity. The problem with the traction given by (3.59) models the generation of elastic waves by explosion.

We seek the solution in the form

$$\boldsymbol{u} = C \operatorname{grad} A \tag{3.61}$$

(see (3.52)), where $C$ is a constant to be determined from the boundary condition (3.59). The vector function (3.61) obviously satisfies the equation (3.60) for $r > \mathfrak{D}$, as well as the limiting absorption principle. Its substitution into the boundary condition requires some algebra. It follows from equations (1.58) and (3.61) and the identity $(\nabla^2 + \kappa^2)A = -4\pi\delta$ that

$$\mathrm{t}^{\mathrm{s}}_m = C\left\{s_m \lambda \nabla^2 + 2\mu s_j \partial^2_{jm}\right\} A|_{r=\mathfrak{D}} = C\left\{-s_m \lambda \kappa^2 + 2\mu s_j \partial^2_{jm}\right\} A|_{r=\mathfrak{D}},$$

where $\partial^2_{jm} := \partial^2/\partial x_j \partial x_m$. As is easy to check,

$$\partial^2_{mj} A = \left[\left(-\kappa^2 - \frac{3i\kappa}{r} + \frac{3}{r^2}\right) s_m s_j + \delta_{mj}\left(\frac{i\kappa}{r} - \frac{1}{r^2}\right)\right] A. \tag{3.62}$$

For the traction on the surface $r = \mathfrak{D}$, after some calculation we find the expression

$$\mathrm{t}^{\mathrm{s}}_m|_{r=\mathfrak{D}} = C\frac{s_m}{\mathfrak{D}^2}\left[-(\lambda + 2\mu)\kappa^2\mathfrak{D}^2 - 4i\mu\kappa\mathfrak{D} + 4\mu\right] A\Big|_{\mathfrak{D}},$$

the comparison of which with (3.59) provides

$$C = -\frac{\mathfrak{p}\mathfrak{D}^3 \exp(-i\kappa\mathfrak{D})}{4\mu - 4i\mu\kappa\mathfrak{D} - (\lambda + 2\mu)\kappa^2\mathfrak{D}^2}. \tag{3.63}$$

We see that the load given by (3.59) generates only the $P$-wave.

Note that the function defined by (3.61) as a solution of the elastodynamics equations outside the cavity, that is, for $r > \mathfrak{D}$, is analytic in $\mathbb{R}^3$ except for the source point $\mathbf{x}^{\mathrm{o}}$. In the whole space $\mathbb{R}^3$, it satisfies the equation[5]

$$\boldsymbol{l}\boldsymbol{u} = -4\pi\nu C \operatorname{grad} \delta(\mathbf{x} - \mathbf{x}^{\mathrm{o}}). \tag{3.64}$$

Let us turn the radius of the cavity $\mathfrak{D}$ to zero (as can be easily seen, the dimensionless small parameter is $\kappa\mathfrak{D} = \frac{\omega\mathfrak{D}}{a}$), simultaneously increasing the amplitude of the surface load in such a way that

$$\mathfrak{p}\mathfrak{D}^3 \to \mathfrak{C} \ \text{ as } \ \mathfrak{D} \to 0, \tag{3.65}$$

with a finite nonzero constant $\mathfrak{C}$. Passing to the limit as $\mathfrak{D} \to 0$ in (3.64) and denoting

$$\lim_{\mathfrak{D}\to 0} \boldsymbol{u} =: \widetilde{\boldsymbol{u}}, \tag{3.66}$$

we get, instead of the boundary value problem in the exterior of the sphere, the problem in the whole space

$$\boldsymbol{l}\widetilde{\boldsymbol{u}} = -4\pi \widetilde{C} \operatorname{grad} \delta(\mathbf{x} - \mathbf{x}^{\mathrm{o}}). \tag{3.67}$$

Here, $\widetilde{C} = -\frac{\nu}{4\mu}\mathfrak{C} = -\frac{\lambda+2\mu}{4\mu}\mathfrak{C}$.

### Rotational load

Instead of (3.59), consider the load

$$\mathrm{t}^{\mathbf{s}}\boldsymbol{u}|_{r=\mathfrak{D}} = -\mathfrak{p}\mathbf{s} \times \boldsymbol{m} \tag{3.68}$$

with an arbitrary nonzero vector $\boldsymbol{m}$. This traction is tangent to the surface of the cavity $r = \mathfrak{D}$. The solution (which is unique under the assumption that the limiting absorption principle holds) can be found in the form

$$\mathbf{u} = \mathrm{C} \operatorname{rot}(\boldsymbol{m}B), \tag{3.69}$$

$B = \exp(i\text{æ}r)/r$. To simplify the subsequent calculation, we temporarily take $\boldsymbol{m} = \mathbf{e}_3$, whence $\boldsymbol{u} = (\mathrm{C}\partial_2 B, -\mathrm{C}\partial_1 B, 0)$. Simple algebra gives

$$\begin{aligned}
\mathrm{t}^{\mathbf{s}}_1|_{r=\mathfrak{D}} &= \mathrm{C}\mu\left[2s_1\partial^2_{12} + s_2\mathscr{D} + s_3\partial^2_{23}\right] B\big|_{r=\mathfrak{D}}, \\
\mathrm{t}^{\mathbf{s}}_2|_{r=\mathfrak{D}} &= \mathrm{C}\mu\left[s_1\mathscr{D} - 2s_2\partial^2_{12} - s_3\partial^2_{13}\right] B\big|_{r=\mathfrak{D}}, \\
\mathrm{t}^{\mathbf{s}}_3|_{r=\mathfrak{D}} &= \mathrm{C}\mu\,(s_1\partial_2 - s_2\partial_1)\,\partial_3 B\big|_{r=\mathfrak{D}},
\end{aligned}$$

with $\mathscr{D} := \partial^2_{22} - \partial^2_{11}$. Using the analog of (3.62) in which $A$ is replaced by $B$ and $\kappa$ by æ, after some calculations we find

$$\mathrm{t}^{\mathbf{s}}_1|_{r=\mathfrak{D}} = C\mu\mathfrak{b}s_2, \quad \mathrm{t}^{\mathbf{s}}_2|_{r=\mathfrak{D}} = -\mathrm{C}\mu\mathfrak{b}s_1, \quad \mathrm{t}^{\mathbf{s}}_3|_{r=\mathfrak{D}} = 0,$$

[5] The left-hand side obviously equals $\operatorname{grad}(\nu\nabla^2 + \rho\omega^2)\boldsymbol{u} = \nu \operatorname{grad}(\nabla^2 + \kappa^2)A$, and to obtain (3.63) we just apply the equation (3.30).

where $\mathfrak{b}$ denotes the constant

$$\mathfrak{b} = \left(-æ^2 - \frac{3iæ}{r} + \frac{3}{r^2}\right) B\bigg|_{r=\mathfrak{D}} = \frac{1}{\mathfrak{D}^2}(-æ^2\mathfrak{D}^2 - 3iæ\mathfrak{D} + 3)B\bigg|_{r=\mathfrak{D}}.$$

The expression for the traction can be written as $\mathbf{t}^{\mathrm{s}}|_{r=\mathfrak{D}} = C\mu\mathfrak{b}\mathbf{s} \times \mathbf{e}_3$. Conversely, replace $\mathbf{e}_3$ by $\boldsymbol{m}$, which gives

$$\mathbf{t}^{\mathrm{s}}|_{r=\mathfrak{D}} = \mathrm{C}\mu\mathfrak{b}\,\mathbf{s} \times \boldsymbol{m}.$$

The comparison with (3.68) enables us to determine the constant $C$:

$$C = -\frac{\mathfrak{p}\mathfrak{D}^3 \exp(-iæ\mathfrak{D})}{\mu(3 - 3iæ\mathfrak{D} - æ^2\mathfrak{D}^2)}. \tag{3.70}$$

The load given by (3.68) generates only the $S$-wave.

Passing as before to the limit under the condition (3.65), we arrive at the following problem in the whole space:

$$\boldsymbol{l}\tilde{\boldsymbol{u}} = -4\pi\tilde{C}\,\mathrm{rot}(\boldsymbol{m}\delta(\mathbf{x} - \mathbf{x}^{\circ})), \tag{3.71}$$

where $\widetilde{C} = \frac{1}{3}\mathfrak{C}$.

### 3.3.2 Concentrated force

A point source given by

$$\boldsymbol{F} = 4\pi\,\boldsymbol{m}\,\delta(\mathbf{x} - \mathbf{x}^{\circ}) \tag{3.72}$$

with a constant vector $\boldsymbol{m} \neq 0$ is called a *concentrated point force (acting at a point* $\mathbf{x}^{\circ}$ *and directed along* $\boldsymbol{m}$*)*. A solution for an arbitrary point source in the time-harmonic case can be obtained from the solution of this problem by replacing $\boldsymbol{m}$ by a proper differential operator with constant vector coefficients (see the beginning of the Section 3.3).

First we choose $\boldsymbol{m}$ in the particular form

$$\boldsymbol{m} = \mathbf{e}_3, \tag{3.73}$$

and solve the equation

$$\nu\,\mathrm{grad}\,\mathrm{div}\,\boldsymbol{u} - \mu\,\mathrm{rot}\,\mathrm{rot}\,\boldsymbol{u} + \rho\omega^2\boldsymbol{u} = -4\pi\mathbf{e}_3\delta(\mathbf{x} - \mathbf{x}^{\circ}). \tag{3.74}$$

Splitting the right-hand side into a sum of potential and solenoidal vectors, which may seem helpful at first glance, does not prove to be convenient. We proceed somewhat differently. Denote

$$\mathrm{div}\,\boldsymbol{u} = \phi, \quad \mathrm{rot}\,\boldsymbol{u} = \boldsymbol{\psi}. \tag{3.75}$$

Applying the divergence operator to both sides of the equation (3.74) and using the elementary identities

$$\operatorname{div}\operatorname{rot}\operatorname{rot}\boldsymbol{u} = 0, \quad \operatorname{div}\operatorname{grad}\operatorname{div}\boldsymbol{u} = \nabla^2 \operatorname{div}\boldsymbol{u} = \nabla^2\phi, \quad \operatorname{div}(\mathbf{e}_3\delta) = \partial_3\delta,$$

$\partial_j := \partial/\partial x_j$, we derive that

$$\nu(\nabla^2 + \kappa^2)\phi = -4\pi\partial_3\delta, \quad \delta := \delta(\mathbf{x} - \mathbf{x}^{\circ})$$

and

$$\phi = \frac{1}{\nu}\partial_3 A. \tag{3.76}$$

Further, applying the rotation operator to both sides of (3.74) and taking into account the relations

$$\operatorname{rot}\operatorname{grad}\operatorname{div}\boldsymbol{u} = 0, \quad \operatorname{rot}\operatorname{rot}\operatorname{rot}\boldsymbol{u} = -\nabla^2 \operatorname{rot}\boldsymbol{u} = -\nabla^2\boldsymbol{\psi},$$

we likewise obtain

$$\mu(\nabla^2 + æ^2)\boldsymbol{\psi} = -4\pi \operatorname{rot}(\mathbf{e}_3\delta),$$

whence

$$\boldsymbol{\psi} = \frac{1}{\mu}\operatorname{rot}(\mathbf{e}_3 B). \tag{3.77}$$

The equation (3.74) provides the expression for $\boldsymbol{u}$:

$$\begin{aligned} \boldsymbol{u} &= -\frac{1}{\rho\omega^2}(\nu \operatorname{grad}\operatorname{div}\boldsymbol{u} - \mu\operatorname{rot}\operatorname{rot}\boldsymbol{u} + 4\pi\delta\mathbf{e}_3) \\ &= -\frac{1}{\rho\omega^2}(\nu\operatorname{grad}\phi - \mu\operatorname{rot}\boldsymbol{\psi} + 4\pi\delta\mathbf{e}_3). \end{aligned} \tag{3.78}$$

It remains to substitute here the explicit expressions (3.76) and (3.77) for $\phi$ and $\boldsymbol{\psi}$. Obviously,

$$\nu\operatorname{grad}\operatorname{div}\boldsymbol{u} = \nu\operatorname{grad}\phi = \partial_3\operatorname{grad}A$$

and

$$\mu\operatorname{rot}\operatorname{rot}\boldsymbol{u} = \mu\operatorname{rot}\boldsymbol{\psi} = \operatorname{rot}\operatorname{rot}(B\mathbf{e}_3) = -\mathbf{e}_3\nabla^2 B + \operatorname{grad}\operatorname{div}(B\mathbf{e}_3).$$

Finally, using the Helmholtz equation for $B$, $\nabla^2 B + æ^2 B = -4\pi\delta$ (which can be obtained from (3.30), (3.34), replacing $k$ by $\kappa$ or æ, and $\frac{e^{ikr}}{r}$ by $A$ or $B$, respectively), we have

$$\mu\operatorname{rot}\boldsymbol{\psi} = \left(æ^2 B + 4\pi\delta\right)\mathbf{e}_3 + \partial_3\operatorname{grad}B.$$

As a consequence, the delta functions on the right-hand side of (3.78) are cancelled, and we arrive at

$$\boldsymbol{u} = \frac{1}{\rho\omega^2}\left[\left(æ^2\mathbf{e}_3 + \partial_3\operatorname{grad}\right)B - \partial_3\operatorname{grad}A\right],$$

or

$$u = \frac{1}{\rho\omega^2}\left[æ^2 \mathbf{e}_3 B + \partial_3 \operatorname{grad}(B - A)\right]. \tag{3.79}$$

For the components of the vector $u$, this reads

$$u_j = \frac{1}{\rho\omega^2}\left[æ^2 \delta_{3j} B + \partial^2_{3j}(B - A)\right]. \tag{3.80}$$

Note that the terms that describe waves $P$ and $S$ (represented, respectively, via $A$ and $B$ and their derivatives) have singularities of order $O(1/r^3)$ at the source point. The total wavefield (3.79) has, as can be easily seen, a weaker singularity of order $O(1/r)$. The matter is that the splitting of the wavefield into waves $P$ and $S$ in the immediate vicinity of the source point is physically meaningless.

### The Green tensor

Let $G^{(p)}$ denote the solution of the equation (3.47) for a concentrated force directed along the $x_p$-axis,

$$\mathbf{F} = 4\pi \mathbf{e}_p \delta(\mathbf{x} - \mathbf{x}^o). \tag{3.81}$$

A component of the vector $G^p_j$ can be obtained from (3.22) upon replacing 3 by $p$. Denote

$$G^p_j =: G_{pj}.$$

Obviously, the last expression is symmetric with respect to the subscripts $p$ and $j$. The $G_{pj}$ are components of a symmetric tensor called the *Green tensor of the operator* $\boldsymbol{l}$ (see (3.48))

$$G_{pj} = G_{jp} = \frac{1}{\rho\omega^2}[æ^2 \delta_{pj} B + \partial^2_{pj}(B - A)]. \tag{3.82}$$

The role played by the Green tensor is analogous to that of the Green function of the Helmholtz operator $\nabla^2 + k^2$.

### Far-field asymptotics

The *far-field area* consists of points the distance of which from the source point is large as compared to the wavelength. In the far-field area, the formulas for the wavefields of point sources are asymptotically simplified. As immediately seen, in derivation of functions of the form $\frac{1}{r}e^{ikr}$ the most important are derivatives of the exponent:

$$\partial_p\left(\frac{e^{ikr}}{r}\right) = ik\partial_p(r)\frac{e^{ikr}}{r} + O\left(\frac{1}{r^2}\right) = iks_p\frac{e^{ikr}}{r} + O\left(\frac{1}{r^2}\right),$$

$$\partial^2_{pq}\left(\frac{e^{ikr}}{r}\right) = -k^2 s_p s_q \frac{e^{ikr}}{r} + O\left(\frac{1}{r^2}\right), \quad kr \gg 1,$$

where $s_j$ are the components of the unit vector

$$\mathbf{s} = \frac{\mathbf{x} - \mathbf{x}^{\mathrm{o}}}{r}, \tag{3.83}$$

directed from the source point to the *observation point* $\mathbf{x}$.

For the center of expansion (3.49) and the center of rotation (3.54), we have

$$\boldsymbol{u} = \frac{i\kappa}{\rho a^2}\mathbf{s}A + O\left(\frac{1}{r^2}\right) \tag{3.84}$$

and

$$\boldsymbol{u} = \frac{i æ}{\rho b^2}[\mathbf{s} \times \boldsymbol{m}]B + O\left(\frac{1}{r^2}\right), \tag{3.85}$$

respectively. In the first case, the polarization of the leading-order term is longitudinal, whereas in the second case it is transverse, quite as for the respective plane waves.

In the particular case of (3.73), for the concentrated force (3.72) after simple calculation we get

$$\boldsymbol{u} = \frac{\mathbf{e}_3 - s_3\mathbf{s}}{\rho b^2}B + \frac{s_3\mathbf{s}}{\rho a^2}A + O\left(\frac{1}{kr^2}\right). \tag{3.86}$$

In the general case of (3.72), this converts to

$$\boldsymbol{u} = \frac{\boldsymbol{m} - (\boldsymbol{m}, \mathbf{s})\mathbf{s}}{\rho b^2}B + \frac{(\boldsymbol{m}, \mathbf{s})\mathbf{s}}{\rho a^2}A + O\left(\frac{1}{kr^2}\right). \tag{3.87}$$

For this source (similarly to the cases of the center of expansion and the center of rotation), the leading terms of asymptotic far-field expansions for both $P$ and $S$ waves, written out in (3.87), are polarized longitudinally and transversely,[6] respectively. If we wrote higher-order terms (which we avoid), it would be seen that in each case both polarizations are present. This holds for the center of rotation as well.

For a wavefield having in the far-field area the asymptotic of the form

$$\begin{gathered}\boldsymbol{u} = \boldsymbol{u}^a + \boldsymbol{u}^b = \frac{e^{i\kappa r}}{r}f(\mathbf{s})\mathbf{s} + \frac{e^{iær}}{r}\boldsymbol{g}(\mathbf{s}) + O\left(\frac{1}{ær^2}\right), \\ \mathbf{s} = \operatorname{grad} r, \quad \boldsymbol{g}(\mathbf{s}) \perp \mathbf{s},\end{gathered} \tag{3.88}$$

where $\mathbf{s} = \operatorname{grad} r$, the scalar function $f(\mathbf{s})$ and the vector function $\mathbf{g}(\mathbf{s})$ are called its *radiation patterns* or *directivities*.

---

[6] The vector multiplier of $B$ in (3.86) is transversely polarized, because the vector $\mathbf{e}_3 - s_3\mathbf{s}$ is orthogonal to the direction of wave propagation $\mathbf{s}$, $(\mathbf{e}_3 - s_3\mathbf{s}, \mathbf{s}) = s_3 - s_3 = 0$.

## 3.4 Point sources in a homogeneous, isotropic, elastic medium. Nonstationary case

We will consider problems of generation of nonstationary elastic waves by immovable point sources. These problems are described by the equations

$$\mathbf{L}\mathbf{u} = \nu \,\mathrm{grad}\,\mathrm{div}\,\mathbf{u} - \mu \,\mathrm{rot}\,\mathrm{rot}\,\mathbf{u} - \rho\ddot{\mathbf{u}} = -\mathbf{F}, \tag{3.89}$$

$\nu := \lambda + 2\mu$, with the right-hand sides of the form $\mathbf{F} = \mathbf{F}(\mathbf{x} - \mathbf{x}^{\mathrm{o}}, t)$, where the vector function $\mathbf{F}$ is localized at a fixed source point $\mathbf{x}^{\mathrm{o}}$. The components of $\mathbf{F}$ are thus finite linear combinations of $\delta(\mathbf{x} - \mathbf{x}^{\mathrm{o}})$ and its derivatives with respect to $x_1$, $x_2$, and $x_3$ with time-dependent coefficients. The dependence of $\mathbf{F}$ on time is assumed to be causal:

$$\mathbf{F}(\mathbf{x} - \mathbf{x}^{\mathrm{o}}, t)\big|_{t<0} \equiv 0, \tag{3.90}$$

and we are interested in the *causal solution*, i.e.,

$$\mathbf{u}(\mathbf{x}, t)\big|_{t<0} \equiv 0. \tag{3.91}$$

The uniqueness of such a solution is proved in Section 3.5.

Similarly to the time-harmonic case, there are several substantial cases here, and we will deal with three of them. In each case, the *source function* $\mathbf{F}(\mathbf{x} - \mathbf{x}_0, t)$ is represented as a product of functions (in general, generalized functions) dependent only on the spatial variables and only on time,

$$\mathbf{F}(\mathbf{x} - \mathbf{x}^{\mathrm{o}}, t) = \widetilde{\mathbf{F}}(\mathbf{x} - \mathbf{x}^{\mathrm{o}})\chi(t). \tag{3.92}$$

The function $\chi(t)$ is called the *time dependence* of the source.

### Center of expansion

First we consider the *nonstationary center of expansion* that is described by a spherically symmetric right-hand side

$$\mathbf{F}(\mathbf{x} - \mathbf{x}^{\mathrm{o}}, t) = 4\pi \,\mathrm{grad}\,\delta(\mathbf{x} - \mathbf{x}^{\mathrm{o}})\chi(t), \tag{3.93}$$

where, in accordance with (3.91),

$$\chi(t)\big|_{t<0} \equiv 0. \tag{3.94}$$

Seeking the solution in the form

$$\mathbf{u} = \mathrm{grad}\,\Phi, \tag{3.95}$$

we find

$$\nu\nabla^2\Phi - \rho\ddot{\Phi} = \nu\left(\nabla^2 - \frac{1}{a^2}\frac{\partial^2}{\partial t^2}\right)\Phi = -4\pi\delta\left(\mathbf{x} - \mathbf{x}^{\mathrm{o}}\right)\chi(t). \tag{3.96}$$

As follows from (3.40), (3.38), and (3.46), the causal solution for $\Phi$ is

$$\Phi = \frac{1}{\nu}\frac{\chi(t-\frac{r}{a})}{r},$$

whence

$$\mathbf{u} = \operatorname{grad}\Phi = \frac{1}{\lambda+2\mu}\operatorname{grad}\left[\frac{\chi\left(t-\frac{r}{a}\right)}{r}\right]. \tag{3.97}$$

### Center of rotation

The *nonstationary center of rotation around* $\boldsymbol{m}$ is described by the source function

$$\mathbf{F}(\mathbf{x}-\mathbf{x}^{\circ},t) = 4\pi \operatorname{rot}\left[\boldsymbol{m}\delta(\mathbf{x}-\mathbf{x}^{\circ})\right]\chi(t), \tag{3.98}$$

where $\boldsymbol{m}$ is an arbitrary constant vector, which is now taken unit for the sake of definiteness. Seeking the solution in the form

$$\mathbf{u} = \operatorname{rot}\boldsymbol{\Psi}, \tag{3.99}$$

we find

$$\mu\nabla^2\boldsymbol{\Psi} - \rho\ddot{\boldsymbol{\Psi}} = \mu\left(\nabla^2 - \frac{1}{b^2}\frac{\partial^2}{\partial t^2}\right)\boldsymbol{\Psi} = -4\pi\mathbf{m}\,\delta\left(\mathbf{x}-\mathbf{x}^{\circ}\right)\chi(t), \tag{3.100}$$

whence the causal solution is

$$\boldsymbol{\Psi} = \frac{\boldsymbol{m}}{\mu}\frac{\chi(t-\frac{r}{b})}{r}.$$

As a result,

$$\mathbf{u} = \operatorname{rot}\boldsymbol{\Psi} = \frac{1}{\mu}\operatorname{rot}\left[\boldsymbol{m}\frac{\chi\left(t-\frac{r}{b}\right)}{r}\right]. \tag{3.101}$$

### Concentrated force

A non-stationary *concentrated force* directed along $\mathbf{e}_3$ corresponds to the source function

$$\mathbf{F}(\mathbf{x}-\mathbf{x}^{\circ},t) = 4\pi\mathbf{e}_3\delta(\mathbf{x}-\mathbf{x}^{\circ})\chi(t). \tag{3.102}$$

Similarly to the time-harmonic case, we do not use potentials.

We introduce the auxiliary notation

$$\Phi = \operatorname{div}\mathbf{u} \quad \text{and} \quad \boldsymbol{\Psi} = \operatorname{rot}\mathbf{u}. \tag{3.103}$$

Applying the operators div and rot to both sides of the equation (3.89), (3.102), we get easily solvable equations

$$\nabla^2\Phi - \frac{1}{a^2}\ddot{\Phi} = -\frac{4\pi}{\nu}\chi(t)\partial_3\delta \quad \text{and} \quad \nabla^2\ddot{\boldsymbol{\Psi}} - \frac{1}{b^2}\ddot{\boldsymbol{\Psi}} = -\frac{4\pi}{\mu}\chi(t)\operatorname{rot}(\delta\mathbf{e}_3). \tag{3.104}$$

Using the result of Section 3.2.3, we obtain

$$\Phi = \frac{1}{\nu}\partial_3\left(\frac{\chi(t-\frac{r}{a})}{r}\right) \quad \text{and} \quad \boldsymbol{\Psi} = \frac{1}{\mu}\,\mathrm{rot}\left(\frac{\chi(t-\frac{r}{b})}{r}\mathbf{e}_3\right). \tag{3.105}$$

The equation (3.89), (3.102) enables us to express $\ddot{\mathbf{u}}$ via the already-known functions:

$$\rho\ddot{\mathbf{u}} = \nu\,\mathrm{grad}\,\mathrm{div}\,\mathbf{u} - \mu\,\mathrm{rot}\,\mathrm{rot}\,\mathbf{u} + 4\pi\mathbf{e}_3\delta\chi(t) = \nu\,\mathrm{grad}\,\Phi - \mu\,\mathrm{rot}\,\boldsymbol{\Psi} + 4\pi\mathbf{e}_3\delta\chi(t), \tag{3.106}$$

$\delta := \delta(\mathbf{x}-\mathbf{x}^\circ)$. It remains to integrate both sides of the latter relation twice with respect to time.

We shall rearrange the right-hand side of (3.106), employing (3.104) and (3.105). First,

$$\nu\,\mathrm{grad}\,\Phi = \partial_3\,\mathrm{grad}\,\frac{\chi\left(t-\frac{r}{a}\right)}{r}.$$

Second, introducing the notation $\frac{\chi(t-\frac{r}{b})}{r} =: \mathrm{x}$, we have

$$\mu\,\mathrm{rot}\,\boldsymbol{\Psi} = \mathrm{rot}\,\mathrm{rot}(\mathrm{x}\mathbf{e}_3) = (-\nabla^2 + \mathrm{grad}\,\mathrm{div})(\mathrm{x}\mathbf{e}_3).$$

Making use of the equation $\left(-\nabla^2 + \frac{1}{b^2}\frac{\partial^2}{\partial t^2}\right)\mathrm{x} = 4\pi\delta\chi(t)$, we find

$$\mu\,\mathrm{rot}\,\boldsymbol{\Psi} = \left(4\pi\delta\chi(t) - \frac{1}{b^2}\ddot{\mathrm{x}}\right)\mathbf{e}_3 + \partial_3\,\mathrm{grad}\,\mathrm{x}.$$

Similarly to the time-harmonic case, the delta functions on the right-hand side of (3.106) cancel, and we arrive at

$$\rho\ddot{\mathbf{u}} = \partial_3\,\mathrm{grad}\,\frac{\chi\left(t-\frac{r}{a}\right) - \chi\left(t-\frac{r}{b}\right)}{r} + \frac{1}{b^2}\ddot{\mathrm{x}}\mathbf{e}_3. \tag{3.107}$$

To integrate this expression twice with respect to time, we will represent the right-hand side of the equation (3.107) as the second-order derivative of a certain function with respect to time. First, we employ the simple formulas

$$\partial^2_{pq}r = \frac{\delta_{pq}}{r} - \frac{x_p x_q}{r^3} = \frac{\delta_{pq} - s_p s_q}{r}, \quad \partial_p\left(\frac{1}{r}\right) = -\frac{x_p}{r^3} = -\frac{s_p}{r^2},$$

$$\partial^2_{pq}\left(\frac{1}{r}\right) = \frac{3x_p x_q}{r^5} - \frac{\delta_{pq}}{r^3} = \frac{-\delta_{pq} + 3s_p s_q}{r^3},$$

to rewrite the right-hand side of (3.107), which involves the differentiation with respect to coordinates as an expression involving only time derivatives. Denoting

$$\chi^a := \chi\left(t-\frac{r}{a}\right), \quad \chi^b := \chi\left(t-\frac{r}{b}\right), \tag{3.108}$$

we get

$$\rho\ddot{\mathbf{u}} = \frac{\ddot{\chi}^a}{a^2} - \frac{\ddot{\chi}^b}{b^2}\frac{s_3\mathbf{s}}{r} + \left(\left(\frac{\dot{\chi}^a}{a} - \frac{\dot{\chi}^b}{b}\right) + \frac{\chi^a - \chi^b}{r}\right)\frac{3\mathbf{s}s_3 - \mathbf{e}_3}{r^2} + \frac{\ddot{\chi}^b}{b^2}\frac{\mathbf{e}_3}{r}$$

$$= \frac{\ddot{\chi}^a}{a^2}\frac{s_3\mathbf{s}}{r} - \frac{\ddot{\chi}^b}{b^2}\frac{s_3\mathbf{s} - \mathbf{e}_3}{r} + \mathsf{y}\frac{3\mathbf{s}s_3 - \mathbf{e}_3}{r^3}. \tag{3.109}$$

Here, $\mathsf{y} = \mathsf{y}(t)$,

$$\mathsf{y} := r\left(\frac{\dot{\chi}^a}{a} - \frac{\dot{\chi}^b}{b}\right) + \chi^a - \chi^b$$

involves the first-order derivatives of $\chi^a$ and $\chi^b$ with respect to time.

Now we make use of the identity noticed by Stokes 1851 [26]:

$$\mathsf{y} = r\left(\frac{\dot{\chi}^a}{a} - \frac{\dot{\chi}^b}{b}\right) + \chi^a - \chi^b = \ddot{\mathsf{z}}, \tag{3.110}$$

where

$$\mathsf{z} = \int_{\frac{r}{a}}^{\frac{r}{b}} \tau\chi(t-\tau)d\tau. \tag{3.111}$$

The right-hand side of (3.109) takes the form of a second-order derivative with respect to time. Thus, we have found

$$\mathbf{u} = \frac{1}{\rho}\left\{\frac{1}{r}\left(\frac{\chi^a}{a^2}(\mathbf{e}_3,\mathbf{s})\mathbf{s} - \frac{\chi^b}{b^2}[(\mathbf{e}_3,\mathbf{s})\mathbf{s} - \mathbf{e}_3]\right) + \frac{3\mathbf{s}s_3 - \mathbf{e}_3}{r^3}\mathsf{z}\right\}. \tag{3.112}$$

Consider the particular case of

$$\chi(t) = \delta(t), \tag{3.113}$$

where

$$\mathbf{u} = \frac{1}{\rho}\left\{\frac{\delta(t-\frac{r}{a})}{a^2 r}(\mathbf{e}_3,\mathbf{s})\mathbf{s} - \frac{\delta(t-\frac{r}{b})}{b^2 r}[(\mathbf{e}_3,\mathbf{s})\mathbf{s} - \mathbf{e}_3] + \frac{3\mathbf{s}s_3 - \mathbf{e}_3}{r^3}\mathsf{z}\right\}. \tag{3.114}$$

In calculating the expression (3.111), we arrived at the function (in the classical sense)

$$\mathsf{z} = t\left[H\left(t - \frac{r}{a}\right) - H\left(t - \frac{r}{b}\right)\right], \tag{3.115}$$

where $H$ is the *Heaviside function* introduced by (3.10).

The first and second terms on the right-hand side of (3.114) have a clear physical interpretation. These are spherical waves in which the perturbation is localized at the wavefronts of $P$- and $S$-waves, which move away from the source point with the appropriate velocities. The corresponding displacements are purely longitudinal and purely transverse, respectively. The third term in (3.115) looks more surprising. It describes a perturbation localized in the spherical annulus between the wavefronts $t = \frac{r}{a}$ and $t = \frac{r}{b}$. It linearly grows with time at each point of the annulus as $\frac{r}{a} < t < \frac{r}{b}$, and it weakens with distance at each time instant as fast as $r^{-3}$.

## 3.5 Conditions at infinity and uniqueness

In this section we discuss methods of determining unique solutions in problems for homogeneous media; the wave field is considered either in the whole space ($\mathbf{x} \in \mathbb{R}^3$) or in the exterior of a bounded domain $\Omega$, $\mathbf{x} \in \mathbb{R}^3 \setminus \Omega$.

### 3.5.1 Limiting absorption principle

We continue the discussion of the *limiting absorption principle* started in Section 3.2.2. Now we are interested in the general case of an anisotropic homogeneous medium, described by the nonhomogeneous equation (1.82). We consider complex values of the frequency (see (3.37)). The solutions are assumed to exponentially decay at infinity with their first derivatives,[7] and then $\operatorname{Im}\omega$ is turned to zero. It remains to establish that the solution of the respective homogeneous problem

$$\boldsymbol{l}\boldsymbol{u} = 0, \tag{3.116}$$

$(\boldsymbol{l}\boldsymbol{u})_j = \partial_m c_{jmpq}\varepsilon_{pq} + \rho\omega^2 u_j$, $\varepsilon_{pq} = \frac{1}{2}\left(\partial_p u_q + \partial_q u_p\right)$, is unique.

**Theorem (on the limiting absorption principle in elastodynamics).** *Let a smooth vector function $\boldsymbol{u}$ satisfy the equation* (3.116) *in* $\mathbb{R}^3$ *for* $\operatorname{Re}\omega > 0$, $\operatorname{Im}\omega > 0$ *and exponentially tend to zero together with its first derivatives, that is,*

$$\max_{|\mathbf{x}|=R} |\boldsymbol{u}| \leqslant \operatorname{const} e^{-\epsilon R}, \quad \max_{|\mathbf{x}|=R} \sqrt{\sum_{j=1}^{3} |\partial_j \boldsymbol{u}|^2} \leqslant \operatorname{const} e^{-\epsilon R} \tag{3.117}$$

*for some* $\epsilon > 0$. *Then* $\boldsymbol{u} \equiv 0$.

We give a proof based on the classical "Green's integration by parts." Let us scalarly multiply $\boldsymbol{l}\boldsymbol{u}$ by $\boldsymbol{u}$ and integrate over the ball $B_R = \{\mathbf{x} : \ |\mathbf{x}| < R\}$:

$$0 = \int_{|\mathbf{x}|<R} (\boldsymbol{l}\boldsymbol{u}, \boldsymbol{u}) d\mathbf{x} = \int_{|\mathbf{x}|<R} \left\{(\partial_m c_{jmpq}\varepsilon_{pq} + \rho\omega^2 u_j)u_j^*\right\} d\mathbf{x}. \tag{3.118}$$

Integration by parts yields

$$\int_{|\mathbf{x}|<R} \left\{-c_{jmpq}\varepsilon_{jm}^*\varepsilon_{pq} + \rho(\operatorname{Re}\omega + i\operatorname{Im}\omega)^2|\boldsymbol{u}|^2\right\} d\mathbf{x} + \int_{|\mathbf{x}|=R} (\mathbf{t}^n\boldsymbol{u}, \boldsymbol{u})\, dS = 0, \tag{3.119}$$

where $(\mathbf{t}^n\boldsymbol{u})_j = \sigma_{js}\cos\widehat{nx_s}$, $\cos\widehat{nx_s}$ are the components of the exterior unit

[7] In general, a rigorous proof of the existence of solutions which decay at infinity for $\operatorname{Im}\omega > 0$ requires more sophisticated techniques than those we employ in the present book (in this connection, see Sanchez-Palencia 1980 [22]). For point sources in isotropic homogeneous media, the existence of such solutions is obvious.

normal to the sphere $|\mathbf{x}| = R$, and $dS$ is its surface element. Turning $R$ to infinity, we apply the condition (3.117), by virtue of which the surface integral tends to zero. The imaginary part of the equation (3.119) now reads

$$2i\rho \operatorname{Re}\omega \operatorname{Im}\omega \int_{\mathbb{R}} |\boldsymbol{u}|^2 d\mathbf{x} = 0, \tag{3.120}$$

and therefore $\boldsymbol{u} \equiv 0$.

### The scalar case

An analogous result can be proved for the scalar case in a likewise but simpler manner. The statement is:

**Theorem (on the limiting absorption principle in the scalar case).** *Under the condition* (3.37), *any solution of the equation* (3.28) *with* $k = k' + ik''$, $k'' > 0$, *which exponentially decays at infinity together with its first derivatives, equals identically zero.*

### On passing to the limit as $\operatorname{Im}\omega \to 0$

The existence of the limit as $\operatorname{Im}\omega \to 0$ for the solutions under discussion in a rather general situation is proved, e.g., in Sanchez-Palencia 1980 [22], see also Vainberg 1989 [28]. In the particular problems we are dealing with in the present book, the existence of such a limit is obvious.

## 3.5.2 ⋆ Radiation conditions

Now we describe another approach to defining the conditions at $|\mathbf{x}| \to \infty$ that forbid arriving waves from infinity and guarantee the uniqueness of the solution. The solution proves to be the same as that specified by the limiting absorption principle. Being less general than the limiting absorption principle, this approach does not require dealing with complex frequencies. We begin with the scalar case.

### Sommerfeld radiation conditions in the scalar case

We shall assume that a function $u$ satisfies the homogeneous Helmholtz equation (3.28) outside a bounded domain $\Omega \subset \mathbb{R}^3$. The classical definition is as follows: a function $u$ satisfies the *Sommerfeld radiation conditions* if

$$\frac{\partial u}{\partial r} - iku = o\left(\frac{1}{r}\right), \quad u = O\left(\frac{1}{r}\right), \quad r \to \infty. \tag{3.121}$$

Here, $r = |\mathbf{x}|$ is the distance from the origin, and $\frac{\partial}{\partial r} = \frac{x_m}{r}\frac{\partial}{\partial x_m}$ is the derivative in a radial direction. The conditions (3.121) mean that $\max\limits_{|\mathbf{x}|=r} \left(r \left|\frac{\partial u}{\partial r} - iku\right|\right) \to 0$

as $r \to \infty$ and the estimate $|u| \leqslant \frac{\text{const}}{r}$ with some constant independent of $\mathbf{x}$ holds for sufficiently large $r$.[8]

### Integral representation of an arbitrary solution of the Helmholtz equation

The aforementioned results are based on an integral representation, and we proceed to deriving it. Let $D \subset \mathbb{R}^3$ be a bounded domain with a smooth boundary $\partial D$, $u(\mathbf{x})$ and $v(\mathbf{x})$ be smooth functions in $D$. Recall the classical Green formula

$$\int_D (u\nabla^2 v - v\nabla^2 u) d\mathbf{x} = \int_D \left[u\left(\nabla^2 + k^2\right) v - v(\nabla^2 + k^2)u\right] d^3\mathbf{x}$$
$$= \int_{\partial D} \left(u \frac{\partial v}{\partial n_{\mathbf{x}}} - v \frac{\partial u}{\partial n_{\mathbf{x}}}\right) dS(\mathbf{x}). \qquad (3.123)$$

Here, $\frac{\partial}{\partial n_{\mathbf{x}}}$ is the differentiation with respect to the outer normal at a point $\mathbf{x} \in \partial D$. Assume that $u$ is a solution of the Helmholtz equation, and $v = \frac{e^{ikr}}{r}$, $r = |\mathbf{x} - \mathbf{y}|$. Taking into account the fact that $(\nabla^2_{\mathbf{x}} + k^2)v = -4\pi\delta(\mathbf{x} - \mathbf{y})$, we find

$$u(\mathbf{x}) = \frac{1}{4\pi} \int_{\partial D} \left(\frac{e^{ik|\mathbf{x}-\mathbf{y}|}}{|\mathbf{x}-\mathbf{y}|} \frac{\partial u(\mathbf{y})}{\partial n_{\mathbf{y}}} - u(\mathbf{y}) \frac{\partial}{\partial n_{\mathbf{y}}} \frac{e^{ik|\mathbf{x}-\mathbf{y}|}}{|\mathbf{x}-\mathbf{y}|}\right) dS(\mathbf{y}). \qquad (3.124)$$

### Integral representation of a function satisfying the radiation conditions

We state two simple lemmas.

**Lemma 1.** *Let functions* $u(\mathbf{x})$ *and* $v(\mathbf{x})$ *(satisfying the Helmholtz equation) satisfy the radiation conditions. Then*

$$\int_{|\mathbf{x}|=R} \left(u \frac{\partial v}{\partial n} - v \frac{\partial u}{\partial n}\right) dS(\mathbf{x}) \to 0 \quad \textit{as} \quad R \to \infty. \qquad (3.125)$$

The proof immediately follows from (3.121) in view of the relation $\frac{\partial}{\partial n} = \frac{\partial}{\partial R}$ and the fact that the area of a sphere of radius $R$ is proportional to $R^2$.

**Lemma 2.** *The solution* $u = \frac{e^{ik|\mathbf{x}-\mathbf{y}|}}{|\mathbf{x}-\mathbf{y}|}$ *of the point source problem* (3.29) *satisfies the radiation conditions* (3.121).

We omit a very simple proof.

---

[8] The two conditions (3.121) can be replaced by a single *integral radiation condition* (see, e.g., Courant and Hilbert 1962 [5])

$$\int_{r=\mathsf{R}} \left|\frac{\partial u}{\partial r} - iku\right|^2 dS \to 0 \text{ as } \mathsf{R} \to \infty, \qquad (3.122)$$

which is applicable also in the case of boundaries that outgo to infinity (in a special manner).

Now let a function $u$ satisfy the radiation conditions outside a bounded smooth domain $\Omega$. Apply (3.124), choosing as $D$ the domain between the sphere $\Sigma_R = \{\mathbf{x} : |\mathbf{x}| = R\}$ of a sufficiently large radius[9] and the complement of $\Omega$. Then $\partial D = \partial\Omega \bigcup \Sigma_R$, and

$$u(\mathbf{x}) = \frac{1}{4\pi}\int_{\partial\Omega}\left(\frac{e^{ikr}}{r}\frac{\partial u}{\partial n} - u\frac{\partial}{\partial n}\frac{e^{ikr}}{r}\right)dS(\mathbf{y}) + \frac{1}{4\pi}\int_{\Sigma_R}\left(\frac{e^{ikr}}{r}\frac{\partial u}{\partial n} - u\frac{\partial}{\partial n}\frac{e^{ikr}}{r}\right)dS(\mathbf{y}). \tag{3.126}$$

Here, $r = |\mathbf{x} - \mathbf{y}|$, and the integral is taken with respect to the coordinates of the point $\mathbf{y}$.

Let $R$ tend to infinity. From Lemmas 1 and 2 it follows that $\int_{\Sigma_R} \to 0$. Passing to the limit and noting that, by the radiation conditions for $u$, the integral over a large sphere vanishes, we get the desired integral representation

$$u(\mathbf{x}) = \frac{1}{4\pi}\int_{\partial\Omega}\left(\frac{e^{ikr}}{r}\frac{\partial u}{\partial n} - u\frac{\partial}{\partial n}\frac{e^{ikr}}{r}\right)dS(\mathbf{y}), \tag{3.127}$$

with $\frac{\partial}{\partial n}$ denoting differentiation with respect to the inner normal to $\partial\Sigma$.

It can be shown without difficulty that the representation (3.127) is valid for any function satisfying the limiting absorption principle.

### Physical meaning of the radiation conditions

**Theorem (on coordinate asymptotics).** *If a function $u$ is a solution of the Helmholtz equation* (3.28) *in the exterior of a bounded domain $\Omega$, and it satisfies the Sommerfeld radiation conditions* (3.121)*, then, as $|\mathbf{x}| \to \infty$, it has the coordinate asymptotics*

$$u(\mathbf{x}) = \frac{e^{ikr}}{r}\mathfrak{h}(\mathbf{s}) + O\left(\frac{1}{r^2}\right), \quad r = |\mathbf{x}|, \quad \mathbf{s} = \frac{\mathbf{x}}{r}. \tag{3.128}$$

Here, $\mathfrak{h}(\mathbf{s})$ is a smooth function dependent only on the direction vector $\mathbf{s}$ and called the *radiation pattern* or *directivity*.

The formula (3.128) is a simple corollary to the integral representation (3.127). Consider there the first term,

$$\frac{1}{4\pi}\int_{\partial\Omega}\frac{e^{ik|\mathbf{x}-\mathbf{y}|}}{|\mathbf{x}-\mathbf{y}|}\frac{\partial u(\mathbf{y})}{\partial n_{\mathbf{y}}}dS(\mathbf{y}), \tag{3.129}$$

[9] $R$ should be so large that $\Omega$ lies inside $\Sigma_R$.

observing that

$$|\mathbf{x}-\mathbf{y}| = \sqrt{(\mathbf{x}-\mathbf{y},\mathbf{x}-\mathbf{y})} = |\mathbf{x}|\left(1-\frac{2(\mathbf{x},\mathbf{y})}{|\mathbf{x}|^2}+\frac{|\mathbf{y}|^2}{|\mathbf{x}|^2}\right)^{\frac{1}{2}}$$
$$= |\mathbf{x}| - \left(\frac{\mathbf{x}}{|\mathbf{x}|},\mathbf{y}\right) + O\left(\frac{1}{|\mathbf{x}|}\right), \tag{3.130}$$
$$\frac{1}{|\mathbf{x}-\mathbf{y}|} = \frac{1}{|\mathbf{x}|}\left(1-\frac{2(\mathbf{x},\mathbf{y})}{|\mathbf{x}|^2}+\frac{|\mathbf{y}|^2}{|\mathbf{x}|^2}\right)^{-\frac{1}{2}} = \frac{1}{|\mathbf{x}|} + O\left(\frac{1}{|\mathbf{x}|^2}\right).$$

The above estimates are uniform for $\mathbf{y} \in \partial\Omega$. Substituting these expressions into (3.129) and using also the relation $e^{ikO\left(\frac{1}{|\mathbf{x}|}\right)} = 1 + O\left(\frac{1}{|\mathbf{x}|}\right)$, it is easy to derive that (3.129) has the asymptotic representation (3.128). We omit a completely analogous consideration of the second term in (3.127).

The *formula of the coordinate asymptotics* (3.128) makes possible a physical interpretation of the corresponding wave process. The Helmholtz equation (3.28) with $k = \frac{\omega}{c}$ arises when considering the time-harmonic solutions of the wave equation $\nabla^2 \mathrm{u} - \frac{1}{c^2}\ddot{\mathrm{u}} = 0$, that is, the solutions described by (3.26). The multiplication of (3.128) by the time dependence $e^{-i\omega t}$ yields

$$\mathrm{u}(\mathbf{x},t) = \frac{e^{-i\omega\left(t-\frac{r}{c}\right)}}{r}\mathfrak{h}\left(\mathbf{s}\right) + O\left(\frac{1}{r^2}\right), \tag{3.131}$$

which, up to terms of order $O\left(\frac{1}{r^2}\right)$, represents a spherical wave propagating away from the origin with the velocity $c$ and decaying with distance inversely as $r$. Its amplitude depends on the direction and is ruled by the *radiation pattern* $\mathfrak{h}\left(\mathbf{s}\right)$.

### Uniqueness under the Sommerfeld radiation conditions

The radiation conditions afford uniqueness in the point source problem stated as follows.

**Theorem.** *Let a generalized function* $u(\mathbf{x})$, $u \in K'(\mathbb{R}^3)$ *be a solution of the equation* (3.27), *where* $\mathscr{F}$ *is also a generalized function from the class* $K'(\mathbb{R}^3)$[10] *and equals zero outside a finite domain. Let* $u(\mathbf{x})$ *satisfy the Sommerfeld conditions* (3.121). *Then* $u(\mathbf{x})$ *is unique.*

The proof is reduced to establishing that any solution of the homogeneous equation (3.28) in $\mathbb{R}^3$, which satisfies the radiation conditions, is necessarily zero. It is quite straightforward. As is well known (see, e.g., Schwartz 1950 [23], Gelfand and Shilov 1968 [9]), any solution of the homogeneous equation (3.28) is infinitely differentiable in the classical sense. We substitute $u$ into the formula (3.127) and assume $\Omega$ to be a ball: $\Omega = \{\mathbf{x} : \ |\mathbf{x}| < R\}$, $\partial\Omega = \{\mathbf{x} : |\mathbf{x}| = R\}$. Turning $R \to \infty$ and making use of (3.125), we deduce that $u(\mathbf{x}) = 0$ for an arbitrary point $\mathbf{x} \in \mathbb{R}^3$, as required.

[10] I.e., $u$ and $\mathscr{F}$ are continuous functionals on compactly supported infinitely differentiable functions (see Gelfand and Shilov 1964 [8]).

The uniqueness theorem is true for the Dirichlet (and Neumann) problems outside a bounded domain $\Omega \subset \mathbb{R}^3$ with a smooth boundary (see, e.g., Smirnov 1964b [25], Tikhonov and Samarsky 1965 [27]).

### Radiation conditions in isotropic elastodynamics

Here we consider a time-harmonic elastodynamic wavefield $\boldsymbol{u}$, which satisfies the homogeneous equation with constant coefficients of the form (3.47) outside a bounded domain. For large $r$, $\boldsymbol{u}$ can be asymptotically split into a sum of $P$ and $S$ waves in accordance with (3.88), where the radiation patterns $f(\mathbf{s})$ and $\boldsymbol{g}(\mathbf{s})$ are smooth functions on the unit sphere $|\mathbf{s}| = 1$. The scalar function $f$ and the vector-valued function $\boldsymbol{g}$ are called the *radiation patterns of $P$- and $S$-waves*, respectively. For uniqueness, one can require the fulfillment of the Sommerfeld radiation conditions for each of the waves.

A different statement of the radiation conditions was proposed by Jones 1984 [15].

To put us on a right track, we replace $u$ by $\boldsymbol{u}$, and $\frac{\partial}{\partial r} = \frac{\partial u}{\partial n}$ by $\mathbf{t}^{\mathrm{s}}(\boldsymbol{u})$ in the first condition (3.121) (see Section 1.4.1). Since at a large distance the leading terms of the asymptotics result from differentiating the exponents in (3.88), we arrive at the condition

$$\mathbf{t}^{\mathrm{s}}(\boldsymbol{u}) - \frac{i\omega(\lambda + 2\mu)}{a}(\boldsymbol{u}, \mathbf{s})\mathbf{s} - \frac{i\omega\mu}{b}\left[\boldsymbol{u} - (\boldsymbol{u}, \mathbf{s})\mathbf{s}\right] = O\left(\frac{1}{r^2}\right). \qquad (3.132)$$

It is natural to add an analog of the second condition (3.121):

$$\boldsymbol{u} = O\left(\frac{1}{r}\right). \qquad (3.133)$$

The *Jones radiation conditions* (3.132)–(3.133) are completely analogous to the Sommerfeld conditions.

**Jones' theorem.** *Let the vector* $\mathbf{F}$ *in* (3.47) *be localized at a point* $\mathbf{y}$. *Then the solution of* (3.47) *in the class* $K'$, *satisfying the Jones conditions* (3.132)–(3.133), is unique.[11]

To prove the theorem, we establish that the respective homogeneous problem has no solution other than identical zero.

Note that since the operator $\boldsymbol{l} = (\lambda + 2\mu)\,\mathrm{grad}\,\mathrm{div} - \mu\,\mathrm{rot}\,\mathrm{rot} + \rho\omega^2\mathbf{I}$ is elliptic, any solution of the homogeneous equation (3.47) from $K'$ is smooth (see, e.g., Gelfand and Shilov 1964, 1967 [8, 9]). We arbitrarily choose a point $\mathbf{y} \in \mathbb{R}^3$, and let $\boldsymbol{G}^{(p)}(\mathbf{x}-\mathbf{y})$ be the wavefields of concentrated forces described in Section 3.3.2. With the use of (3.87), it is easy to observe that they satisfy the Jones conditions. Apply the reciprocity relation (1.95) to the vectors $\boldsymbol{u}$

[11] Actually, Jones proved (see Jones 1984 [15]) a more general statement.

and $G^{(p)}$, taking the ball $|\mathbf{x}| \leqslant R$ as the domain of integration:

$$\int_{|\mathbf{x}|\leqslant R} \Big\{(\boldsymbol{u} \cdot \boldsymbol{l}\boldsymbol{G}^{(p)}) - (\boldsymbol{G}^{(p)} \cdot \boldsymbol{l}\boldsymbol{u})\Big\}\, d\mathbf{x} = \int_{|\mathbf{x}|=R} \Big\{(\boldsymbol{u} \cdot \mathbf{t}^{\mathbf{s}}\boldsymbol{G}^{(p)}) - (\boldsymbol{G}^{(p)} \cdot \mathbf{t}^{\mathbf{s}}\boldsymbol{u})\Big\}\, dS, \quad (3.134)$$

$\mathbf{s} = \frac{\mathbf{x}}{|\mathbf{x}|}$. Turning $R$ to infinity and taking into account the Jones conditions for $\boldsymbol{u}$ and $\boldsymbol{G}^{(p)}$, with a little difficulty we see that the integrand on the right-hand side has order $O(\frac{1}{R^3})$. Thus, the right-hand side tends to zero. The left-hand side equals $-4\pi u_p(\mathbf{y})$ for $p = 1, 2, 3$. In view of the arbitrariness of the point $\mathbf{y}$, we got that $\boldsymbol{u} \equiv 0$, which completes the proof.

### 3.5.3 Uniqueness theorem in the nonstationary case

Let the displacement vector $\mathbf{u}(\mathbf{x}, t)$ satisfy the equation (3.89) with the source (3.92), where $\widetilde{\mathbf{F}}(\mathbf{x} - \mathbf{x}^{\circ})$ has the form

$$\widetilde{\mathbf{F}}(\mathbf{x} - \mathbf{x}^{\circ}) = \mathbf{m}(\partial_1, \partial_2, \partial_3)\delta(\mathbf{x} - \mathbf{x}^{\circ}), \quad (3.135)$$

and $\mathbf{m}$ is a polynomial the coefficients of which are constant vectors. The time dependence of the source $\chi(t)$ is a generalized function (we note that generalized functions dependent on different variables can be multiplied). The initial condition is (3.91).

We present a proof of the uniqueness of the Cauchy problem (3.89), (3.91) based on a good old idea due to Holmgren 1901 [11], which can be stated as follows: the solvability of a linear problem implies the uniqueness of the conjugate problem. It is the mainstream approach to proving the uniqueness of solutions of a Cauchy problem (see, e.g., the monograph by Gelfand and Shilov 1967 [9] and references therein).

**Theorem.** *The Cauchy problem* (3.89), (3.91) *cannot have two different nonzero solutions in the class of generalized vector-valued functions* $\mathbf{u} \in K'(\mathbb{R}^4)$.[12]

To follow the proof, it is recommended the reader be familiar with §§ 1, 2 and Appendix 2 of Chapter 1 in the monograph by Gelfand and Shilov 1967 [9]. We borrowed our notation from there.

*A proof.* Let $\mathbf{f} \in K(\mathbb{R}^4)$, and let $\mathbf{w}$ be a solution of the Cauchy problem "in the backward direction":

$$\mathbf{L}\mathbf{w} = \mathbf{f}, \quad \mathbf{w}|_{t>t_0} = 0. \quad (3.136)$$

[12] We employ this notation for vectors whose components are continuous functionals on the set $K(\mathbb{R}^4)$ of finite infinitely differentiable functions of four variables $x_1$, $x_2$, $x_3$, and $t$. A vector belongs to $K'$ or $K$ if its components belong to the corresponding space.

We assume that $\mathbf{f}|_{t>t_0} = 0$; otherwise, it is arbitrary. A solution is sought for $t \leqslant t_0$. We aim at showing that the relation

$$\mathbf{Lu} = 0, \quad \mathbf{u}|_{t<0} = 0, \quad \mathbf{u} \in K', \tag{3.137}$$

implies $\mathbf{u} = 0$.

Let $\eta \in C^\infty(\mathbb{R}^1)$ be a *cut-off function* defined for all values of $t$ such that

$$\eta(t) = \begin{cases} 1 & \text{for } t \geqslant -\frac{1}{2}, \\ 0 & \text{for } t < -1. \end{cases}$$

The function $\eta(t)\mathbf{w}(\mathbf{x}, t)$ is infinitely differentiable (since $\mathbf{f} \in K(\mathbb{R}^4)$), and by the finiteness of the velocity of propagation of perturbation (see Section 2.5), it is compactly supported. Thus, $\eta\mathbf{w} \in K(\mathbb{R}^4)$.

We will temporarily use the notation $(\boldsymbol{\varphi}, \boldsymbol{\psi})$ for the value of a vector-valued generalized function $\boldsymbol{\varphi} \in K'$ on a vector $\boldsymbol{\psi} \in K$. The expression $(\mathbf{u}, \mathbf{L}(\eta\mathbf{w}))$ equals $(\mathbf{u}, \mathbf{f})$, because the function $\eta$ differs from 1 where $\mathbf{w} = 0$. As follows from the definition of the action of a differential operator on generalized functions,

$$(\mathbf{u}, \mathbf{L}(\eta\mathbf{w})) = (\mathbf{Lu}, \eta\mathbf{w}).$$

Since $\mathbf{Lu} = 0$, we have

$$(\mathbf{u}, \mathbf{f}) = (\mathbf{u}, \mathbf{L}(\eta\mathbf{w})) = (\mathbf{Lu}, \eta\mathbf{w}) = 0,$$

that is, $(\mathbf{u}, \mathbf{f}) = 0$ for an arbitrary $\mathbf{f} \in K(\mathbb{R}^4)$. This exactly means that $\mathbf{u} = 0$, which proves the theorem.

---

## 3.6 ⋆ Comments to Chapter 3

Initially, the delta function was introduced in a formal manner by O. Heaviside, and P. A. M. Dirac. Owing to S. L. Sobolev and L. Schwartz, it became a rigorously defined mathematical object within the theory of generalized functions (or distributions). Good expositions of the basics of the theory of generalized functions, which is doubtless a splendid advance in mathematics of the 20th century, can be found, e.g., in Gelfand and Shilov 1964 [8], Jones 1982 [14], and Vladimirov 1984 [29].

The above material concerning the wavefields of point sources is quite traditional. The set of three concentrated forces wavefields with time dependence (3.113), directed along the coordinate axes, is an elastodynamic analog of the fundamental solution of the Cauchy problem for hyperbolic equations and systems (see Gelfand and Shilov 1967 [9]). The corresponding wavefields were dealt with, e.g., in Babich 1957 [2] and Hudson 1980 [12]. More complicated multipole point sources, attracting attention in seismology, were considered, e. g., in Aki and Richards 1980 [1] and in Dahlen and Tromp 1998 [6]. Wavefield described by the expression (3.111) and conditions of its vanishing between the wavefronts were discussed in detail by Kiselev and Tagirdzhanov 2015 [17].

For general homogeneous anisotropic media, the wavefield of a point source cannot be represented by a finite sum of elementary functions.[13] In the time-harmonic case, the employment of Fourier integrals (e.g., Musgrave 1970 [20], Every 1986 [7]) is natural. In the nonstationary case, the Radon transform is helpful (e.g., Yeatts 1984 [30], Norris 2007 [21], among others).

For solutions of problems of generation of elastic waves by loads of the form (3.59) and (3.68) applied to surfaces of small spherical cavities in a smoothly inhomogeneous medium, the results described by (3.64) and (3.71) remain true (see Kiselev 1982 [16]), though one could expect that terms of the form $\mathbf{A}\delta(\mathbf{x}-\mathbf{x}^\circ)$ would appear on the right-hand sides.

The limiting absorption principle was first stated, as far as we know, by Ignatowsky 1905 [13]. Essentially, the proof of uniqueness in Section 3.5.1 does not require that the solution and its derivatives vanish at infinity. An easily generalizable to elastodynamics uniqueness theorem in the class of moderately growing generalized functions is presented for the scalar case, e.g., by Vladimirov 1984 [29].

## References to Chapter 3

[1] Aki, K. and Richards, P. G. 1980. *Quantitative seismology: Theory and methods.* San Francisco: W. H. Freeman and Company. *Аки К., Ричардс П.* Количественная сейсмология. Т. 1. М.: Мир, 1983.

[2] Babich, V. M. 1957. Sobolev – Kirchhoff method in dynamics of an inhomogeneous elstic medium. *Vestn. Leningrad University. Ser. Mat. Mech. Astron.* 3(13):146–60. *Бабич В. М.* Метод С. Л. Соболева – Кирхгофа в динамике неоднородной упругой среды. Вестн. ЛГУ. Сер. мат. мех. астрон., 1957, № 3(13). С. 146–160.

[3] Blake Jr., F. G. 1952. Spherical wave propagation in solid media. *J. Acoust. Soc. Amer.* 24(2):211–5.

[4] Burridge, R., Chadwick, P. and Norris, A. N. 1993. Fundamental elastodynamic solutions for anisotropic media with ellipsoidal slowness surfaces. *Proc. Roy. Soc. London A.* 440(1910):655–81.

[13] Exceptional cases where it can be done are discussed by Burridge, Chadwick, and Norris 1993 [4].

[5] Courant, R. and Hilbert, D. 1962. *Methods of mathematical physics. Vol. 2.* New York: Interscience. *Курант Р.* Уравнения с частными производными. М.: Мир, 1964.

[6] Dahlen, F. A. and Tromp, J. 1998. *Theoretical global seismology.* Princeton University Press.

[7] Every, A. G. 1986. Formation of phonon-focusing caustics in crystals. *Phys. Rev. B.* 34:2852–62.

[8] Gelfand, I. M. and Shilov, G. E. 1964. *Generalized functions. Volume I: Properties and operations.* New York: Academic Press. *Гельфанд И. М., Шилов Г. Е.* Обобщенные функции и действия над ними (Обобщенные функции. Вып. 1). М.: Добросвет, 2000.

[9] Gelfand, I. M. and Shilov, G. E. 1967. *Generalized functions, Volume 3: Theory of differential equations.* New York: Academic Press. *Гельфанд И. М., Шилов Г. Е.* Некоторые вопросы теории дифференциальных уравнений (Обобщенные функции. Вып. 3). М.: Физматгиз, 1958.

[10] Gurvich, I. I. 1968. On the theory of a spherical emitter of shear waves. *Inz. AN SSSR. Fizika Zemli.* № 1:44–52. *Гурвич И. И.* К теории сферического излучателя поперечных сейсмических волн. Изв. АН СССР. Физика Земли, 1968. № 1. С. 44–52.

[11] Holmgren, F. 1901. Über Systeme von linearen partiellen Differentialgleichungen. *Stockh. Öfversigt Kongl. Veten-Akad. Förth.* 58:91–103.

[12] Hudson, J. A. 1980. *Excitation and propagation of elastic waves.* Cambridge: Cambridge University Press.

[13] Ignatowsky, W. 1905. Reflexion elektromagnetisches Wellen an einem Draht. *Ann. Physik.* 323(13):495–22.

[14] Jones, D. S. 1982. *The theory of generalised functions.* Cambridge: Cambridge University Press.

[15] Jones, D. S. 1984. A uniqueness theorem in elastodynamics. *Quart. J. Mech. Appl. Math.* 37:121–42.

[16] Kiselev, A. P. 1982. Small spherical emitter in an inhomogeneous elastic medium, *Mech. Solids.* 17(4):106–14. *Киселев А. П.* Малый сферический излучатель в неоднородной упругой среде. Изв. АН СССР. Механика твердого тела, 1982. № 4. С. 119–126.

[17] Kiselev, A. P. and Tagirdzanov, A. M. 2015. Paradoxical properties of non-stationary wavefields of point sources in isotropic, elastic medium. *Acoust. Phys.* 61:388–91.

[18] Klein, F. 1937. Lectures on developmant om mathematics in XIX century. Moscow: GONTI. *Клейн Ф.* Лекции о развитии математики в XIX столетии. Т. 1. М.–Л.: ГОНТИ, 1937.

[19] Kolmogorov, A. N. and Fomin, S. V. 1970. *Introduction to real analysis.* New York: Dover. *Колмогоров А. Н., Фомин С. В.* Элементы теории функций и функционального анализа. М.: Наука, 1976.

[20] Musgrave, M. J. P. L. 1970. *Crystal acoustics.* San Francisco: Holden-Day.

[21] Norris, A. N. 2007. Wavefront singularities associated with the conical point in elastic solids with cubic symmetry. *Wave Motion.* 44:513–27.

[22] Sanchez-Palencia, E. 1980. *Non-homogeneous media and vibration theory.* Berlin: Springer. *Санчес-Паленсия Э.* Неоднородные среды и теория колебаний. М.: Мир, 1984.

[23] Schwartz, L. 1950. *Théorie des distributions.* Tome 1. Paris: Hermann.

[24] Smirnov, V. I. 1964a. *A course of higher mathematics. Volume 2.* Oxford: Pergamon. *Смирнов В. И.* Курс высшей математики. Т. 2. М.: Наука, 1981.

[25] Smirnov, V. I. 1964b. *A course of higher mathematics. Volume 4.* Reading MA: Addison-Wesley. *Смирнов В. И.* Курс высшей математики. Т. 4. Ч. 2. М.: Наука, 1981.

[26] Stokes, G. G. 1851. On the dynamical theory of diffraction. *Trans. Cambridge Phil. Soc.* 9(1):1–62.

[27] Tikhonov, A. N. and Samarsky, A. A. 1965. *Equations of mathematical physics.* Interscience Publishers. *Тихонов А. Н., Самарский А. А.* Уравнения математической физики. М.: Изд. МГУ, 1999.

[28] Vainberg, B. R. 1989. *Asymptotic methods in equations of mathematical physics.* New York: Gordon and Breach. *Вайнберг Б. Р.* Асимптотические методы в уравнениях математической физики. М.: Изд-во МГУ, 1982.

[29] Vladimirov, V.S. 1984. *Equations of mathematical physics.* Moscow: Mir. *Владимиров В. С.* Уравнения математической физики. М.: Наука, 1971.

[30] Yeatts, F. R. 1984. Elastic radiation from a point force in an anisotropic medium. *Phys. Rev. B.* 29:1674–84.

# Chapter 4

## Ray Method for Volume Waves in Isotropic Media

In this chapter (as well as in Chapter 5) we address an important class of wave processes in which a wave approximately behaves as a plane wave with its characteristics smoothly varying from point to point. The frequency is treated as a large parameter. Such expansions are known as *high-frequency asymptotics*. A detailed analysis allows us not only to take into account the change in the amplitude along the rays but also to describe, using higher-order terms, polarization anomalous from the point of view of the theory of homogeneous plane waves. Here, we consider the case of isotropic media.

## 4.1 Ray ansatz and transport equations

### 4.1.1 Ray ansatz and local plane waves

#### Discussion of ray ansatz

For successfully constructing asymptotic formulas, it is crucial to properly guess the form of the sought-for expansion. We seek the displacement vector in the form[1]

$$\mathbf{u} = e^{i\omega(\tau(\mathbf{x})-t)} \left\{ \boldsymbol{u}^0(\mathbf{x}) + \frac{\boldsymbol{u}^1(\mathbf{x})}{-i\omega} + \dots \right\}, \quad \omega \to \infty, \tag{4.1}$$

or

$$\mathbf{u} = e^{i\omega(\tau(\mathbf{x})-t)} \sum_{m=0}^{\infty} \frac{\boldsymbol{u}^m(\mathbf{x})}{(-i\omega)^m}, \quad \omega \to \infty. \tag{4.2}$$

Here, the scalar function $\tau = \tau(\mathbf{x})$ called the *eikonal* and the vectors $\boldsymbol{u}^0$, $\boldsymbol{u}^1$, ... independent of $\omega$ are to be determined. Generally speaking, it is assumed that the vectors $\boldsymbol{u}^0$, $\boldsymbol{u}^1$, ... are complex. They are called the *vector amplitudes* of zeroth, first, etc., order, while the expression (4.2) is called the

[1] It is often required to multiply the ansatz (4.1) by some function of $\omega$, for example, by $(-i\omega)^{-\xi}$.

*ray ansatz.* The surfaces in $\mathbb{R}^3$ on which the *phase* $\omega(\tau - t)$ is constant,

$$\tau(\mathbf{x}) - t = \text{const}, \tag{4.3}$$

are called *wavefront sets* or simply *wavefronts.* The function $\tau$ will be mainly real except for a few specified cases.

Homogeneous plane waves have the form (4.2) with

$$\omega\tau(\mathbf{x}) = \mathbf{k}\cdot\mathbf{x} \quad \text{and} \quad \boldsymbol{u}^0 = \text{const}, \quad \boldsymbol{u}^1 = \boldsymbol{u}^2 = \ldots = 0.$$

For spherical waves (see Chapter 3, with the footnote to (4.1) taken into account), we have

$$\omega\tau(\mathbf{x}) = |\mathbf{k}||\mathbf{x}| \quad \text{and} \quad \boldsymbol{u}^0 = \frac{1}{|\mathbf{x}|}\mathbf{A}\left(\frac{\mathbf{x}}{|\mathbf{x}|}\right).$$

Further consideration in this chapter can be regarded as a perturbation theory for homogeneous plane waves (except for Section 4.5.2, where the model solution is an inhomogeneous wave). This perturbable wave is called a *local plane wave.*

Saying that the dimensional parameter $\omega$ plays the role of a large parameter, we mean that

$$\omega \to \infty, \tag{4.4}$$

or, which is the same,

$$\frac{1}{\omega} \to 0. \tag{4.5}$$

Dimensionless small parameters characterizing the inhomogeneity of a medium can be introduced, such as

$$\frac{c|\operatorname{grad}\mu|}{\omega\mu} \ll 1, \quad \frac{c|\operatorname{grad}\lambda|}{\omega\lambda} \ll 1, \quad \frac{c|\operatorname{grad}\rho|}{\omega\rho} \ll 1, \tag{4.6}$$

where $c$ is the minimal velocity of propagation in the domain under consideration. At a distance of order of the wavelength $\frac{2\pi c}{\omega}$, a medium is regarded as approximately homogeneous. In the case of a homogeneous medium, the quantities (4.6) equal zero identically. Whenever the conditions (4.6) are satisfied, we say that the medium is *slowly varying.*

We assume also that the parameters characterizing the wave process, namely,

$$\frac{\omega\mathfrak{R}_{1,2}}{c} \gg 1, \tag{4.7}$$

are large, where $\mathfrak{R}_{1,2}$ are the principal radii of curvatures (i.e., the quantities inverse to the principal curvatures of a wavefront, see Section A.4).[2] Only under these conditions, can a wave show the desired similarity to the plane wave.

[2]This condition arises, e.g., in Section 4.7.

The inequalities (4.6) and (4.7) are required in the construction of the theory in the leading order; for higher-order approximations, it is necessary to impose analogous dimensionless conditions on higher derivatives of parameters of a medium and conditions on a wavefield.

In general, the convergence of the ray series (4.2) is not assumed. The series (4.2) belongs to the class of the so-called asymptotic series, the convergence of which is not required.[3] The requirement that the order of smallness of a term with respect to the powers of the corresponding parameter unrestrictedly grow with increasing its number is necessary. In such a case, a series is said to have an *asymptotic nature* (for details, see Section 4.10).

The ray ansatz (4.2) can be regarded as an analog of the Green – Liouville expansion for a one-dimensional equation of the second order (called the WKB by physicists). Alternatively, this ansatz can be devised when analyzing far fields of point sources (see Chapter 3). Expansion (4.2) is the most important example of a high-frequency asymptotic in the theory of elasic waves.

### Local plane wave and local plane vector

Now we introduce a simple, in essence, notion of crucial importance for further consideration, namely, that of local plane wave. It naturally appears as we fix a point $\mathbf{x} \in \mathbb{R}^3$ and expand in its neighborhood the eikonal $\tau(\mathbf{x}+\widetilde{\mathbf{x}})$, ($|\widetilde{\mathbf{x}}|$ is small) with respect to $\widetilde{x}_1$, $\widetilde{x}_2$, $\widetilde{x}_3$, $\widetilde{\mathbf{x}} = (\widetilde{x}_1, \widetilde{x}_2, \widetilde{x}_3)$, up to linear terms

$$\tau(\mathbf{x}+\widetilde{\mathbf{x}}) \approx \tau(\mathbf{x}) + \left.\frac{\partial \tau}{\partial x_j}\right|_{\mathbf{x}} \widetilde{x}_j. \tag{4.8}$$

In the amplitude, we retain only the leading term at the point $\mathbf{x}$

$$\sum_{m=0}^{\infty} \frac{\boldsymbol{u}^m(\mathbf{x}+\mathbf{x}')}{(-i\omega)^m} \approx \boldsymbol{u}^0(\mathbf{x}). \tag{4.9}$$

We thus obtain an approximate equality of the ray ansatz (4.2) and a plane wave, which is a solution of the form (2.4) with the *local wave vector*

$$\boldsymbol{k} = \boldsymbol{k}(\mathbf{x}) = \omega \operatorname{grad} \tau(\mathbf{x}), \tag{4.10}$$

and the vector amplitude

$$\boldsymbol{h} = \boldsymbol{h}(\mathbf{x}) = e^{i\omega\tau(\mathbf{x})} \boldsymbol{u}^0. \tag{4.11}$$

The expression

$$\widetilde{\mathbf{u}} := \boldsymbol{h}(\mathbf{x}) e^{i\boldsymbol{k}\cdot\widetilde{\mathbf{x}} - i\omega t} \tag{4.12}$$

is called the *local plane wave*. It depends on the variables $\widetilde{\mathbf{x}}$ and describes a plane wave in a homogeneous medium with coordinates $\widetilde{\mathbf{x}} \in \mathbb{R}^3$ and "frozen"

[3] The nonconvergence of expansions of cylinder functions in inverse powers of the large argument is well known (see, e.g., Olver 1974 [51]).

parameters $\lambda = \lambda(\mathbf{x})$, $\mu = \mu(\mathbf{x})$, and $\rho = \rho(\mathbf{x})$.[4] Replacing the ansatz (4.1) by its leading term at the point $\mathbf{x} + \widetilde{\mathbf{x}}$, we arrive at the approximate relation

$$e^{i\omega\tau(\mathbf{x}+\widetilde{\mathbf{x}})-i\omega t}\boldsymbol{u}^0(\mathbf{x}+\widetilde{\mathbf{x}}) \approx \boldsymbol{h}e^{i\boldsymbol{k}\cdot\widetilde{\mathbf{x}}-i\omega t}, \tag{4.13}$$

which is valid in a small neighborhood of $\mathbf{x}$.

The velocity of the wavefront, calculated according to the formula (2.9), that is, of the surface $\tau(\mathbf{x}) = t + \text{const}$, is equal to $\frac{1}{|\operatorname{grad}\tau|}\frac{\operatorname{grad}\tau}{|\operatorname{grad}\tau|}$; the related local plane wave at the point $\mathbf{x}$ has the same phase velocity. The corresponding slowness vector is

$$\boldsymbol{p} = \frac{\boldsymbol{k}}{\omega} = \frac{\omega \operatorname{grad}\tau}{\omega} = \operatorname{grad}\tau. \tag{4.14}$$

The result of the above discussion can be stated as follows: the ray ansatz is a perturbed local plane wave with the direction, amplitude, and phase gradually varying from point to point. We will return to a local plane wave when we shall consider the energy of wave processes.

### 4.1.2 Recurrent system

Elastodynamics equations describing time-harmonic waves $\mathbf{u}(\mathbf{x}, t) = e^{-i\omega t}\boldsymbol{u}(\mathbf{x}; \omega)$ in isotropic media have the form

$$\boldsymbol{l}\boldsymbol{u} \equiv \mathfrak{L}\boldsymbol{u} + \rho\omega^2\boldsymbol{u} = 0, \tag{4.15}$$

where the operator $\mathfrak{L}$ was introduced by (1.60). Substituting in (4.15) the expression

$$\boldsymbol{u} = e^{i\omega\tau}\boldsymbol{U}, \tag{4.16}$$

we obtain

$$\mathfrak{L}\boldsymbol{u} + \rho\omega^2\boldsymbol{u} = (i\omega)^2 e^{i\omega\tau}\left(\mathfrak{N} + \frac{\mathfrak{M}}{i\omega} + \frac{\mathfrak{L}}{(i\omega)^2}\right)\boldsymbol{U} = 0. \tag{4.17}$$

Here, $\mathfrak{N}$ is an operator acting on vectors in the three-dimensional space according to the formula

$$\mathfrak{N}\boldsymbol{u} := (\lambda + \mu)(\boldsymbol{u}\cdot\boldsymbol{p})\boldsymbol{p} + (\mu\boldsymbol{p}\cdot\boldsymbol{p} - \rho)\boldsymbol{u}, \tag{4.18}$$

where $\boldsymbol{p}$ is the slowness vector (4.14), $\mathfrak{M}$ is a matrix differential operator of the first order,

$$\begin{aligned}\mathfrak{M}\boldsymbol{u} := {} & (\lambda + \mu)[\boldsymbol{p}\operatorname{div}\boldsymbol{u} + \operatorname{grad}(\boldsymbol{u}\cdot\boldsymbol{p})] + \mu[\boldsymbol{u}\operatorname{div}\boldsymbol{p} + 2(\boldsymbol{p}\cdot\operatorname{grad})\boldsymbol{u}] \\ & + (\boldsymbol{u}\cdot\boldsymbol{p})\operatorname{grad}\lambda + (\operatorname{grad}\mu\cdot\boldsymbol{u})\boldsymbol{p} + \boldsymbol{u}(\operatorname{grad}\mu\cdot\boldsymbol{p}).\end{aligned} \tag{4.19}$$

Note that for real $\tau$, the coefficients of the operators $\mathfrak{N}$, $\mathfrak{M}$, and $\mathfrak{L}$ are real.

---

[4]The concept of local plane wave is related to a comparatively "advanced mathematics." An auxiliary space $\mathbb{R}^3$ is assigned to a point $\mathbf{x} \in \mathbb{R}^3$, where the Lamé parameters and the volume density are "frozen," the coordinates on this $\mathbb{R}^3$ are denoted by $\widetilde{\mathbf{x}}$. Actually, we deal with the tangent space at the point $\mathbf{x}$. A set of tangent spaces at all the points $\mathbf{x}$ is the *tangent bundle* (see, e.g., Novikov and Taimanov 2006 [50]).

Now, in agreement with (4.2), we expand $\boldsymbol{U}$ in inverse powers of the large parameter,

$$\boldsymbol{U}(\mathbf{x};\omega) = \left\{\boldsymbol{u}^0(\mathbf{x}) + \frac{\boldsymbol{u}^1(\mathbf{x})}{-i\omega} + \dots\right\},$$

substitute this into (4.17), and equate to zero the coefficients at $\omega^2$ and $\omega^1$, and then also those at lower powers of $\omega$, etc. We obtain the recurrent equations

$$\mathfrak{N}\boldsymbol{u}^0 = 0, \tag{4.20}$$

$$\mathfrak{N}\boldsymbol{u}^1 = \mathfrak{M}\boldsymbol{u}^0, \tag{4.21}$$

$$\mathfrak{N}\boldsymbol{u}^2 = \mathfrak{M}\boldsymbol{u}^1 - \mathfrak{L}\boldsymbol{u}^0, \tag{4.22}$$

$$\mathfrak{N}\boldsymbol{u}^{m+2} = \mathfrak{M}\boldsymbol{u}^{m+1} - \mathfrak{L}\boldsymbol{u}^m, \quad m = 1, 2, \dots \tag{4.23}$$

The equations (4.21)–(4.23) are called the *transport equations.*

Since the operator $\mathfrak{N}$ does not contain differentiation, each of the transport equations is a system of three linear algebraic equations. The matrix of this system $\|\mathfrak{N}_{mn}\| = \|(\lambda(\mathbf{x})+\mu(\mathbf{x}))p_m p_n + (\mu(\mathbf{x})p_l p_l - \rho(\mathbf{x}))\delta_{mn}\|$ had already appeared in Chapter 2 where we dealt with plane waves in homogeneous media. Its occurrence here inevitably results from the relationship between the ansatz (4.1) and the local plane waves, which was discussed in Section 4.1.1.

### 4.1.3 Waves $P$ and $S$

As we already know (see Section 2.3), the equation (4.20) has nontrivial solutions in two cases. First, zero is an eigenvalue of the matrix $\mathfrak{N}$, where

$$(\operatorname{grad}\tau)^2 = \frac{1}{a^2}, \tag{4.24}$$

which corresponds to the local velocity of a wave $P$,

$$a = a(\mathbf{x}) = \sqrt{\frac{\lambda(\mathbf{x}) + 2\mu(\mathbf{x})}{\rho(\mathbf{x})}}. \tag{4.25}$$

The corresponding eigensubspace of the matrix $\mathfrak{N}$ is one-dimensional, and

$$\boldsymbol{u}^0(\mathbf{x}) \parallel \operatorname{grad}\tau(\mathbf{x}). \tag{4.26}$$

This means that both real and imaginary parts of the vector $\boldsymbol{u}^0$ are parallel to $\operatorname{grad}\tau$. Second, a nontrivial solution exists under the condition

$$(\operatorname{grad}\tau)^2 = \frac{1}{b^2} \tag{4.27}$$

where

$$b = b(\mathbf{x}) = \sqrt{\frac{\mu(\mathbf{x})}{\rho(\mathbf{x})}} \tag{4.28}$$

is the local velocity of a wave $S$. In the latter case, the eigensubspace of the matrix $\mathfrak{N}$ is two-dimensional, and the polarization of the zeroth order is transversal,

$$\boldsymbol{u}^0(\mathbf{x}) \perp \operatorname{grad}\tau(\mathbf{x}), \tag{4.29}$$

i.e., $\operatorname{Re}\boldsymbol{u}^0 \perp \operatorname{grad}\tau$ and $\operatorname{Im}\boldsymbol{u}^0 \perp \operatorname{grad}\tau$.

---

## 4.2 Eikonal equation and rays

In this section, we present important results related to solving the *eikonal equation* with $c > 0$ standing for a local velocity

$$(\operatorname{grad}\tau)^2 = \frac{1}{c^2}, \quad c = c(\mathbf{x}). \tag{4.30}$$

We have arrived at such an equation in both above cases. Note that for a constant velocity $c =$ const, it has particular solutions of the form

$$\tau(\mathbf{x}) = \frac{1}{c}\mathbf{x}\cdot\mathbf{s}, \quad |\mathbf{s}| = 1 \quad \text{and} \quad \tau(\mathbf{x}) = \frac{1}{c}|\mathbf{x}|,$$

which appeared in the description of plane and spherical waves.

As known (see, e.g., Babich and Buldyrev 2007 [6]), the classical theory of the eikonal equation is based on variational calculus. The main objects of this theory are rays. The concept of ray emerged in optics in antiquity (if not even earlier). First we just describe the key constructions, and then derive them in the context of variational theory.

### 4.2.1 Fermat functional and rays

Consider the *Fermat functional* defined by the integral

$$\int_{M_0}^{M} \frac{ds}{c} \tag{4.31}$$

along the curves $\ell$ connecting points $M_0$ and $M$, $M_0, M \in \mathbb{R}^3$. Here, $ds = ds(\mathbf{x}) = \sqrt{(dx_1)^2 + (dx_2)^2 + (dx_3)^2}$ is the differential of the arc length and $c = c(\mathbf{x}) > 0$. The prescribed function $c(\mathbf{x})$ is assumed to be smooth. The kinematic meaning of the integral (4.31) is the travel time of a wave with the velocity $c$ along a given *comparison curve* $\ell$ from $M_0$ to $M$.

Let a curve $\ell$ connecting $M_0$ and $M$ be such that

$$\delta\int_{M_0}^{M} \frac{ds}{c} = 0. \tag{4.32}$$

In accordance with the terminology of variational calculus, such an $\ell$ is called an *extremal of the Fermat functional* (4.31). An extremal with a specified direction is called a *ray*. We will always choose the direction in which $\tau$ grows,

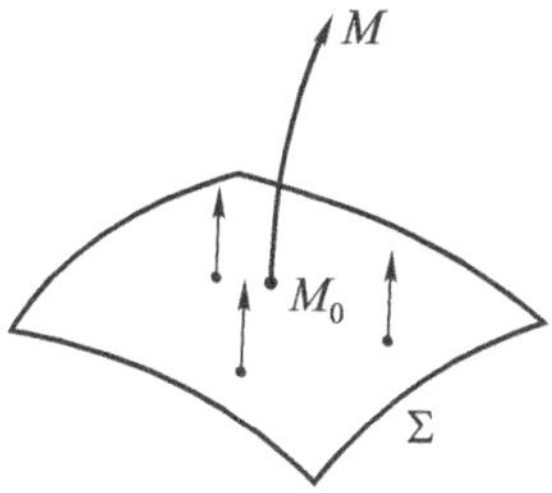


**Figure 4.1**: Ray launched from a surface

which is in agreement with the definitions given above for the rays of plane waves.

Rays in the two-dimensional case are defined similarly.

The condition (4.32) is necessary for the travel time along the corresponding curve connecting $M_0$ and $M$ to be minimal as compared with that along close curves (for $M$ sufficiently close to $M_0$, this condition is also sufficient). The relation (4.32) is called the *Fermat principle.*

Let a segment of the ray connecting points $M_0$ and $M$ be parametrically described as $x_j = x_j(\sigma)$, $j = 1, 2, 3$, $\sigma_0 \leqslant \sigma \leqslant \sigma_1$. The equation (4.32) takes the form

$$\delta \int_{\sigma_0}^{\sigma_1} \Phi d\sigma, \quad \Phi = \frac{1}{c(\mathbf{x}(\sigma))} \sqrt{\sum\nolimits_{j=1}^{3} (x'_j)^2}, \quad {}' := \frac{d}{d\sigma}. \tag{4.33}$$

The vanishing of the variation of the integral $\int_{\sigma_0}^{\sigma_1}$ is equivalent to the Euler equations

$$\frac{\partial \Phi}{\partial x_j} - \frac{d}{d\sigma}\frac{\partial \Phi}{\partial x'_j} = 0, \quad j = 1, 2, 3. \tag{4.34}$$

The system (4.34) can easily be rewritten in vector form

$$\frac{d}{ds}\left(\frac{\mathbf{s}}{c}\right) - \operatorname{grad}\left(\frac{1}{c}\right) = 0, \tag{4.35}$$

where the parameter along the curve is its length $s$, and $\mathbf{s}$ is the unit tangent vector. As is easily seen from (4.35), for a constant velocity $c$ the rays are straight lines.

### 4.2.2 Solving the eikonal equation with the help of rays

Recall that a surface $\Sigma \subset \mathbb{R}^3$ on which the function $\tau$ is constant is called a *wavefront.* We take the direction of $\operatorname{grad}\tau(\mathbf{x})$ as that of propagation of a wave at a point $\mathbf{x}$. Let the value $\tau|_\Sigma = \text{const}$ be known. Launch from each point $M_0 \in \Sigma$ the ray along the normal to $\Sigma$ (see Fig. 4.1). Let $M$ be a point on such a ray. Define the value $\tau(M)$ by the formula

$$\tau(M) = \tau(M_0)|_\Sigma \pm \int_{M_0}^{M} \frac{ds}{c}, \quad M_0 \in \Sigma, \tag{4.36}$$
$$\tau(M_0) = \tau|_\Sigma = \text{const},$$

where integration is performed along the ray, the sign "+" is taken if the ray $M_0M$ is directed along the normal to $\Sigma$, otherwise the sign "−" is to be taken. Near $\Sigma$ the function (4.36) satisfies the eikonal equation. We will prove it at the end of Section 4.2.3.

### 4.2.3 Cauchy problem for the eikonal equation

#### Classical real Cauchy problem for the eikonal equation

Description of wave phenomena often requires solving the eikonal equation (4.30) with the eikonal's value prescribed on some smooth surface $\Sigma$,

$$\tau|_\Sigma = \tau^\circ, \tag{4.37}$$

where $\tau^\circ = \tau^\circ(\mathbf{x})$ is a smooth function. Furthermore, let us choose the direction of a normal to $\Sigma$ and accordingly specify the sign of the normal derivative $\frac{\partial \tau}{\partial n}$. This is the classical Cauchy problem for the eikonal equation. Such a problem arises, for example, in the case where $\Sigma$ bounds an elastic medium or interfaces two elastic media, with an incident wave defined by the ray expansion.

Let us proceed to solving the above problem. First of all, the condition (4.37) allows one to find $\operatorname{grad}\tau$ on $\Sigma$. Indeed, differentiating $\tau^\circ$ along $\Sigma$, we find the components of $\operatorname{grad}\tau$ that are tangent to $\Sigma$ and will be denoted by $\operatorname{grad}_\Sigma \tau$. The eikonal equation (4.30) enables one to determine the normal component of $\operatorname{grad}\tau$ by the formula

$$\left.\frac{\partial \tau}{\partial n}\right|_\Sigma = \pm\sqrt{\left.\frac{1}{c^2}\right|_\Sigma - (\operatorname{grad}_\Sigma \tau^\circ)^2}. \tag{4.38}$$

Note that the sign of the expression (4.38) must be specified.

Here, we deal with real $\tau^\circ$ and assume that[5]

$$|\operatorname{grad}_\Sigma \tau^\circ| < \frac{1}{c}. \tag{4.39}$$

Thus, $\tau|_\Sigma$ and $\operatorname{sgn}\frac{\partial \tau}{\partial n}\big|_\Sigma$ completely define all the components of $\operatorname{grad}\tau$ on $\Sigma$.

Now we launch from each point $M_0 \in \Sigma$ a ray in the direction of $\operatorname{grad}\tau$ and define on this ray the eikonal in accordance with formula (4.36)

$$\tau(M) = \tau^\circ(M_0) + \int_{M_0}^{M} \frac{ds}{c}, \tag{4.40}$$

[5]This condition is satisfied in the case of lacking of the total internal reflection.

where the integral is taken along the ray. Here, we choose the plus sign in (4.49) aiming at the growth of the eikonal in the direction of grad $\tau$.

With the help of a variational consideration, we will get convinced that the function $\tau$ so constructed is a solution of the problem. This solution is unique (the exact meaning of the latter statement and its proof can be found, e.g., in Courant and Hilbert 1962 [14], Smirnov 1964b [59].

## General form of the first variation

Let us recall some classical results of variational calculus (see, e.g., Smirnov 1964b [59], Gelfand and Fomin 1963 [27]), which will be necessary in the sequel. Consider a general functional of the form

$$I = \int_{M_0}^{M} \Phi d\sigma, \tag{4.41}$$

where the integral is taken along a smooth curve connecting $M_0$ and $M$, $\sigma$ is a parameter along the curve, $\Phi = \Phi(\mathbf{x}, \mathbf{x}')$, $' = \frac{d}{d\sigma}$. Fix a curve $\ell$ connecting $M_0$ and $M$, and consider an increment of the functional (4.41) allowing the variation of not only the initial curve but also the starting and ending points $M_0$ and $M$.[6] Calculation shows (see, e.g., Gelfand and Fomin 1963 [27], Smirnov 1964b [59]) that the variation of the functional (4.41), i.e., the principal linear part of its increment has the form

$$\delta I = \int_{M_0}^{M} \left( \frac{\partial \Phi}{\partial x_i} - \frac{d}{d\sigma} \frac{\partial \Phi}{\partial x_i'} \right) \delta x_i d\sigma + \left( \Phi - x_i' \frac{\partial \Phi}{\partial x_i'} \right) \delta\sigma \Big|_{M_0}^{M} + \frac{\partial \Phi}{\partial x_i'} \delta x_i \Big|_{M_0}^{M}, \tag{4.42}$$

Here, $\delta x_i|_{M_0,M}$ are variations of the positions of the starting and ending points $M_0$ and $M$, and $\delta\sigma|_{M_0,M}$ is an increment of the parameter in such a variation.

If $\ell$ is an extremal, then the integral in (4.42) vanishes. Further, for a functional of the specific form $\Phi = |\mathbf{x}'_\sigma| \phi(\mathbf{x})$ (the Fermat functional is exactly of such a form with $\phi(\mathbf{x}) = \frac{1}{c(\mathbf{x})}$) direct calculation[7] shows that

$$\Phi - x_i' \frac{\partial \Phi}{\partial x_i'} = 0. \tag{4.43}$$

Consider the last term on the right-hand side of (4.42). If at the point $M$ the following condition holds:

$$\frac{\partial \Phi}{\partial x_i'} \delta x_i \Big|_{M} = 0, \tag{4.44}$$

then it is said that at $M$ the *transversality condition* holds. If the point $M$

[6] Here, comparison curves are merely curves close to $\ell$ and do not necessarily connect $M_0$ and $M$.

[7] The calculation employs the first-degree homogeneity of $\Phi$ with respect to $\mathbf{x}'$ and the Euler homogeneity relation (see Section 2.4.4).

under variation moves along some surface, then the transversality condition at this point reads as follows:

$$\frac{\partial\Phi}{\partial x_i'}\delta x_i = \frac{1}{c}\frac{x_i'}{\sqrt{\sum_{i=1}^{3}(x_i')^2}}\delta x_i = \frac{\mathrm{s}_i}{c}\delta x_i = \frac{1}{c}(\mathbf{s},\delta\mathbf{x}) = 0,$$

where $\mathbf{s} = (s_1, s_2, s_3)$ is a unit vector tangent to the ray, $\delta\mathbf{x} = (\delta x_1, \delta x_2, \delta x_3)$. The transversality condition has thus been reduced to the orthogonality of the ray to this surface.

**Why does expression (4.40) solve the Cauchy problem?**

The fulfillment of the condition (4.37) is obvious. We will vary the position of the point $M$ in the formula (4.40) assuming $\ell$ to be the ray connecting $M_0$ and $M$. In our case, the comparison curves are rays. Then

$$\delta\tau = \delta\tau^{\text{o}} + \left.\frac{\partial\Phi}{\partial x_i'}\delta x_i\right|_{M_0}^{M} = \delta\tau^{\text{o}} + \left.\frac{\partial\Phi}{\partial x_i'}\delta x_i\right|_{M} - \left.\frac{\partial\Phi}{\partial x_i'}\delta x_i\right|_{M_0}, \tag{4.45}$$

where $\delta\tau^{\text{o}}$ is an increment of the initial data under changing $M_0$ as a result of the variation of $M$.

The terms related to $M_0$ cancel because the equality $\delta\tau^{\text{o}}|_{M_0} = \frac{\partial\Phi}{\partial x_i'}\delta x_i|_{M_0}$ holds. To prove it, note that

$$\left.\frac{\partial\Phi}{\partial x_i'}\delta x_i\right|_{M_0} = \left.\frac{\mathrm{s}_i}{c}\delta x_i\right|_{M_0} = (\operatorname{grad}\tau)_i\delta x_i|_{M_0}\,. \tag{4.46}$$

The vector $\delta\mathbf{x} = (\delta x_1, \delta x_2, \delta x_3)$ is tangent to the surface $\Sigma$ on which $\tau = \tau^{\text{o}}$, therefore $\left.\frac{\partial\Phi}{\partial x_i'}\delta x_i\right|_{M_0}$ is the increment of $\tau|_\Sigma$ when moving along $\Sigma$ by an infinitely small vector $\delta\mathbf{x}$, that is, $\left.\frac{\partial\Phi}{\partial x_i'}\delta x_i\right|_{M_0} = \delta\tau^{\text{o}}$. The formula (4.45) takes the form $\delta\tau = \frac{\partial\Phi}{\partial x_j'}\delta x_j$. The substitution of the expression (4.33) for $\Phi$ gives

$$\delta\tau = \frac{\partial\Phi}{\partial x_j'}\delta x_j = \frac{1}{c}\frac{x_i'}{\sqrt{\sum_{i=1}^{3}(x_i')^2}}\delta x_j. \tag{4.47}$$

Now the $\delta x_i$ are arbitrary, therefore (4.47) leads to the relation $\frac{\partial\tau}{\partial x_i} = \frac{1}{c}\mathrm{s}_i$, or

$$\operatorname{grad}\tau = \frac{\mathbf{s}}{c}, \tag{4.48}$$

where $\mathbf{s}$ is a unit vector tangent to the ray. Thus, the construction (4.40) solves the Cauchy problem for the eikonal equation (4.30).

A similar consideration proving that the formula (4.40) is a solution of the eikonal equation for the central field of rays with center $M_0$ (see Section 4.2.4) is omitted.

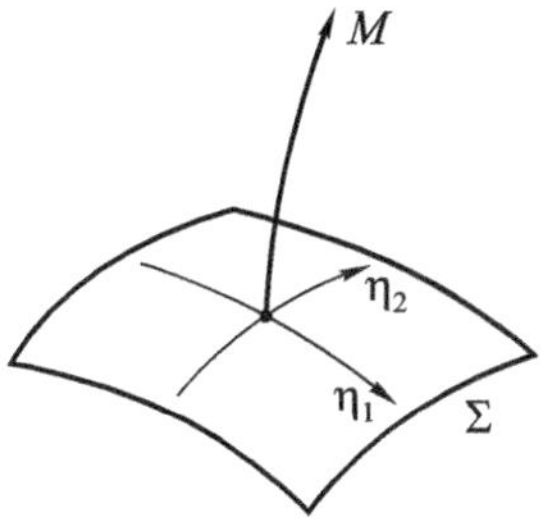


**Figure 4.2**: Ray coordinates associated with the surface $\Sigma$

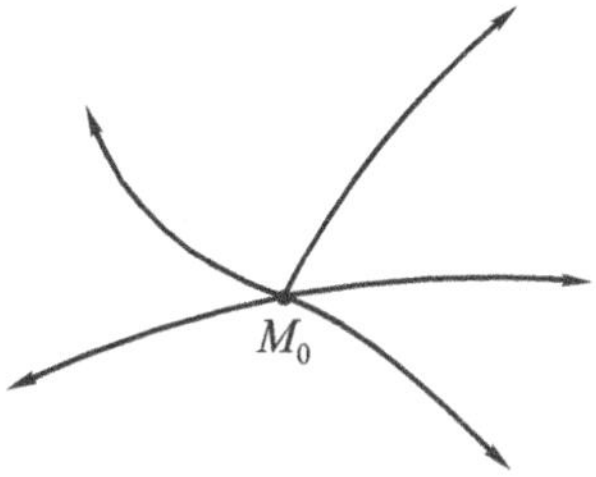


**Figure 4.3**: Central field of rays

### 4.2.4 Ray coordinates and field of rays

We are going to introduce special coordinates best suited for solving transport equations.

Let some regular coordinates $(\eta_1, \eta_2)$, which are, generally speaking, nonorthogonal, be introduced on the surface $\Sigma$ (see Section 4.2.2). We define a ray launched from the point $M_0 \in \Sigma$ by its coordinates $\eta_1$ and $\eta_2$ (see Fig. 4.2). The assignment of $\tau(M)$ uniquely determines the position of a point on this ray. We will describe points in a neighborhood of the surface $\Sigma$ by the regular coordinates $(\eta_1, \eta_2, \tau)$. Such coordinates are called *ray coordinates*, and the related family of rays is called a field of rays or *ray field*.

Very important is the degenerate case of the previous construction of solutions of the eikonal equation, where the initial surface $\Sigma$ reduces to a single point, which we denote by $M_0$. Let us launch from $M_0$ the rays in all directions and assume that

$$\tau(M) = \tau(M_0) + \int_{M_0}^{M} \frac{ds}{c}, \tag{4.49}$$

where the integral is taken along rays. The previous consideration remains in force, and the expression (4.49) also satisfies the eikonal equation (4.30). The family of rays launched from a fixed point $M_0$ is called the *central field of rays*, and $M_0$ is called its *center* (see Fig. 4.3).

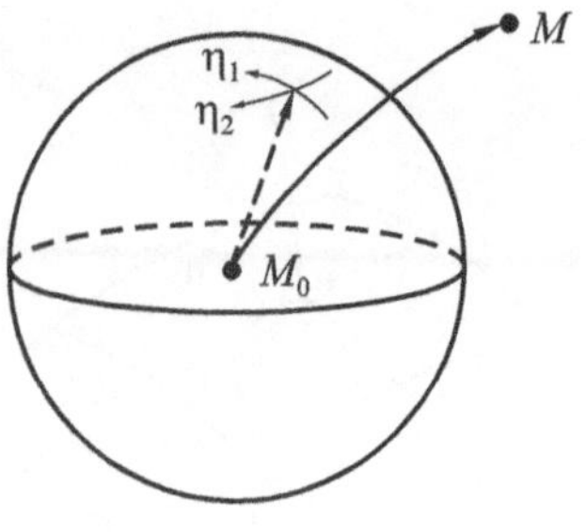


**Figure 4.4**: Ray coordinates associated with a central field of rays

For the central field of rays (see Section 4.2.4), the role of coordinates $\eta_1$ and $\eta_2$ can be played by the coordinates of the end of the unit vector $\mathbf{s}|_{M_0}$ which is tangent to the ray launched from $M_0$ to $M$ (see Fig. 4.4).

In both cases, $\eta_1$ and $\eta_2$ define a ray, and $\tau$ defines the position of a point $M$ on the ray, and the coordinate system $(\eta_1, \eta_2, \tau)$ is called *ray coordinates.*

### 4.2.5 ⋆ Complex eikonal

In the case of the total internal reflection, the condition (4.39) should be replaced by

$$|\operatorname{grad}_\Sigma \tau^{\mathrm{o}}| > \frac{1}{c}, \tag{4.50}$$

and there is no real solution of the eikonal equation.

#### Eikonal equation in the coordinates $(q^1, q^2, n)$

First, introduce coordinates suitable near the surface $\Sigma$. These coordinates are important for studying the Cauchy problem in the case of (4.50). They will also appear when considering the Rayleigh wave in Chapter 8. We restrict the consideration to a small neighborhood of $\Sigma$. Introduce on $\Sigma$ some regular coordinates $q_\alpha$, $\alpha = 1, 2$, and take the distance $q_3 = n$ from $\Sigma$ as the third coordinate. In the area considered, we put $n > 0$. We fix a sufficiently small $\epsilon = \text{const} > 0$, and let $\Sigma_\epsilon : \{n = \epsilon\}$ be *a surface parallel to* $\Sigma$. It can be proved that $\Sigma_\epsilon$ is a smooth surface.

Let us show that a normal to $\Sigma$ is a normal to $\Sigma_\epsilon$ as well. Indeed, in vector form the equation of $\Sigma_\epsilon$ reads

$$\mathbf{r}_\epsilon = \mathbf{r}_0(q_1, q_2) + \epsilon \boldsymbol{n}(q_1, q_2), \tag{4.51}$$

where $\mathbf{r}_\epsilon$ and $\mathbf{r}_0$ are radius vectors of a point on $\Sigma_\epsilon$ and $\Sigma$, respectively, and $\boldsymbol{n}$ is a unit normal to $\Sigma$. Consider the scalar products

$$\left(\frac{\partial \mathbf{r}_\epsilon}{\partial q_\alpha}, \boldsymbol{n}\right) = \left(\frac{\partial \mathbf{r}_0}{\partial q_\alpha}, \boldsymbol{n}\right) + \epsilon \left(\frac{\partial \boldsymbol{n}}{\partial q_\alpha}, \boldsymbol{n}\right) = 0, \quad \alpha = 1, 2.$$

Further, the second term on the right-hand side equals zero, because $\boldsymbol{n}^2 = 1$, and therefore $\frac{\partial \boldsymbol{n}^2}{\partial q_\alpha} = 2(\frac{\partial \boldsymbol{n}}{\partial q_\alpha}, \boldsymbol{n}) = 0$. The first term vanishes, because both vectors $\frac{\partial \mathbf{r}_0}{\partial q_1}$ and $\frac{\partial \mathbf{r}_0}{\partial q_2}$ lie in the tangent plane to $\Sigma$. Thus, it is established that $\boldsymbol{n}$ is orthogonal to $\frac{\partial \mathbf{r}_\epsilon}{\partial q_1}$ and $\frac{\partial \mathbf{r}_\epsilon}{\partial q_2}$, and therefore to $\Sigma_\epsilon$.

A squared differential of arc length in the coordinates $(n, q_1, q_2)$ equals

$$\sum\nolimits_{p=1}^{3} (dx_p)^2 = (d(\mathbf{r}_0 + n\boldsymbol{n}))^2 = \sum\nolimits_{\alpha,\beta=1}^{2} g_{\alpha_\beta} dq_\alpha dq_\beta + dn^2 = g_{ij} dq_i dq_j,$$

$q_3 = n$. By virtue of the orthogonality of $\frac{\partial(\mathbf{r}_0 + n\boldsymbol{n})}{\partial q_\alpha}$ to $\boldsymbol{n}$, we have $g_{13} = g_{23} = g_{31} = g_{32} = 0$, and the matrix $\|g_{ij}\|$ and its inverse $\|g^{ij}\|$ are of the form

$$\|g_{ij}\| = \begin{pmatrix} g_{11} & g_{12} & 0 \\ g_{21} & g_{22} & 0 \\ 0 & 0 & 1 \end{pmatrix}, \quad \|g^{ij}\| = \begin{pmatrix} g^{11} & g^{12} & 0 \\ g^{21} & g^{22} & 0 \\ 0 & 0 & 1 \end{pmatrix},$$

where $\begin{pmatrix} g^{11} & g^{12} \\ g^{21} & g^{22} \end{pmatrix} = \begin{pmatrix} g_{11} & g_{12} \\ g_{21} & g_{22} \end{pmatrix}^{-1}$, $g_{\alpha_\beta} = \mathrm{g}_{\alpha\beta}|_\Sigma + O(n)$, and $\mathrm{g}_{\alpha\beta}|_\Sigma := \left(\frac{\partial \mathbf{r}_0}{\partial q_\alpha}, \frac{\partial \mathbf{r}_0}{\partial q_\beta}\right)$, $\alpha, \beta = 1, 2$.

**Solution of the eikonal equation near $\Sigma$**

As follows from the results of the Appendix, the following relation holds:

$$(\operatorname{grad}\tau)^2 = g^{ij} \frac{\partial \tau}{\partial q_i} \frac{\partial \tau}{\partial q_j} = \sum\nolimits_{\alpha,\beta=1}^{2} g^{\alpha\beta} \frac{\partial \tau}{\partial q_\alpha} \frac{\partial \tau}{\partial q_\beta} + \left(\frac{\partial \tau}{\partial n}\right)^2, \tag{4.52}$$

and the eikonal equation in the coordinates $q_1$, $q_2$, $n$ has the form

$$(\operatorname{grad}\tau)^2 \equiv \left(\frac{\partial \tau}{\partial n}\right)^2 + (\operatorname{grad}_\Sigma \tau)^2 = \frac{1}{c^2}, \tag{4.53}$$

where

$$(\operatorname{grad}_\Sigma \tau)^2 := \sum\nolimits_{\alpha,\beta=1}^{2} g^{\alpha\beta} \frac{\partial \tau}{\partial q_\alpha} \frac{\partial \tau}{\partial q_\beta}. \tag{4.54}$$

The operator $(\operatorname{grad}_\Sigma)^2$ is sometimes called the first differential Beltrami parameter. From (4.53) it follows that

$$\frac{\partial \tau}{\partial n} = \pm i \sqrt{(\operatorname{grad}_\Sigma \tau)^2 - \frac{1}{c^2}}. \tag{4.55}$$

We choose the sign "+" to ensure the decay of the factor $e^{i\omega\tau}$ with distance from the boundary, in accordance with what we did in the theory of plane waves. We arrive at the Cauchy problem

$$\frac{\partial \tau}{\partial n} = i \sqrt{(\operatorname{grad}_\Sigma \tau)^2 - \frac{1}{c^2}}, \qquad \tau|_{n=0} = \tau|_\Sigma = \tau^\circ. \tag{4.56}$$

Since the wavefield with such an eikonal rapidly decays as the distance from the boundary grows, only a small neighborhood of $\Sigma$ is of interest in applications.

A solution of this problem "in small" can easily be constructed as an expansion in powers of $n$:

$$\tau = \tau\big|_{n=0} + \frac{\partial \tau}{\partial n}\bigg|_{n=0} n + \frac{1}{2}\frac{\partial^2 \tau}{\partial n^2}\bigg|_{n=0} n^2 + \dots \tag{4.57}$$

The leading-order terms are as follows:

$$\tau = \tau^{\circ} + in\sqrt{(\operatorname{grad}_{\Sigma} \tau^{\circ})^2 - \frac{1}{c^2}\bigg|_{n=0}} + O(n^2). \tag{4.58}$$

We restrict the discussion to such a "poor" solution of the Cauchy problem, because in this case we can hardly expect a more complete one. The variational theory presented in the previous sections is substantially real, and its complex counterpart is not yet available. The correctness (for example, in the Hadamard's sense) of the problem (4.56) is doubtful.

## 4.3 Solving transport equations. The wave $P$

### 4.3.1 Zeroth-order approximation. Consistency condition and the Umov equation

Considering the case of a wave $P$, we imply that $\tau$ is a solution of the eikonal equation (4.24) corresponding to the velocity of waves $P$, and mark other quantities describing the wavefield by the symbol $^P$. Let us write (4.26) in the form

$$\boldsymbol{u}^{P0} = \phi^0 \operatorname{grad} \tau = \frac{\phi^0}{a}\mathbf{s}, \tag{4.59}$$

where $\phi^0 = \phi^0(\mathbf{x})$ is an arbitrary scalar function for the present. This is all that can be obtained from (4.20).

The equation for the following approximation (4.21) is a nonhomogeneous system of linear algebraic equations for the components of the vector $\boldsymbol{u}^1$. Since the corresponding homogeneous system has a nontrivial solution, the system (4.21) is, generally speaking, inconsistent.

#### Consistency condition

Recall an important result of linear algebra that is known as the finite-dimensional *Fredholm alternative.* Consider a system of linear equations

$$\mathbf{AX} = \mathbf{f},$$

where $\mathbf{A}$ is an $m \times m$ matrix , $\mathbf{A} = \begin{pmatrix} \mathsf{a}_{11} & \cdots & \mathsf{a}_{1m} \\ \vdots & \ddots & \vdots \\ \mathsf{a}_{m1} & \cdots & \mathsf{a}_{mm} \end{pmatrix}$, $\mathbf{X} = \begin{pmatrix} \mathsf{x}_1 \\ \vdots \\ \mathsf{x}_m \end{pmatrix}$, $f = \begin{pmatrix} \mathsf{f}_1 \\ \vdots \\ \mathsf{f}_m \end{pmatrix}$. The system is consistent if and only if the right-hand side $\mathbf{f}$ is orthogonal to any solution $\mathbf{Y}$ of the system

$$\mathbf{A}^{+}\mathbf{Y} = 0$$

with the Hermitian conjugate matrix $\mathbf{A}^{+} = \begin{pmatrix} \mathsf{a}^*_{11} & \cdots & \mathsf{a}^*_{m1} \\ \vdots & \ddots & \vdots \\ \mathsf{a}^*_{1m} & \cdots & \mathsf{a}^*_{mm} \end{pmatrix}$, that is, the consistency condition is as follows:

$$(\mathbf{f}, \mathbf{Y}) := \sum_{n=1}^{n=m} \mathsf{f}_n \mathsf{Y}^*_n = 0, \tag{4.60}$$

(see, e.g., Smirnov 1964c [60]).[8]

In our case, the role of the matrix $\mathbf{A}$ is played by $\mathfrak{M}^P = \mathfrak{M}^{P+}$, and $\mathbf{Y}$ is a vector proportional to grad $\tau$; see (4.59). Thus, the consistency condition of the equation (4.21) can be written as

$$(\mathfrak{M}^P \boldsymbol{u}^{P0}, \boldsymbol{u}^{P0}) = 0. \tag{4.61}$$

**Continuity equation for the energy fluid or the Umov hydrodynamic equation**

Equating to zero the real part of $(\mathfrak{M}^P \boldsymbol{u}^{P0}, \boldsymbol{u}^{P0})$, after some algebra with the use of formula (4.59) we obtain an important relation

$$\operatorname{Re}(\mathfrak{M}^P \boldsymbol{u}^{P0}, \boldsymbol{u}^{P0}) = \operatorname{div}\left\{(\lambda + 2\mu)|\boldsymbol{u}^{P0}|^2 \operatorname{grad}\tau\right\} = 0. \tag{4.62}$$

The latter is just a necessary consistency condition of equation (4.21), because the condition (4.61) is equivalent to the system of equations (4.62) added by

$$\operatorname{Im}\big(\mathfrak{M}^P \boldsymbol{u}^{P0}, \boldsymbol{u}^{P0}\big) = 0. \tag{4.63}$$

From (4.62) and (4.63) we are going to deduce a simple expression for the function $\phi^0$; see (4.59).

Now we give to the equation (4.62) an interpretation of crucial importance. The equation can be written in the form

$$\begin{gathered} \operatorname{div}(\bar{\tilde{\mathfrak{e}}}\boldsymbol{v}^{gr}) = 0, \\ \bar{\tilde{\mathfrak{e}}} := \frac{\omega^2}{2}\rho|\tilde{\boldsymbol{u}}|^2 = \frac{\omega^2}{2}\rho(\mathbf{x})|\tilde{\boldsymbol{u}}(\mathbf{x})|^2. \end{gathered} \tag{4.64}$$

Here, $\bar{\tilde{\mathfrak{e}}}$ is the density of energy averaged over the period of the corresponding

[8]If the standard scalar product in $\mathbb{C}^n$ defined by (4.60) is replaced by another one, then $\mathbf{A}^{+}$ must be understood as the corresponding conjugate matrix.

local plane wave $\widetilde{\boldsymbol{u}}$, see (4.12), and $\boldsymbol{v}^{gr}(\mathbf{x})$ is the *group velocity of the wave P*, which equals the group velocity of $\widetilde{\boldsymbol{u}}$:

$$\boldsymbol{v}^{gr}(\mathbf{x}) = a(\mathbf{x})\frac{\operatorname{grad}\tau}{|\operatorname{grad}\tau|}. \tag{4.65}$$

In order to get out of the inconvenient factor of proportionality $\frac{\omega^2}{2}$, we introduce the *reduced density of energy averaged over time* of a local plane wave:

$$\overline{\overline{\mathcal{E}}} := \frac{2}{\omega^2}\overline{\overline{\mathfrak{e}}} = \rho|\widetilde{\boldsymbol{u}}|^2 = \rho(\mathbf{x})|\widetilde{\boldsymbol{u}}(\mathbf{x})|^2 = \rho(\mathbf{x})|\boldsymbol{u}^0(\mathbf{x})|^2. \tag{4.66}$$

From (4.64)–(4.65) it follows that

$$\operatorname{div}\left(\overline{\overline{\mathcal{E}}}\boldsymbol{v}^{gr}\right) = 0. \tag{4.67}$$

The equation (4.67) and its numerous counterparts arising in subsequent chapters have the form of a continuity equation, which is well known in hydrodynamics (see, e.g., Landau and Lifshitz 1987 [43], Kochin, Kibel, and Roze 1964 [40]). This equation gives a mathematical description of the conservation of mass of the *energy fluid*. The role of the density of its mass is played by $\overline{\overline{\mathcal{E}}}$. The hydrodynamic interpretation of the equation (4.67) agrees with the fact that the vector $\boldsymbol{v}^{gr}$ defined by the formula (4.65) is directed along the ray, and the following equalities hold:

$$\frac{dx_j}{d\tau} = v_j^{gr}, \quad j = 1, 2, 3, \quad \text{or} \quad \frac{d\mathbf{x}}{d\tau} = \boldsymbol{v}^{gr}. \tag{4.68}$$

Indeed,

$$\boldsymbol{v}^{gr} = a\mathbf{s} = a\frac{d\mathbf{x}}{ds} = a\frac{d\mathbf{x}}{d\tau}\frac{d\tau}{ds} = a\frac{1}{a}\frac{d\mathbf{x}}{d\tau} = \frac{d\mathbf{x}}{d\tau}. \tag{4.69}$$

The vector $\boldsymbol{v}^{gr}$ is tangent to the ray. Thus, rays play the role of lines of flow of the energy fluid (more information on the lines of flow in hydrodynamics is given, e.g., in Kochin, Kibel, and Roze 1964 [40]). This fluid flows with the velocity $\boldsymbol{v}^{gr}$ together with the wave, which completely agrees with ideas due to Rayleigh and Reynolds (1870s) on the relationship between the group velocity and (the process of) propagation of energy (see, e.g., Sommerfeld 1950 [62]).

Formula (4.67) gives an exact meaning to the pioneer idea by Umov 1874, 1950 [63, 64] on plane wave energy flowing with the velocity of an elastic wave and satisfying the continuity equation; see also Section 4.10. Equality (4.67), which is of great importance in the theory of high-frequency elastic waves, is called the *Umov equation*.

### 4.3.2 Zeroth-order approximation. Formulas for $\overline{\overline{\mathcal{E}}}$ and $\boldsymbol{u}^{P0}$

#### A ray tube and a formula for $\operatorname{div}(\mathcal{A}\boldsymbol{v}^{gr})$

Derive a general formula of vector analysis, which facilitates finding expressions for $\overline{\overline{\mathcal{E}}}$ and $\boldsymbol{u}^{P0}$, as well as for many other quantities in the sequel.

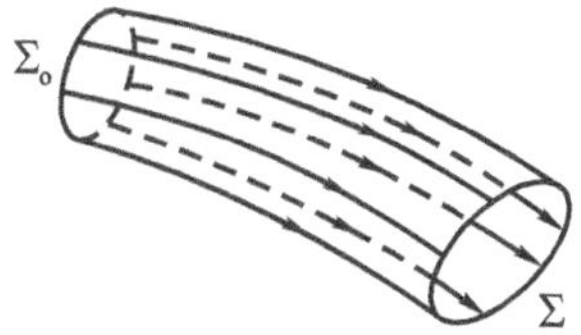


**Figure 4.5**: Ray tube

Consider a formula for $\operatorname{div}(\mathcal{A}\boldsymbol{v}^{gr})$, where $\mathcal{A}$ is an arbitrary smooth, and generally speaking, complex function. In similarity to other important mathematical statements, this formula admits fairly many different proofs.[9] We give a derivation that is the most transparent from the physical point of view. It is based on the concept of a *ray tube*. Let us define this term.

Let $\Sigma_o$ be an area on some surface that is not tangent to the vector field $\boldsymbol{v}^{gr}$ at any point. The rays intersecting $\Sigma_o$ form a geometric object, which we will call a *ray tube* (e.g., Smirnov 1964a [58]); see Fig. 4.5.

Regarding rays as lines of flow of the energy fluid, one can expect that no energy passes through the lateral boundary of the ray tube. Thus, a ray tube is similar to a waveguide such that the energy flow does not change from one cross-section to another. This qualitative consideration will be confirmed by calculation. Let $\Sigma_o$ and $\Sigma$ be two cross-sections of a ray tube, which are transverse to the rays and non-intersecting each other; see Fig. 4.5. Integrating the expression $\operatorname{div}(\mathcal{A}\boldsymbol{v}^{gr})$ over a domain $\Omega$ cut by $\Sigma_o$ and $\Sigma$ out from the ray tube ($\mathcal{A}$ is an arbitrary smooth function), and applying the Ostrogradsky–Gauss divergence formula, we obtain

$$\int_\Omega \operatorname{div}(\mathcal{A}\boldsymbol{v}^{gr})d\mathbf{x} = \int_{\partial\Omega} \mathcal{A}v_n^{gr}\,dS, \qquad v_n^{gr} := (\boldsymbol{n}, \boldsymbol{v}^{gr}), \tag{4.70}$$

where $\boldsymbol{n}$ is a unit outward normal to $\Omega$, and $dS$ is a surface element of $\partial\Omega$. Let the normals to the cross-sections $\Sigma_o$ and $\Sigma$ be oriented in the direction of energy fluid flow (i.e., in the direction of its velocity vector $\boldsymbol{v}^{gr} = \frac{\partial \mathbf{x}}{\partial \tau}$); we denote these normals by $\boldsymbol{N}$. For the sake of definiteness, let $\boldsymbol{N}$ be an outward normal to $\Omega$ on $\Sigma$, and an inward one on $\Sigma_o$. Taking into account the fact that $(\boldsymbol{v}^{gr}, \boldsymbol{n}) = 0$ on the lateral surface of the tube, we write formula (4.70) as

$$\int_\Omega \operatorname{div}(\mathcal{A}\boldsymbol{v}^{gr})d\mathbf{x} = \int_\Sigma \mathcal{A}v_N^{gr}\,dS - \int_{\Sigma_o} \mathcal{A}v_N^{gr}\,dS. \tag{4.71}$$

We shall derive the required formula for $\operatorname{div}(\mathcal{A}\boldsymbol{v}^{gr})$ from (4.71) by taking the ray tube very narrow and the cross-sections very close to each other. Let the ray tube be formed by the rays of which the ray coordinates $\eta'_{1,2}$ satisfy

[9] For example, Gauss published four proofs of the fundamental theorem of algebra (see, e.g., Fine and Rosenberger 1991).

the inequalities

$$\eta_1 \leqslant \eta_1' \leqslant \eta_1 + d\eta_1, \quad \eta_2 \leqslant \eta_2' \leqslant \eta_2 + d\eta_2. \tag{4.72}$$

Let the cross-sections be close wavefronts defined by $\Sigma_o : \tau = \text{const}$ and $\Sigma : \tau = \text{const} + d\tau$, and let the domain $\Omega$ be

$$\eta_1 \leqslant \eta_1' \leqslant \eta_1 + d\eta_1, \quad \eta_2 \leqslant \eta_2' \leqslant \eta_2 + d\eta_2, \quad \tau \leqslant \tau' \leqslant \tau + d\tau. \tag{4.73}$$

The left-hand side of formula (4.72), up to higher-order infinitesimal terms, equals

$$\int_\Omega \operatorname{div}(\mathcal{A}\boldsymbol{v}^{gr})d\mathbf{x} \approx \operatorname{div}(\mathcal{A}\boldsymbol{v}^{gr})|_{(\eta_1,\eta_2,\tau)}\, |\Omega|, \tag{4.74}$$

where $|\Omega|$ is the volume of the domain $\Omega$. The domain is almost a parallelepiped built on the vectors $\frac{\partial \mathbf{x}}{\partial \eta_1}d\eta_1$, $\frac{\partial \mathbf{x}}{\partial \eta_2}d\eta_2$, and $\frac{\partial \mathbf{x}}{\partial \tau}d\tau = \boldsymbol{v}^{gr}d\tau$. The volume of a parallelepiped built on any vectors $\boldsymbol{a}$, $\boldsymbol{b}$, and $\boldsymbol{c}$ equals

$$|([\boldsymbol{a} \times \boldsymbol{b}], \boldsymbol{c})| = \left| \det \begin{pmatrix} a_1 & b_1 & c_1 \\ a_2 & c_2 & c_2 \\ a_3 & b_3 & c_3 \end{pmatrix} \right|.$$

Consequently,

$$\int_\Omega \operatorname{div}(\mathcal{A}\boldsymbol{v}^{gr})d\mathbf{x} \approx \operatorname{div}(\mathcal{A}\boldsymbol{v}^{gr})\mathscr{D}d\eta_1 d\eta_2 d\tau, \tag{4.75}$$

where

$$\mathscr{D} = \left| \frac{D(x,y,z)}{D(\eta_1,\eta_2,\tau)} \right| = \left| \det \begin{pmatrix} \frac{\partial x}{\partial \eta_1} & \frac{\partial x}{\partial \eta_2} & \frac{\partial x}{\partial \tau} \\ \frac{\partial y}{\partial \eta_1} & \frac{\partial y}{\partial \eta_2} & \frac{\partial x}{\partial \tau} \\ \frac{\partial z}{\partial \eta_1} & \frac{\partial z}{\partial \eta_2} & \frac{\partial z}{\partial \tau} \end{pmatrix} \right|. \tag{4.76}$$

Consider now the right-hand side of the formula (4.71). Since $v_N^{gr} = \left(\frac{\partial \mathbf{x}}{\partial \tau}, \boldsymbol{N}\right)$ on $\Sigma_o$ and $\Sigma$, up to higher-order terms, we have

$$\begin{aligned} &\int_\Sigma \mathcal{A}v_N^{gr}dS - \int_{\Sigma_o} \mathcal{A}v_N^{gr}dS \\ &\quad\approx \mathcal{A}|_{(\eta_1,\eta_2,\tau+d\tau)} \left(\frac{\partial \mathbf{x}}{\partial \tau}, \boldsymbol{N}\right) \left| \left[\frac{\partial \mathbf{x}}{\partial \eta_1} \times \frac{\partial \mathbf{x}}{\partial \eta_2}\right] \right| d\eta_1 d\eta_2 \Big|_{\tau+d\tau} \\ &\quad\quad - \mathcal{A}|_{(\eta_1,\eta_2,\tau)} \left(\frac{\partial \mathbf{x}}{\partial \tau}, \boldsymbol{N}\right) \left| \left[\frac{\partial \mathbf{x}}{\partial \eta_1} \times \frac{\partial \mathbf{x}}{\partial \eta_2}\right] \right| d\eta_1 d\eta_2 \Big|_{\tau}. \end{aligned}$$

We notice that $(\frac{\partial \mathbf{x}}{\partial \tau}, \boldsymbol{N}) \left| \left[\frac{\partial \mathbf{x}}{\partial \eta_1} \times \frac{\partial \mathbf{x}}{\partial \eta_2}\right] \right| d\eta_1 d\eta_2 d\tau$ is the volume of the parallelepiped built on the vectors $\frac{\partial \mathbf{x}}{\partial \eta_1}d\eta_1$, $\frac{\partial \mathbf{x}}{\partial \eta_2}d\eta_2$, and $\frac{\partial \mathbf{x}}{\partial \tau}d\tau$, and therefore $(\frac{\partial \mathbf{x}}{\partial \tau}, \boldsymbol{N}) \left| \left[\frac{\partial \mathbf{x}}{\partial \eta_1} \times \frac{\partial \mathbf{x}}{\partial \eta_2}\right] \right| = \mathscr{D}$. We have used the fact that $\left(\frac{\partial \mathbf{x}}{\partial \tau}, \boldsymbol{N}\right) \geqslant 0$, because the vectors $\frac{\partial \mathbf{x}}{\partial \tau} = \boldsymbol{v}^{gr}$ and $\boldsymbol{N}$ are directed to the same side.

Eventually, for the right-hand side of (4.71) we obtain the expression

$$\mathcal{A}\mathscr{D}|_{(\eta_1,\eta_2,\tau+d\tau)}\,d\eta_1 d\eta_2 - \mathcal{A}\mathscr{D}|_{(\eta_1,\eta_2,\tau)}\,d\eta_1 d\eta_2 \approx \frac{\partial(\mathcal{A}\mathscr{D})}{\partial\tau} d\eta_1 d\eta_2 d\tau. \tag{4.77}$$

Equating the right-hand sides of (4.75) and (4.77), we arrive at the required essential formula

$$\operatorname{div}(\mathcal{A}\boldsymbol{v}^{gr}) = \frac{1}{\mathscr{D}}\frac{\partial(\mathcal{A}\mathscr{D})}{\partial\tau}. \tag{4.78}$$

Recall that the smooth function $\mathcal{A}$ is arbitrary and, generally speaking, complex.

## Back to the Umov equation (4.67)

We put $\mathcal{A} = \overline{\overline{\mathcal{E}}}$ in (4.78), and by virtue of the Umov equation (4.67), we obtain

$$\operatorname{div}\left(\overline{\overline{\mathcal{E}}}\boldsymbol{v}^{gr}\right) = \frac{1}{\mathscr{D}}\frac{\partial(\overline{\overline{\mathcal{E}}}\mathscr{D})}{\partial\tau} = 0, \tag{4.79}$$

whence

$$\overline{\overline{\mathcal{E}}}\mathscr{D} = \Psi(\eta_1,\eta_2), \tag{4.80}$$

where $\Psi(\eta_1,\eta_2)$ is an arbitrary smooth nonnegative function that is constant on each ray.

Formula (4.80) has a clear hydrodynamic interpretation. The quantity $\frac{\omega^2}{2}\overline{\overline{\mathcal{E}}}\mathscr{D}d\tau\eta_1 d\eta_2$ is an amount of energy flowing during the time $d\tau$ through a cross-section $\tau = \text{const}$ of the ray tube (playing the role of a waveguide) and it remains constant along the tube (up to higher-order terms). The counterparts of relation (4.80) are widely known in hydrodynamics (see Landau and Lifshitz 1987 [43], Kochin, Kibel, and Roze 1964 [40]), where a fluid having the density $\rho$ and velocity $\boldsymbol{v}$ flows along the corresponding lines of flow (which are the hydrodynamic counterparts of rays).

## Formulas for $\boldsymbol{u}^{P0}$

An easy matter is to derive an expression for $|\boldsymbol{u}^{P0}|$ from the formula (4.80). The expression comprises an arbitrary function that is constant on each ray. Indeed, from (4.59) and (4.80) it follows that

$$\frac{\rho}{a^2}|\phi^0|^2\mathscr{D} = \Psi(\eta_1,\eta_2) \quad \Leftrightarrow \quad |\phi^0| = \frac{a\sqrt{\Psi(\eta_1,\eta_2)}}{\sqrt{\rho\mathscr{D}}}.$$

We introduce an object important for the ray method the *geometrical spreading* (of rays) $J$, by the formula

$$J = \frac{1}{a}\mathscr{D}. \tag{4.81}$$

We thus have $|\phi^0| = \frac{\sqrt{\Psi(\eta_1,\eta_2)}}{\sqrt{\rho a J}}$ and

$$|\boldsymbol{u}^{P0}| = \frac{\psi(\eta_1,\eta_2)}{\sqrt{\rho a J}}. \tag{4.82}$$

Here, $\psi(\eta_1,\eta_2) = \sqrt{\Psi(\eta_1,\eta_2)}$ is an arbitrary smooth nonnegative function. The geometrical spreading will be discussed in the next section.

Let $\mathbf{x}$ and $\mathbf{x}^{\circ}$ be points on the same ray. From formulas (4.81) and (4.82) it follows that

$$|\boldsymbol{u}^{P0}(\mathbf{x})| = |\boldsymbol{u}^{P0}(\mathbf{x}^{\circ})| \frac{\sqrt{\rho(\mathbf{x}^{\circ})\mathscr{D}(\mathbf{x}^{\circ})}}{\sqrt{\rho(\mathbf{x})\mathscr{D}(\mathbf{x})}} = |\boldsymbol{u}^{P0}(\mathbf{x}^{\circ})| \frac{\sqrt{\rho(\mathbf{x}^{\circ})a(\mathbf{x}^{\circ})J(\mathbf{x}^{\circ})}}{\sqrt{\rho(\mathbf{x})a(\mathbf{x})J(\mathbf{x})}}. \tag{4.83}$$

To fully describe $\phi^0$, we invoke the condition (4.63) not yet used. It is apparent that if $\phi^0$ is real, this condition holds automatically. In general, we put

$$\phi^0 = |\phi^0| e^{i\upsilon}, \quad \upsilon := \arg \phi^0. \tag{4.84}$$

After some algebra employing the explicit expression (4.19) of the operator $\mathfrak{M}^P$, the equality (4.63) takes the form

$$\begin{aligned} \operatorname{Im}(\mathfrak{M}^P \boldsymbol{u}^{P0}, \boldsymbol{u}^{P0}) &= 2(\lambda + 2\mu) |\boldsymbol{u}^{P0}|^2 (\operatorname{grad} \upsilon, \operatorname{grad} \tau) \\ &\equiv 2\rho |\boldsymbol{u}^{P0}|^2 \frac{\partial \upsilon}{\partial \tau} = 0, \end{aligned} \tag{4.85}$$

that is, $\upsilon = \arg \phi^0$ remains constant along the ray. From (4.82) and (4.85) we get the final expression for $\boldsymbol{u}^{P0}$:

$$\boldsymbol{u}^{P0} = \frac{\psi^0(\eta_1,\eta_2)}{\sqrt{\rho a J}} \mathbf{s} = \sqrt{\frac{a}{\rho J}} \psi^0(\eta_1,\eta_2) \operatorname{grad} \tau, \tag{4.86}$$

where $\psi^0(\eta_1,\eta_2)$ is an arbitrary smooth complex function. The function $\psi^0$ is called the *diffraction coefficient* of the zeroth-order approximation. In literature, the term *"initial data"* (for the transport equation) is also commonly used as the synonym for the term "diffraction coefficient."

### 4.3.3 Ray coordinates and the geometrical spreading

In the previous section (see (4.81)) we have encountered for the first time with the geometrical spreading of rays, the characteristic of a field of rays that is very important for the ray method. In brief, the geometrical spreading shows how the area of a cross-section of a narrow ray tube varies along a ray. Let us consider this more thoroughly.

The geometrical spreading is determined up to an arbitrary smooth factor constant along a ray. This factor does not affect the right-hand side of (4.83). Notice that if $\eta_1$ and $\eta_2$ are ray coordinates, then the vectors $\frac{\partial \mathbf{x}}{\partial \eta_1}$ and $\frac{\partial \mathbf{x}}{\partial \eta_2}$

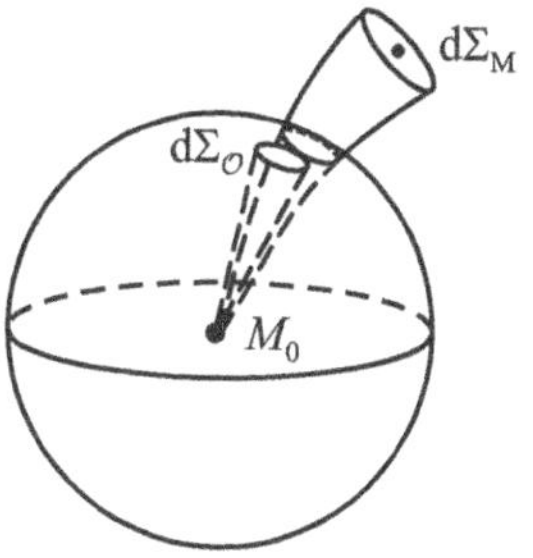


**Figure 4.6**: Defining the geometrical spreading for a central field of rays

are orthogonal to the rays. The formula (4.81) implies that $J = \left|\frac{\partial \mathbf{x}}{\partial \eta_1} \times \frac{\partial \mathbf{x}}{\partial \eta_2}\right|$. Thus, $J$ depends on the choice of the coordinates $\eta_1$ and $\eta_2$. Notice also that $J d\eta_1 d\eta_2 = \left|\frac{\partial \mathbf{x}}{\partial \eta_1} \times \frac{\partial \mathbf{x}}{\partial \eta_2}\right| d\eta_1 d\eta_2$ is the area of a cross-section of the ray tube which is orthogonal to the ray (4.72) — of course, up to infinitely small terms of higher order.

The coordinates $(\eta_1, \eta_2)$ for a given field of rays can be introduced in several different ways. Let $(\eta_1, \eta_2)$ and $(\widehat{\eta}_1, \widehat{\eta}_2)$ be two coordinate systems, and

$$J = \left|\frac{\partial \mathbf{x}}{\partial \eta_1} \times \frac{\partial \mathbf{x}}{\partial \eta_2}\right| \quad \text{and} \quad \widehat{J} = \left|\frac{\partial \mathbf{x}}{\partial \widehat{\eta}_1} \times \frac{\partial \mathbf{x}}{\partial \widehat{\eta}_2}\right|. \tag{4.87}$$

It can be easily shown that on each ray the ratio

$$\frac{J}{\widehat{J}} = \left|\det \begin{pmatrix} \frac{\partial \widehat{\eta}_1}{\partial \eta_1} & \frac{\partial \widehat{\eta}_2}{\partial \eta_1} \\ \frac{\partial \widehat{\eta}_1}{\partial \eta_2} & \frac{\partial \widehat{\eta}_2}{\partial \eta_2} \end{pmatrix}\right|$$

is constant. This agrees with determining the geometrical spreading up to a factor constant along a ray.

The case of a central field of rays has its own specifics which we are now going to discuss.

### Geometrical spreading of a central field of rays

Let the rays of a central field be launched from a point $M_0$. In accordance with Section 4.2.4, we parameterize the rays by unit vectors $\mathbf{s}$ tangent to them at the point $M_0$. The locus of the ends of these vectors is the unit sphere centered at $M_0$ and denoted by $\Sigma_{\mathcal{O}}$. Let $d\Sigma_{\mathcal{O}}$ be a small area on $\Sigma_{\mathcal{O}}$ surrounding the end $\mathcal{O}'$ of the unit vector tangent to the ray $MM_0$ at the point $M_0$. We denote the boundary of the area $d\Sigma_{\mathcal{O}}$ by $\mathcal{C}$. The rays, to which the unit tangents end on $\mathcal{C}$, form the ray tube of the central field of rays; see Fig. 4.6. Let us cross the tube by a wavefront $\tau = \text{const}$ containing the point $M$. The ray tube cuts out of the wavefront an area $d\Sigma_M$.

*The geometrical spreading corresponding to a ray of a central field* is the limit as $d\Sigma_{\mathcal{O}}$ is collapsing into a point $\mathcal{O}'$ on the sphere:

$$J = \lim_{d\Sigma_M \to \mathcal{O}'} \frac{|d\Sigma_M|}{|d\Sigma_{\mathcal{O}}|}. \tag{4.88}$$

Introduce the coordinates $\eta_1, \eta_2$ in the area $d\Sigma_{\mathcal{O}}$ on the unit sphere. It can be easily seen that

$$J = r^2 + O(r^3) \quad \text{as} \quad r \to 0, \quad r := |M_0M|. \tag{4.89}$$

Let us characterize a ray launched from $M_0$ by the coordinates $\eta_1$ and $\eta_2$, and the position of a point $M$ on the ray by the eikonal $\tau = \int_{M_0}^{M} \frac{ds}{c}$. Nearby a ray, the coordinate system $\eta_1, \eta_2, \tau$ is regular. One can easily derive the following formula for $J$:

$$J = \frac{|\mathbf{x}_{\eta_1} \times \mathbf{x}_{\eta_2}|_M}{|\mathbf{x}_{\eta_1} \times \mathbf{x}_{\eta_2}|_{M_0}}.$$

Obviously, this expression does not depend on the choice of the coordinates $\eta_1, \eta_2$.

In the case of a central field of rays we can also expect that the geometrical spreading is determined up to a factor constant on each ray. Consider a central field of rays in a homogeneous medium (with the velocity $c$), for which the eikonal is:

$$\tau = \frac{r}{c}. \tag{4.90}$$

The classical angles of the *spherical coordinate system* are often taken as the ray coordinates of the end of a unit vector on the sphere:

$$\eta_1 = \vartheta, \quad \eta_2 = \varphi, \tag{4.91}$$

where $\vartheta$ and $\varphi$ are defined by the relations

$$x = r\sin\vartheta\cos\varphi, \quad y = r\sin\vartheta\sin\varphi, \quad z = r\cos\varphi, \tag{4.92}$$

$0 \leqslant \vartheta \leqslant \pi$, $0 \leqslant \varphi < 2\pi$. Instead of (4.88), the geometrical spreading becomes

$$J = \frac{1}{c}\left|\frac{D(x,y,z)}{D(\tau,\eta_1,\eta_2)}\right| = \frac{1}{c}\left|\frac{D(x,y,z)}{D\left(\frac{r}{c},\vartheta,\varphi\right)}\right|$$
$$= r^2 \left|\det\begin{pmatrix} \sin\vartheta\cos\varphi & \sin\vartheta\sin\varphi & \cos\vartheta \\ -\cos\vartheta\cos\varphi & \cos\vartheta\sin\varphi & -\sin\vartheta \\ -\sin\vartheta\sin\varphi & \sin\vartheta\cos\varphi & 0 \end{pmatrix}\right|.$$

Direct calculation gives

$$J = r^2 \sin\vartheta.$$

Despite its traditionality, this choice has a deficiency that at the North and

South poles, $\vartheta = 0$ and $\vartheta = \pi$ (as well as on the meridian $\phi = 0$), the coordinates lose their regularity.

We can alternatively choose the coordinates as follows:

$$\eta_1 = \cos\vartheta, \quad \eta_2 = \varphi, \tag{4.93}$$

which gives $J = \frac{1}{c}\left|\frac{D(x,y,z)}{D(\frac{r}{c},\cos\vartheta,\varphi)}\right|$. In this case, the calculation of the determinant leads to the expression

$$J = r^2. \tag{4.94}$$

For the coordinates (4.93), the points $\eta_1 = \pm 1$ are singular. It is known (see, e.g., Novikov and Taimanov 2006 [50]) that no coordinates regular at all points exist on a sphere.

### 4.3.4 Anomalous polarization

We have ensured the consistency of equation (4.21). Let us proceed with solving it. We split the sought-for solution into two terms: parallel and perpendicular to $\operatorname{grad}\tau$:

$$\boldsymbol{u}^{P1} = \boldsymbol{u}^{P1\|} + \boldsymbol{u}^{P1\perp}, \quad \text{where} \quad \boldsymbol{u}^{P1\|} \parallel \operatorname{grad}\tau \quad \text{and} \quad \boldsymbol{u}^{P1\perp} \perp \operatorname{grad}\tau. \tag{4.95}$$

The term $\boldsymbol{u}^{P1\perp}$ is called the *additional* or *anomalously polarized* component of the wave $P$. It describes transverse polarization, which is absent for the homogeneous plane wave $P$.

The zero subspace of the matrix $\mathfrak{N}^P$ is spanned on the vector $\operatorname{grad}\tau$. On its orthogonal complement, equation (4.21) is uniquely solvable, and one can easily derive that

$$\boldsymbol{u}^{P1\perp} = \frac{a^2\mathfrak{M}^P\boldsymbol{u}^{P0}}{\rho(b^2 - a^2)}. \tag{4.96}$$

After some transformation of the right-hand side, we find

$$\boldsymbol{u}^{P1\perp} = -\boldsymbol{P}(\boldsymbol{B}\phi^0 + \operatorname{grad}\phi^0), \tag{4.97}$$

where $\phi^0 = a^2\boldsymbol{u}^{P0}\cdot\operatorname{grad}\tau = a\boldsymbol{u}^{P0}\cdot\mathrm{s}$, $\mathrm{s} = a\operatorname{grad}\tau$ (see (4.59)), $\boldsymbol{B}$ denotes a vector characterizing the inhomogeneity of the medium,

$$\boldsymbol{B} = \frac{1}{1-\gamma^2}\left\{4\gamma^2\frac{\operatorname{grad} b}{b} + (2\gamma^2 - 1)\frac{\operatorname{grad}\rho}{\rho}\right\}, \quad \gamma := \frac{b}{a}, \tag{4.98}$$

and $\boldsymbol{P}$ is an operator of projecting to a plane that is orthogonal to the ray of the wave $P$,

$$\boldsymbol{P}\boldsymbol{X} = \boldsymbol{X} - (\boldsymbol{X},\mathrm{s})\,\mathrm{s}.$$

For homogeneous media, the expression (4.98) vanishes. Obviously, $|\phi^0| = a|\boldsymbol{u}^{P0}|$. The term $\boldsymbol{P}(\operatorname{grad}\phi^0)$ in (4.97) characterizes the variation of the intensity of the wave in the direction perpendicular to the ray. This term equals zero virtually in all interesting cases (except for homogeneous plane waves and spherical waves with the pattern independent of angles).

### 4.3.5 ⋆ First longitudinally polarized correction

The consistency condition for equation (4.22)

$$(\mathfrak{M}^P \boldsymbol{u}^{P1} - \mathfrak{L}\boldsymbol{u}^{P0}) \cdot \operatorname{grad}\tau = 0,$$

with (4.95) taken into account, reads as

$$\left(\mathfrak{M}^P \boldsymbol{u}^{P1\|}, \operatorname{grad}\tau\right) = \left(\mathfrak{L}\boldsymbol{u}^{P0} - \mathfrak{M}^P \boldsymbol{u}^{P1\perp}, \operatorname{grad}\tau\right). \tag{4.99}$$

We put

$$\boldsymbol{u}^{P1\|} = \phi^1 \operatorname{grad}\tau. \tag{4.100}$$

Using (4.59) and (4.100), we represent the left-hand side of (4.99) in the form

$$\left(\mathfrak{M}^P \boldsymbol{u}^{P1\|}, \operatorname{grad}\tau\right) = \left(\mathfrak{M}^P \left(\phi \boldsymbol{u}^{P0}\right), \operatorname{grad}\tau\right), \quad \phi := \frac{\phi^1}{\phi^0}. \tag{4.101}$$

We further transform this expression exploiting the equation $(\mathfrak{M}^P \boldsymbol{u}^{P0}, \operatorname{grad}\tau) = 0$, which is a form of the consistency condition (4.21). It is useful to note that if $\phi = \text{const}$, then the expression (4.101) equals zero. If $\phi \neq \text{const}$, only the terms that include the derivatives of $\phi$ contribute to the expression (4.101); see formula (4.19) in this regard. After some algebra, we get

$$\left(\mathfrak{M}^P \left(\phi \boldsymbol{u}^{P0}\right), \operatorname{grad}\tau\right) = 2(\lambda + 2\mu)(\operatorname{grad}\tau)^2 \phi^0 \left(\operatorname{grad}\phi, \operatorname{grad}\tau\right) = 2\frac{\rho}{a^2}\phi^0 \frac{\partial \phi}{\partial \tau}. \tag{4.102}$$

Essentially, we used here the classical method of variation of constants (see, e.g., Smirnov 1964a [58]). The "constant" to be "varied" is $\phi = \frac{\phi^1}{\phi^0}$.

Substituting (4.102) into (4.99), we arrive at an ordinary first-order differential equation, which is immediately integrated. We finally obtain

$$\boldsymbol{u}^{P1\|} = \sqrt{\frac{a}{\rho J}} \left\{ \psi^1 - \int^{\tau} \frac{a^2}{2\rho} \sqrt{\frac{\rho J}{a}} \left[\mathfrak{M}^P \boldsymbol{u}^{P1\perp} - \mathfrak{L}\boldsymbol{u}^{P0}\right] \operatorname{grad}\tau d\tau \right\} \operatorname{grad}\tau, \tag{4.103}$$

where $\psi^1 = \psi^1(\eta_1, \eta_2)$ is an arbitrary smooth function called the *diffraction coefficient of the first approximation.*

### 4.3.6 ⋆ Higher-order approximations

Going on in a similar manner, we denote

$$\boldsymbol{u}^{Pm} = \boldsymbol{u}^{Pm\|} + \boldsymbol{u}^{Pm\perp}, \tag{4.104}$$

where

$$\boldsymbol{u}^{Pm\|} \parallel \operatorname{grad}\tau, \quad \boldsymbol{u}^{Pm\perp} \perp \operatorname{grad}\tau.$$

The transverse component of $\boldsymbol{u}^{Pm\perp}$ is uniquely expressed via $\boldsymbol{u}^{P0}, \ldots, \boldsymbol{u}^{P,m-1}$:

$$\boldsymbol{u}^{Pm\perp} = \frac{a^2}{\rho(a^2 - b^2)} \left\{ \mathfrak{M}^P \boldsymbol{u}^{P,m-1} - \mathfrak{L}\boldsymbol{u}^{P,m-2} \right\}. \tag{4.105}$$

Here $\boldsymbol{u}^{P,-1} = \boldsymbol{u}^{P,-2} = 0$. The expression for the longitudinal component,

$$\boldsymbol{u}^{Pm\|} = \sqrt{\frac{a}{\rho J}} \left\{ \psi^m - \int^{\tau} \frac{a^2}{2\rho} \sqrt{\frac{\rho J}{a}} \left[ \mathfrak{M}^P \boldsymbol{u}^{P,m,\perp} - \mathfrak{L}\, \boldsymbol{u}^{P,m-1} \right] \operatorname{grad} \tau d\tau \right\} \operatorname{grad} \tau, \tag{4.106}$$

includes an arbitrary smooth function of two variables $\psi^m = \psi^m(\eta_1, \eta_2)$, called the *diffraction coefficient* of the $m$-th order for the wave $P$.

---

## 4.4 Solving transport equations. The wave $S$

Consideration of the wave $S$ is more tricky as compared to the case of the wave $P$. This is caused by the two-dimensionality of the zero subspace of the operator $\mathfrak{N}$, which makes possible linear, elliptic, and circular polarization in the zeroth approximation. The quantities describing the wavefield will be marked in the current section by $^S$.

Let $\tau$ satisfy the eikonal equation (4.27) related to the velocity of the wave $S$; therefore, the polarization of $\boldsymbol{u}^{S0}$ is transverse; see (4.29). We assume that a respective field of rays is chosen and the ray coordinates $(\eta_1, \eta_2, \tau)$ are introduced.

### 4.4.1 Zeroth-order approximation. A preliminary consideration. The Rytov law

**An absolute value of $\boldsymbol{u}^{S0}$**

We start with a simple question on the absolute value of the amplitude of the zeroth approximation. Similarly to the case of the wave $P$, the consistency condition (4.20) leads to the equation

$$(\mathfrak{M}^S \boldsymbol{u}^{S0}, \boldsymbol{u}^{S0}) = 0. \tag{4.107}$$

As above, the equation (4.107) implies a formula similar to (4.82) for the absolute value of $\boldsymbol{u}^{S0}$,

$$|\boldsymbol{u}^{S0}| = \frac{\chi(\eta_1, \eta_2)}{\sqrt{\rho b J}}, \tag{4.108}$$

where $\chi(\eta_1, \eta_2)$ is an arbitrary non-negative smooth function of two variables, and $J$ is the geometrical spreading of rays of a wave $S$.

Formulas (4.29) and (4.108) do not describe $\boldsymbol{u}^{S0}$ completely.

**Reducing the consistency condition to a system of ordinary differential equations along a ray**

Let $\mathbf{n}$ and $\boldsymbol{\nu}$ be a unit principal normal and binormal to a ray at a point $(\tau, \eta_1, \eta_2)$. The vectors $\mathbf{n}$ and $\boldsymbol{\nu}$ smoothly depend on a point on the ray if the curvature of the ray does not vanish. If thc ray is a straight line, then $\mathbf{n}$ and $\boldsymbol{\nu}$ are any pair of vectors orthogonal to the ray and to each other. It is assumed that the unit vectors $\mathbf{s} = b\nabla\tau$, $\mathbf{n}$, and $\boldsymbol{\nu}$ form a right triplet, $\mathbf{s} \times \mathbf{n} = \boldsymbol{\nu}$. Since $(\mathbf{s}, \mathbf{n}) = (\mathbf{s}, \boldsymbol{\nu}) = 0$ in all cases, the transverse character of the zeroth-order approximation (4.29) means that $\boldsymbol{u}^{S0}$ is a linear combination of $\mathbf{n}$ and $\boldsymbol{\nu}$:

$$\boldsymbol{u}^{S0} = u_{\mathrm{n}}\mathbf{n} + u_{\nu}\boldsymbol{\nu}. \tag{4.109}$$

The consistency condition for equation (4.22) in the case of the wave $S$ can be written in the form

$$(\mathfrak{M}^S \boldsymbol{u}^{S0}, \mathbf{n}) = (\mathfrak{M}^S \boldsymbol{u}^{S0}, \boldsymbol{\nu}) = 0. \tag{4.110}$$

We shall employ the Frenet formulas (see, e.g., Smirnov 1964a [58])

$$\frac{d\mathbf{s}}{ds} = K\mathbf{n}, \quad \frac{d\mathbf{n}}{ds} = -K\mathbf{s} - T\boldsymbol{\nu}, \quad \frac{d\boldsymbol{\nu}}{ds} = T\mathbf{n}, \tag{4.111}$$

where $K$ and $T$ are the *curvature* and *torsion* of the ray, respectively. Also we use the fact that $\frac{d}{d\tau} = b\frac{d}{ds}$, where $s$ is the arc length. After some algebra, equations (4.110) become

$$\begin{aligned} &2\frac{du_{\mathrm{n}}}{d\tau} + \left[b^2\nabla^2\tau + \frac{1}{\rho}(\nabla\mu, \nabla\tau)\right] u_{\mathrm{n}} + 2Tbu_{\nu} = 0, \\ &2\frac{du_{\nu}}{d\tau} + \left[b^2\nabla^2\tau + \frac{1}{\rho}(\nabla\mu, \nabla\tau)\right] u_{\nu} - 2Tbu_{\mathrm{n}} = 0. \end{aligned} \tag{4.112}$$

These equations hold for real $\boldsymbol{u}^{S0}$, as well as for complex $u_{\mathrm{n}}$ and $u_{\nu}$.

We have arrived at a more complicated, vector problem than in the case of the wave $P$. The complication is due to the terms $2Tbu_{\nu}$ and $-2Tbu_{\mathrm{n}}$. First we consider an interesting (and worth attention) case where they vanish.

**Case of $T \equiv 0$**

If $T \equiv 0$, the ray is a plane curve (see, e.g., Smirnov 1964a [58]). The system (4.112) splits into two independent equations which implies that

$$\operatorname{div}\left\{\mu|u_{\mathrm{n}}|^2 \operatorname{grad}\tau\right\} = 0, \quad \operatorname{div}\left\{\mu|u_{\nu}|^2 \operatorname{grad}\tau\right\} = 0. \tag{4.113}$$

Verbally repeating the consideration given in Section 4.3.1, we obtain

$$u_{\mathrm{n}} = \frac{\chi_1(\eta_1, \eta_2)}{\sqrt{\rho b J}}, \quad u_\nu = \frac{\chi_2(\eta_1, \eta_2)}{\sqrt{\rho b J}}, \tag{4.114}$$

where $J$ is the geometrical spreading, and $\chi_1(\eta_1, \eta_2)$ and $\chi_2(\eta_1, \eta_2)$ are arbitrary smooth complex functions.

### Case of real $\boldsymbol{u}^{S0}$

The case of real $\boldsymbol{u}^{S0}$ is simpler than the general case and has some specific features; it is therefore reasonable to consider it separately. Since $u_{\mathrm{n}}$ and $u_\nu$ are assumed to be real, the following relations hold, with the real function $\theta = \theta(\tau)$ appropriately chosen:

$$\begin{gathered} u_{\mathrm{n}} = \Phi \cos\theta, \quad u_\nu = \Phi \sin\theta, \\ \Phi = \sqrt{u_{\mathrm{n}}^2 + u_\nu^2} = \left|\boldsymbol{u}^{S0}\right|. \end{gathered} \tag{4.115}$$

Substituting (4.115) into equations (4.112), we get

$$\begin{gathered} \left(2\frac{d\Phi}{d\tau} + \mathrm{H}\Phi\right)\cos\theta + 2\left(-\frac{d\theta}{d\tau} + Tb\right)\Phi\sin\theta = 0, \\ \left(2\frac{d\Phi}{d\tau} + \mathrm{H}\Phi\right)\sin\theta + 2\left(\frac{d\theta}{d\tau} - Tb\right)\Phi\cos\theta = 0, \end{gathered} \tag{4.116}$$

where $\mathrm{H} := b^2\nabla^2\tau + \frac{1}{\rho}(\nabla\mu, \nabla\tau)$. Equations (4.116) are equivalent to the relations

$$2\frac{d\Phi}{d\tau} + \left[b^2\nabla^2\tau + \frac{1}{\rho}(\nabla\mu, \nabla\tau)\right]\Phi = 0 \tag{4.117}$$

and

$$\frac{d\theta}{d\tau} - Tb = 0. \tag{4.118}$$

Multiply the relation (4.117) by $\Phi$. After some calculation, we obtain

$$\operatorname{div}\left\{\mu\Phi^2 \operatorname{grad}\tau\right\} = 0. \tag{4.119}$$

By virtue of $\Phi^2 = u_{\mathrm{n}}^2 + u_\nu^2$, the relation (4.119) agrees with formulas (4.113).

The important formula (4.118) is called the *Rytov law*. The equation (4.118), with (4.115) and (4.109) taken into account, means that the vector $\boldsymbol{u}^{S0}$ lying in the plane orthogonal to the ray rotates with the velocity proportional to the torsion of the ray as the wave propagates along the ray.

The vector of physical displacement in the zeroth approximation is

$$\mathrm{Re}\left[e^{i\omega(\tau-t)}\boldsymbol{u}^{S0}\right] = \boldsymbol{u}^{S0}\cos\omega(\tau - t). \tag{4.120}$$

In accordance with Section (2.3.5), the polarization is linear; at each point of the ray a particle oscillates along the vector $\boldsymbol{u}^{S0}$.

## 4.4.2 Rytov law. The case of complex $\boldsymbol{u}^{S0}$

Let us make in the system (4.112) the substitution

$$u_{\rm n} = \frac{1}{\sqrt{\rho b J}} v_{\rm n}, \quad u_\nu = \frac{1}{\sqrt{\rho b J}} v_\nu.$$

Taking into account that the functions (4.114) satisfy the system (4.112) as $T \equiv 0$, for $v_{\rm n}$ and $v_\nu$ we obtain simpler equations

$$\frac{dv_{\rm n}}{d\tau} = Tbv_\nu, \quad \frac{dv_\nu}{d\tau} = -Tbv_{\rm n}. \tag{4.121}$$

An easy matter is to find the fundamental system of solutions, that is, a pair of linearly independent solutions of this system of equations. Such solutions may be the vectors

$$\begin{pmatrix} v_{\rm n}^1 \\ v_\nu^1 \end{pmatrix} = \begin{pmatrix} \cos\theta(\tau) \\ \sin\theta(\tau) \end{pmatrix} \quad \text{and} \quad \begin{pmatrix} v_{\rm n}^2 \\ v_\nu^2 \end{pmatrix} = \begin{pmatrix} -\sin\theta(\tau) \\ \cos\theta(\tau) \end{pmatrix}, \tag{4.122}$$

where, in accordance with (4.118),

$$\theta(\tau) = \int_{\tau^o}^{\tau} Tb d\tau, \tag{4.123}$$

and $\tau^o$ is fixed. The check that the vectors (4.122) satisfy the system (4.121) is a routine exercise, and their linear independence is obvious.

Thus, any solution of the system (4.112) can be represented in the form

$$\begin{pmatrix} u_{\rm n} \\ u_\nu \end{pmatrix} = \frac{1}{\sqrt{\rho b J}} \boldsymbol{\Pi} \begin{pmatrix} \chi_1 \\ \chi_2 \end{pmatrix}, \tag{4.124}$$

where $\boldsymbol{\Pi}$ is a matrix the columns of which are the vectors (4.122):

$$\boldsymbol{\Pi} = \begin{pmatrix} \cos\theta & -\sin\theta \\ \sin\theta & \cos\theta \end{pmatrix}. \tag{4.125}$$

The result can be written as follows:

$$\boldsymbol{u}^{S0} = \frac{1}{\sqrt{\rho b J}} \boldsymbol{\Pi} \boldsymbol{\chi}^0, \tag{4.126}$$

where the *vector diffraction coefficient* has the form $\boldsymbol{\chi}^0(\eta_1, \eta_2) = \chi_1(\eta_1, \eta_2)\mathbf{n} + \chi_2(\eta_1, \eta_2)\boldsymbol{\nu}$. The functions $\chi_1(\eta_1, \eta_2)$ and $\chi_2(\eta_1, \eta_2)$ are, generally speaking, complex.

### Physical displacement

Consider first the wavefield at a point $\mathbf{x}^o$ with the ray coordinates $(\tau^o, \eta_1, \eta_2)$. As $\tau = \tau^o$, $\boldsymbol{\Pi}$ becomes the identity matrix, $\boldsymbol{\Pi}|_{\tau=\tau^o} = \begin{pmatrix} 1 & 0 \\ 0 & 1 \end{pmatrix}$, and

$$\mathrm{Re}[e^{i\omega(\tau - t)} \boldsymbol{u}^{S0}]\Big|_{\mathbf{x}^o} = \mathrm{Re}\left[ \frac{1}{\sqrt{\rho b J}}\Big|_{\mathbf{x}^o} (\chi_1 \mathbf{n} + \chi_2 \boldsymbol{\nu}) e^{i\omega(\tau^o - t)} \right]. \tag{4.127}$$

The vector of physical displacement at a point $\mathbf{x}^o + \widetilde{\mathbf{x}}$ close to $\mathbf{x}^o$, see Section 4.1.1, is approximately equal to

$$\operatorname{Re}\left[e^{i\omega(\tau(\mathbf{x}+\widetilde{\mathbf{x}})-t)}\boldsymbol{u}^{S0}\right] \approx \operatorname{Re}\left[\frac{e^{i\omega\tau}}{\sqrt{\rho b J}}\boldsymbol{\chi}^0\bigg|_{\mathbf{x}^o} e^{i\omega(\boldsymbol{k}\cdot\widetilde{\mathbf{x}}-t)}\right], \tag{4.128}$$

where $\frac{e^{i\omega\tau}}{\sqrt{\rho b J}}\boldsymbol{\chi}^0\Big|_{\mathbf{x}^o} e^{i\omega(\boldsymbol{k}\cdot\widetilde{\mathbf{x}}-t)}$ is a local plane wave, $\boldsymbol{k} = \omega \operatorname{grad}\tau|_{\mathbf{x}^o} = \frac{\omega}{b}\mathbf{s}\big|_{\mathbf{x}}$ is the related local wave vector. Putting

$$\boldsymbol{\chi}^0 = \chi_1\mathbf{n} + \chi_2\boldsymbol{\nu} = \boldsymbol{P} + i\boldsymbol{Q}, \quad \boldsymbol{P} = \operatorname{Re}\boldsymbol{\chi}^0, \quad \boldsymbol{P} = \operatorname{Im}\boldsymbol{\chi}^0, \tag{4.129}$$

one can repeat the construction of Section 2.3.5 for a local plane wave $S$ occurring in (4.128). Let us introduce the angle $\psi$ by

$$\boldsymbol{\chi}^0 = e^{i\psi}(\boldsymbol{P}' + i\boldsymbol{Q}'), \quad \operatorname{Im}\psi = 0, \quad \boldsymbol{P}' \perp \boldsymbol{Q}',$$

assuming that the relations (2.50)–(2.52) hold. As a result, the displacement (4.127) takes the form

$$\operatorname{Re}[e^{i\omega(\tau-t)}\boldsymbol{u}^{S0}] \approx \frac{1}{\sqrt{\rho b J}}\bigg|_{\mathbf{x}^o}\left[\boldsymbol{P}'\cos(\Theta-\omega t) - \boldsymbol{Q}'\sin(\Theta-\omega t)\right], \tag{4.130}$$

where $\Theta = \omega\tau(\mathbf{x}^o) + \boldsymbol{k}\cdot\widetilde{\mathbf{x}} + \psi$. On the ray itself, $\widetilde{\mathbf{x}} = 0$, and

$$\Theta = \omega\tau(\mathbf{x}^o) + \psi. \tag{4.131}$$

At the point $\mathbf{x}^o$, linear, elliptic, or circular polarization is possible. If $\tau \neq \tau^o$, then

$$\operatorname{Re}[e^{i\omega(\tau-t)}\boldsymbol{u}^{S0}] \approx \frac{1}{\sqrt{\rho b J}}\bigg|_{\mathbf{x}} \boldsymbol{\Pi}\left[\boldsymbol{P}'\cos(\Theta-\omega t) - \boldsymbol{Q}'\sin(\Theta-\omega t)\right], \tag{4.132}$$

where $\boldsymbol{\Pi}$ is the rotation matrix (4.125). Hence, the type of polarization (whether it is linear, or elliptic, or circular) is kept along the ray.

The rotation of the vectors $\boldsymbol{P}'$ and $\boldsymbol{Q}'$ as the result of applying the matrix $\boldsymbol{\Pi}$ is also called the *Rytov law*.

### 4.4.3 Anomalous polarization

Now the equation (4.21) is consistent. We seek its solution in the form

$$\boldsymbol{u}^{S1} = \boldsymbol{u}^{S1\parallel} + \boldsymbol{u}^{S1\perp}, \quad \boldsymbol{u}^{S1\parallel} \parallel \operatorname{grad}\tau, \quad \boldsymbol{u}^{S1\perp} \perp \operatorname{grad}\tau. \tag{4.133}$$

The term $\boldsymbol{u}^{S1\parallel}$ describing the longitudinal component is called *additional* or *anomalously polarized component* of the wave $S$.

The zero subspace of the matrix $\mathfrak{N}$ is orthogonal to $\operatorname{grad}\tau$. On the orthogonal complement to $\operatorname{grad}\tau$, the equation (4.21) is uniquely solvable, and with little difficulty we obtain

$$\boldsymbol{u}^{S1\parallel} = \frac{b^2\mathfrak{M}^S\boldsymbol{u}^{S0}}{\rho(a^2-b^2)} = \frac{\gamma^2\mathfrak{M}^S\boldsymbol{u}^{S0}}{\rho(1-\gamma^2)}, \quad \gamma = \frac{b}{a}. \tag{4.134}$$

Let us transform the right-hand side of (4.134). Using the fact that $\boldsymbol{u}^{S1\|} \parallel \operatorname{grad}\tau$ and $\boldsymbol{u}^{S10} \perp \operatorname{grad}\tau$, we have

$$\mathfrak{M}^S \boldsymbol{u}^{S0} = \{(\operatorname{grad}\mu, \boldsymbol{u}^{S0}) + (\lambda+\mu)\operatorname{div}\boldsymbol{u}^{S0} + 2\mu(\mathbf{s}, (\mathbf{s}, \operatorname{grad})\boldsymbol{u}^{S0})\}\mathbf{s}$$

with $\mathbf{s} = b\operatorname{grad}\tau$. From the Frenet formulas (4.111) it follows that the last term in curly brackets equals $-2\mu K(\boldsymbol{u}^{S0}, \mathbf{n})$, whence

$$\boldsymbol{u}^{S1\|} = b\left\{(\boldsymbol{A}, \boldsymbol{u}^{S0}) + \operatorname{div}\boldsymbol{u}^{S0}\right\}\mathbf{s}, \tag{4.135}$$

where

$$\boldsymbol{A} = \frac{\gamma^2}{1-\gamma^2}\left(2\frac{\operatorname{grad}b}{b} + \frac{\operatorname{grad}\rho}{\rho} - 2K\mathbf{n}\right), \quad \gamma = \frac{b}{a}. \tag{4.136}$$

The first term in curly brackets in (4.135) characterizes the inhomogeneity of the medium while the second characterizes the variation of the intensity of the wave in the direction perpendicular to that of its propagation. This term is nonzero in all interesting cases, except for the case of homogeneous plane waves. For homogeneous media, the vector (4.136) is zero, and (4.135) is reduced to the simple relation

$$\boldsymbol{u}^{S1\|} = b\,\mathbf{s}\operatorname{div}\boldsymbol{u}^{S0}. \tag{4.137}$$

### 4.4.4 ⋆ First transversely polarized correction

Consider the equation of the first approximation (4.22). The matrix $\mathfrak{N}$ is degenerate on the two-dimensional subspace orthogonal to $\operatorname{grad}\tau$. The consistency condition of the equation (4.22) can be written in the form

$$(\mathfrak{M}^S\boldsymbol{u}^{S1} - \mathfrak{L}\boldsymbol{u}^{S0})\cdot\mathbf{n} = (\mathfrak{M}^S\boldsymbol{u}^{S1} - \mathfrak{L}\boldsymbol{u}^{S0})\cdot\boldsymbol{\nu} = 0. \tag{4.138}$$

Let

$$\boldsymbol{u}^{S1\perp} = u_{\mathrm{n}}\mathbf{n} + u_\nu\boldsymbol{\nu} \tag{4.139}$$

in (4.133). Now the equations (4.138) read

$$\begin{aligned} &2\frac{du_{\mathrm{n}}}{d\tau} + \left[b^2\nabla^2\tau + \frac{1}{\rho}(\nabla\mu, \nabla\tau)\right]u_{\mathrm{n}} + 2Tbu_\nu = \mathrm{F}_{\mathrm{n}},\\ &2\frac{du_\nu}{d\tau} + \left[b^2\nabla^2\tau + \frac{1}{\rho}(\nabla\mu, \nabla\tau)\right]u_\nu - 2Tbu_{\mathrm{n}} = \mathrm{F}_\nu, \end{aligned} \tag{4.140}$$

$$\mathrm{F}_{\mathrm{n}} = \frac{1}{\rho}(\mathfrak{L}\boldsymbol{u}^{S0} - \mathfrak{M}^S\boldsymbol{u}^{S1})\cdot\mathbf{n}, \quad \mathrm{F}_\nu = \frac{1}{\rho}(\mathfrak{L}\boldsymbol{u}^{S0} - \mathfrak{M}^S\boldsymbol{u}^{S1})\cdot\boldsymbol{\nu}.$$

From Section 4.4.2 we know that the fundamental matrix of the corresponding homogeneous system (4.140) has the form $\frac{1}{\sqrt{\rho b J}}\mathbf{\Pi}$; see (4.124), (4.125). In accordance with the classical method of variation of constants, we seek a solution of the form

$$\begin{pmatrix} u_{\mathrm{n}} \\ u_{\nu} \end{pmatrix} = \frac{1}{\sqrt{\rho b J}} \mathbf{\Pi} \boldsymbol{\Upsilon}, \quad \boldsymbol{\Upsilon} = \boldsymbol{\Upsilon}(\tau) = \begin{pmatrix} \Upsilon_{\mathrm{n}} \\ \Upsilon_{\nu} \end{pmatrix},$$

and for the column $\boldsymbol{\Upsilon}$ obtain

$$\frac{2}{\sqrt{\rho b J}} \mathbf{\Pi} \frac{d\boldsymbol{\Upsilon}}{d\tau} = \mathbf{F}, \quad \mathbf{F} = \begin{pmatrix} \mathrm{F}_{\mathrm{n}} \\ \mathrm{F}_{\nu} \end{pmatrix},$$

wherefrom $\boldsymbol{\Upsilon}$ can be found by simple calculation. The result is as follows:

$$\boldsymbol{u}^{S1\|} = \frac{1}{\sqrt{\rho b J}} \mathbf{\Pi} \left\{ \boldsymbol{\chi}^{1} + \int^{\tau} \frac{\sqrt{\rho b J}}{2\rho} \mathbf{\Pi}^{-1} \left[ \mathfrak{L} \boldsymbol{u}^{S0} - \mathfrak{M}^{S} \boldsymbol{u}^{S1\|} \right] d\tau \right\}, \tag{4.141}$$

where the *diffraction coefficient* $\boldsymbol{\chi}^1 = \boldsymbol{\chi}^1(\eta_1, \eta_2)$ is an arbitrary smooth vector, which can be regarded as orthogonal to the ray.

### 4.4.5 ⋆ Higher-order approximations

Denoting

$$\boldsymbol{u}^{Sm} = \boldsymbol{u}^{Sm\|} + \boldsymbol{u}^{Sm\perp}, \quad m \geqslant 2, \tag{4.142}$$

where

$$\boldsymbol{u}^{Sm\|} \parallel \operatorname{grad} \tau, \quad \boldsymbol{u}^{Sm\perp} \perp \operatorname{grad} \tau,$$

one can easily find that the longitudinal component of $\boldsymbol{u}^{Sm}$ is uniquely expressed via $\boldsymbol{u}^{S0}$, ..., $\boldsymbol{u}^{S,m-1}$:

$$\boldsymbol{u}^{Sm\|} = \frac{\gamma^2}{\rho(1-\gamma^2)} \left\{ \mathfrak{M}^{S} \boldsymbol{u}^{S,m-1} - \mathfrak{L} \boldsymbol{u}^{S,m-2} \right\}. \tag{4.143}$$

The transversal component

$$\boldsymbol{u}^{Sm\perp} = \frac{1}{\sqrt{\rho b J}} \mathbf{\Pi} \left\{ \boldsymbol{\chi}^{m} + \int^{\tau} \frac{\sqrt{\rho b J}}{2\rho} \mathbf{\Pi}^{-1} \left[ \mathfrak{L} \boldsymbol{u}^{S,m-2} - \mathfrak{M}^{S} \boldsymbol{u}^{S,m-1,\|} \right] d\tau \right\} \tag{4.144}$$

involves the *diffraction coefficient*, that is, a smooth vector $\boldsymbol{\chi}^m = \boldsymbol{\chi}^m(\eta_1, \eta_2) \perp \operatorname{grad} \tau$.

---

## 4.5 Reflection of the wave defined by a ray expansion

For seismic application, the above theory of propagation in a smoothly inhomogeneous medium might be insufficient. Seismology and especially seismic exploration typically deal with sharp interfaces. The corresponding theory, with many examples, can be found in Červený 2001 [13] and Popov 2002 [53]). We restrict ourselves to a brief description of the scheme for a relatively simple problem of reflection from a free boundary. With this example, we will illustrate important features of the general theory.

### 4.5.1 Ansatz and the statement of the problem of determining reflected and converted waves

#### Ansatz

Let a wave $\boldsymbol{u}^{\mathrm{i}}$, e.g., a wave $P$, which is defined by the ray expansion

$$\mathbf{u}^{\mathrm{i}} = e^{i\omega(\tau^{\mathrm{i}}(\mathbf{x})-t)}\left\{\boldsymbol{u}^{\mathrm{i}0}(\mathbf{x}) + \frac{\boldsymbol{u}^{\mathrm{i}1}(\mathbf{x})}{-i\omega} + \dots\right\}, \quad \omega \to \infty, \tag{4.145}$$

fall on a smooth traction-free boundary $\mathscr{B}$ of an isotropic medium. We can naturally expect that the incident wave (4.145) gives rise to monotype $PP$ and converted $PS$ waves which we denote (not quite in correspondence with the nomenclature given in Chapter 2) by $\mathbf{u}^{\mathrm{m}}$ and $\mathbf{u}^{\mathrm{c}}$. We assume that they can also be described by the ray expansions

$$\mathbf{u}^{\mathrm{m}} = e^{i\omega(\tau^{\mathrm{m}}(\mathbf{x})-t)}\left\{\boldsymbol{u}^{\mathrm{m}0}(\mathbf{x}) + \frac{\boldsymbol{u}^{\mathrm{m}1}(\mathbf{x})}{-i\omega} + \dots\right\}, \tag{4.146}$$

$$\mathbf{u}^{\mathrm{c}} = e^{i\omega(\tau^{\mathrm{c}}(\mathbf{x})-t)}\left\{\boldsymbol{u}^{\mathrm{c}0}(\mathbf{x}) + \frac{\boldsymbol{u}^{\mathrm{c}1}(\mathbf{x})}{-i\omega} + \dots\right\}. \tag{4.147}$$

As $\omega \to \infty$, each of the expressions (4.145)–(4.147) asymptotically satisfies the equation (4.15). The total wavefield

$$\mathbf{u} = \mathbf{u}^{\mathrm{i}} + \mathbf{u}^{\mathrm{m}} + \mathbf{u}^{\mathrm{c}} \tag{4.148}$$

satisfies the traction-free boundary condition on $\mathscr{B}$,

$$\mathbf{t}^{n}\mathbf{u}|_{\mathscr{B}} = 0. \tag{4.149}$$

The waves $\mathbf{u}^{\mathrm{m}}$ and $\mathbf{u}^{\mathrm{c}}$ are uniquely determined by the incident wave from:
a) the boundary condition on $\mathscr{B}$;
b) the nondegeneracy condition;
c) the requirement that monotype and converted waves propagate away from the boundary.

#### Nondegeneracy condition

The nondegeneracy condition is that the incident wave (4.145) falls on the boundary $\mathscr{B}$, and its rays are nontangent to $\mathscr{B}$. This can be analytically expressed as follows. Let $\boldsymbol{n}$ be a normal to $\mathscr{B}$, directed toward the elastic medium. The nondegeneracy condition has the form

$$\left.\frac{\partial\tau^{\mathrm{i}}}{\partial n}\right|_{\mathscr{B}} < 0. \tag{4.150}$$

Taking into account the fact that the rays of the incident wave are along the vector $\nabla\tau^{\mathrm{i}}$, one can easily see that (4.150) means that these rays are nontangent to and directed to $\mathscr{B}$.

### Conditions on eikonals. Snell's law

In order that the total wavefield (4.148) satisfy the boundary condition (4.149) in a higher order of $1/\omega$, it is necessary that the eikonals coincide on $\mathscr{B}$:

$$\tau^{\mathrm{m}}\big|_{\mathscr{B}} = \tau^{\mathrm{c}}\big|_{\mathscr{B}} = \tau^{\mathrm{i}}\big|_{\mathscr{B}}. \tag{4.151}$$

The requirement that the reflected waves go away from the boundary is reduced to the conditions

$$\left.\frac{\partial\tau^{\mathrm{m}}}{\partial n}\right|_{\mathscr{B}} > 0, \quad \text{and} \quad \left.\frac{\partial\tau^{\mathrm{c}}}{\partial n}\right|_{\mathscr{B}} > 0. \tag{4.152}$$

Since the incidence wave is a wave $P$,

$$(\nabla\tau^{\mathrm{i,m}})^2 = \frac{1}{a^2}.$$

Together with the condition (4.150) and the first of the conditions (4.152), this gives

$$\left.\frac{\partial\tau^{\mathrm{m}}}{\partial n}\right|_{\mathscr{B}} = -\left.\frac{\partial\tau^{\mathrm{i}}}{\partial n}\right|_{\mathscr{B}} > 0. \tag{4.153}$$

Formula (4.153) and the equality of the tangent components $\nabla\tau^{\mathrm{i}}\big|_{\mathscr{B}}$ and $\nabla\tau^{\mathrm{m}}\big|_{\mathscr{B}}$, see (4.151), imply the classical law of reflection of monotype waves, that is, "the angle of incidence equals the angle of reflection."

For the eikonal of the converted wave, which is in our case the wave $S$, the tangent component of $\nabla\tau^{\mathrm{c}}\big|_{\mathscr{B}}$ again coincides with that of $\nabla\tau^{\mathrm{i}}\big|_{\mathscr{B}}$; however, the moduli of the vectors $\nabla\tau^{\mathrm{i}}\big|_{\mathscr{B}}$ and $\nabla\tau^{\mathrm{c}}\big|_{\mathscr{B}}$ are different:

$$\left|\nabla\tau^{\mathrm{i}}\right| = \frac{1}{a} < \left|\nabla\tau^{\mathrm{c}}\right| = \frac{1}{b}. \tag{4.154}$$

Simple consideration leads to the Snell law $\frac{\sin\alpha^{\mathrm{i}}}{\sin\beta^{\mathrm{c}}} = \frac{a}{b}$ for incidence and conversion angles, which appeared in the theory of plane waves.

### Constructing the eikonals of reflected and converted waves

Knowing $\nabla\tau^{\mathrm{i}}|_{\mathscr{B}}$, one can construct, as described in Section 4.2.3, the field of rays of the reflected monotype wave and find $\tau^{\mathrm{m}}$. The same is also true for the converted wave $\tau^{\mathrm{c}}$.

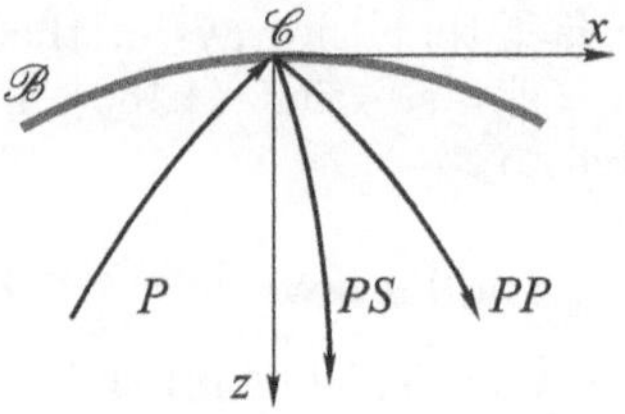


**Figure 4.7**: Geometry of reflection from a curvilinear boundary for the case of the incident wave $P$

### 4.5.2 Constructing the wavefield in higher orders

#### Solving transport equations and determining diffraction coefficients

After finding $\tau^{\rm m}$ and $\tau^{\rm c}$, we can proceed to constructing $\boldsymbol{u}^{\rm m0}$ and $\boldsymbol{u}^{\rm c0}$. Substitute the sum (4.148) into the boundary conditions (4.149) and consider terms of order $O(\omega)$. The boundary can be treated "in small" as a plane, while incident and reflected waves are, in essence, local plane waves. This allows one to employ the results of Section 2.6 for finding $\boldsymbol{u}^{\rm m0}\big|_{\mathscr{B}}$ and $\boldsymbol{u}^{\rm c0}\big|_{\mathscr{B}}$, which determine the initial data for the transport equations, and thus the vectors $\boldsymbol{u}^{\rm m0}$ and $\boldsymbol{u}^{\rm c0}$.

Having found $\boldsymbol{u}^{\rm m0}$ and $\boldsymbol{u}^{\rm c0}$, we can proceed to finding $\boldsymbol{u}^{\rm m1}$ and $\boldsymbol{u}^{\rm c1}$. Consideration of the boundary conditions in order $O(1)$ gives the initial data for the transport equations, which enables one to determine the amplitudes of the first-order approximation outside the boundary. Then one can proceed to finding $\boldsymbol{u}^{\rm m2}$ and $\boldsymbol{u}^{\rm c2}$, etc.

Discuss the construction of $\boldsymbol{u}^{\rm m0}$ and $\boldsymbol{u}^{\rm c0}$ in more detail. To uniquely determine these vector functions at each point, the corresponding diffraction coefficients must be known; in our example of incidence of the wave $P$, these are $\psi^{\rm m0}$ and $\chi^{\rm c0}$. Let the ray of an incident wave intersect a free boundary $\mathscr{B}$ at a point $\mathscr{C}$. We introduce a local Cartesian coordinate system $(x, y, z)$ as follows. Let the unit vector $\mathbf{e}_z$ of the axis $z$ coincide with the unit vector of the internal normal to $\mathscr{B}$ at the point $\mathscr{C} \in \mathscr{B}$. We denote by $\nabla_{\mathscr{B}}\tau^{\rm i}(\mathscr{C})$ the projection of the vector $\nabla\tau^{\rm i}(\mathscr{C})$ to the plane tangent to $\mathscr{B}$ at the point $\mathscr{C} \in \mathscr{B}$. If $\nabla_{\mathscr{B}}\tau^{\rm i}(\mathscr{C}) \neq 0$, then we take $\mathbf{e}_x = \frac{\nabla_{\mathscr{B}}\tau^{\rm i}(\mathscr{C})}{|\nabla_{\mathscr{B}}\tau^{\rm i}(\mathscr{C})|}$ as the unit vector of the axis $x$. If $\nabla_{\mathscr{B}}\tau^{\rm i}(\mathscr{C}) = 0$, then we take any unit vector orthogonal to $\mathbf{e}_z$ as that of the axis $\mathbf{e}_x$. We define the unit vector $\mathbf{e}_y$ as $\mathbf{e}_y = \mathbf{e}_z \times \mathbf{e}_x$ so that $\mathbf{e}_x$, $\mathbf{e}_y$ and $\mathbf{e}_z$ form a right-hand triple. In local coordinates near $\mathscr{C}$, see Fig. 4.7, the wavefield coincides in the zeroth-order approximation with that considered in Chapter 2 in the problem of incidence of a plane wave on the free boundary of a homogeneous half-space. In the vicinity of the point $\mathscr{C}$, consideration can be performed in the plane $(x, z)$.[10]

[10]Let us elucidate this point. The values of eikonals of incident and reflected waves on $\mathscr{B}$ are equal (otherwise the boundary conditions are not satisfied). Therefore the tangent values

For the case of incidence of the wave $P$ at the point $\mathscr{C}$ of the boundary, the vector amplitude equals

$$\boldsymbol{u}^{i0} = \sqrt{\frac{a}{\rho J^{\mathrm{i}}}} \psi^{i0}(\eta_1^{\mathrm{i}}, \eta_2^{\mathrm{i}}) \operatorname{grad} \tau^{\mathrm{i}}, \tag{4.155}$$

where $J^{\mathrm{i}}$ is the corresponding geometrical spreading. Then for a monotype reflected and for converted waves we have, respectively,

$$\boldsymbol{u}^{\mathrm{m}0} = \sqrt{\frac{a}{\rho J^{\mathrm{m}}}} \psi^{\mathrm{m}0}(\eta_1^{\mathrm{m}}, \eta_2^{\mathrm{m}}) \operatorname{grad} \tau^{\mathrm{m}}, \quad \boldsymbol{u}^{\mathrm{c}0} = \frac{\boldsymbol{\chi}^{\mathrm{c}0}(\eta_1^{\mathrm{c}}, \eta_2^{\mathrm{c}})}{\sqrt{\rho b J^{\mathrm{c}}}}. \tag{4.156}$$

Notation will be explained below.

The incident wave uniquely determines the geometrical spreadings $J^{\mathrm{m}}$ and $J^{\mathrm{c}}$ of monotype reflected and converted waves. For the two-dimensional case, such a procedure is described in detail in Section 4.8. As to the respective diffraction coefficients, in a small neighborhood of the point $\mathscr{C}$ the process of reflection – conversion coincides, in the zeroth-order approximation, with that for a plane boundary of a homogeneous half-space, and at $\mathscr{C}$ we obtain

$$\begin{gathered} \frac{\psi^{\mathrm{m}0}(\eta_1^{\mathrm{m}}, \eta_2^{\mathrm{m}})}{\sqrt{J^{\mathrm{m}}}} = A^a \frac{\psi^{\mathrm{i}0}(\eta_1^{\mathrm{i}}, \eta_2^{\mathrm{i}})}{\sqrt{J^{\mathrm{i}}}}, \\ \frac{\boldsymbol{\chi}^{\mathrm{c}0}(\eta_1^{\mathrm{c}}, \eta_2^{\mathrm{c}})}{\sqrt{\rho b J^{\mathrm{c}}}} = B^a \sqrt{\frac{a}{\rho J^{\mathrm{i}}}} \psi^{\mathrm{i}0}(\eta_1^{\mathrm{i}}, \eta_2^{\mathrm{i}}) [\operatorname{grad} \tau^{\mathrm{c}} \times \mathbf{e}_y]. \end{gathered} \tag{4.157}$$

Here, $A^a$ and $B^a$ are reflection and conversion coefficients (see Chapter 2). Finally,

$$\begin{gathered} \psi^{\mathrm{m}0}(\eta_1^{\mathrm{m}}, \eta_2^{\mathrm{m}}) = A^a \sqrt{\frac{J^{\mathrm{m}}}{J^{\mathrm{i}}}} \Bigg|_{\mathscr{B}} \psi^{\mathrm{i}0}(\eta_1^{\mathrm{i}}, \eta_2^{\mathrm{i}}), \\ \boldsymbol{\chi}^{\mathrm{c}0}(\eta_1^{\mathrm{c}}, \eta_2^{\mathrm{c}}) = B^a \sqrt{\frac{a J^{\mathrm{c}}}{b J^{\mathrm{i}}}} \Bigg|_{\mathscr{B}} [\operatorname{grad} \tau^{\mathrm{c}} \times \mathbf{e}_y] \psi^{\mathrm{i}0}(\eta_1^{\mathrm{i}}, \eta_2^{\mathrm{i}}). \end{gathered} \tag{4.158}$$

Reflection, conversion, and transmission of volume waves on nonplane boundaries in inhomogeneous media are discussed in much detail from the seismic point of view (see, e.g., in Červený 2001 [13], Popov 2002 [53]).

## On generalizations

Similarly, the case where the incident wave (4.145) is the wave $S$ can be examined. Here one can face the phenomenon of total internal reflection, where the eikonal equation for $\tau^{\mathrm{c}}$ has no real solutions. The problem of reflection – transmission at the interface between two media is handled similarly, but it is more cumbersome.

---

of $\nabla\tau^{\mathrm{i}}$ and $\nabla\tau^{\mathrm{m}}$ (generalizing the horizontal slownesses which have appeared in Chapter 2) coincide on $\mathscr{B}$. In consequence of the nontangency of the rays of the incident wave to $\mathscr{B}$, $(\nabla\tau^{\mathrm{i}} - \nabla\tau^{\mathrm{m}})|_{\mathscr{B}} \neq 0$, whence $(\nabla\tau^{\mathrm{i}} - \nabla\tau^{\mathrm{m}})|_{\mathscr{B}} = \gamma\mathbf{n}$ with some $\gamma \neq 0$, that is, $\nabla\tau^{\mathrm{m}}$ lies in one plane with $\nabla\tau^{\mathrm{i}}$ and $\mathbf{n}$. Similarly, one can establish that $\nabla\tau^{\mathrm{c}}$ lies in the same plane.

The condition (4.150) is of ultimate importance. In the vicinity of the points of the boundary where $\frac{\partial \tau^{\mathrm{i}}}{\partial n}\big|_{\mathscr{B}} = 0$, smooth $\tau^{\mathrm{m}}$ and $\tau^{\mathrm{c}}$ cannot, generally speaking, be found. The condition $\frac{\partial \tau^{\mathrm{i}}}{\partial n}\big|_{\mathscr{B}} = 0$ means that the ray of the incident wave is tangent to the boundary. Shade and penumbra zones as well as other peculiarities of the wavefield, may exist. We shall not analyze sophisticated collisions that occur in this case (some comprehension of which for the two-dimensional scalar case can be gained from the book of Babič and Kirpičnikova 1979 [8]).

---

## 4.6 ⋆ Riemannian geometry in ray theory

### 4.6.1 Riemannian geometry and Fermat principle

The systematic presentation of the following basic material can be found in a textbook [16] by Dubrovin, Fomenko, and Novikov 1984, also see the Appendix.

#### Riemannian structure

Let $\Omega \subset \mathbb{R}^m$ be a domain in the $m$-dimensional space. The points of the domain are denoted by $\mathbf{x} = (x_1, \ldots, x_m)$. Let *a Riemannian metric* on $\Omega$ be defined:

$$g_{ik}(\mathbf{x}) dx_i dx_k, \quad g_{ik} = g_{ki}, \quad 1 \leqslant i, k \leqslant m. \tag{4.159}$$

Here, the $g_{ik}(\mathbf{x})$ are smooth functions of $\mathbf{x}$, and the matrix $\|g_{ik}\|$ is positive definite at each point $\mathbf{x}$. This matrix is called a *metric tensor*, or just a *metric*. Summation over repeated indices is assumed.

*A Riemannian length* of a parametrically defined curve segment between points $M_0, M_1 \in \Omega$, $M_0 = \mathbf{x}(\sigma_0)$, $M_1 = \mathbf{x}(\sigma_1)$, $\ell = x_1(\sigma), \ldots, x_m(\sigma)$, $\sigma_0 \leqslant \sigma \leqslant \sigma_1$, is defined as

$$\int_{\sigma_0}^{\sigma_1} \sqrt{g_{ik}(\mathbf{x}) dx_{\mathrm{i}} dx_k} = \int_{\sigma_0}^{\sigma_1} \sqrt{g_{ik}(\mathbf{x}) \frac{dx_{\mathrm{i}}}{d\sigma} \frac{dx_k}{d\sigma}} d\sigma. \tag{4.160}$$

The integral does not depend on the parameterization because of the homogeneity of the expression $\sqrt{g_{ik} x'_{\mathrm{i}} x'_k}$, $x'_j = \frac{dx_j}{d\sigma}$, with respect to $x'_j$ (see any variational calculus textbook, for example, Smirnov 1964b [59], Gelfand and Fomin 1963 [27]). The extremals of the integrals (4.160) are called *geodesic lines* or *geodesics*. They are defined by the condition

$$\delta \int_{M_0}^{M_1} \sqrt{g_{ik}(\mathbf{x}) dx_i dx_k} = 0. \tag{4.161}$$

If $\ell$ is such a curve that the integral (4.160) has a minimal value (i.e., the Riemannian length of any other curve connecting $M_0$ and $M_1$ is greater), then $\ell$ is a geodesic for sure.

### Semigeodesic coordinates

Consider a problem of finding a curve $\ell$ of minimal Riemannian length, connecting a fixed point $M_0$ with some smooth surface $\Sigma$. This curve is a geodesic, and at each point $M \in \Sigma$ the *transversality condition*

$$g_{ik}(M)x_i'\delta x_k = 0, \quad x_i' := \frac{dx_i}{d\sigma} \tag{4.162}$$

holds. Here, $\delta x_k$ are the coordinates of the vectors from the tangent plane to $\Sigma$ at the point $M$ where $\ell$ meets $\Sigma$. The transversality condition is also called orthogonality in the corresponding Riemannian metrics (4.159).

We define now the so-called semigeodesic coordinate system that is of a great concern in Riemannian geometry. Let the points of the surface $\Sigma$ be characterized by their coordinates $\eta_1, \dots, \eta_{m-1}$. Let us launch a geodesic orthogonal to $\Sigma$ at $M_o \in \Sigma$, and let $\eta_1, \dots, \eta_{m-1}$ be the coordinates of $M_o$. The points $M$ on $\ell$ are described by the Riemannian length $\tau = \int_{M_o}^{M} \sqrt{g_{ij}dx_i dx_j}$ of the part $M_oM$ of the geodesic, where the integral is taken along $\ell$. The coordinates $(\tau, \eta_1, \dots, \eta_{m-1})$ are called *semigeodesic coordinates.* An example of such coordinates is the coordinate system $(q^1, q^2, n)$ associated with the surface $\Sigma \subset \mathbb{R}^3$ used in Section 4.2.5.

### Conformally Euclidean metric

A metric is called *conformally Euclidean* if there is a smooth function $c(\mathbf{x}) > 0$ such that for any $\mathbf{x} \in \Omega$

$$g_{ik}(\mathbf{x})dx_i dx_k = \frac{1}{c^2(\mathbf{x})} \sum_{j=1}^{m} (dx_j)^2, \tag{4.163}$$

i.e.,

$$g_{ik}(\mathbf{x}) = \frac{\delta_{ik}}{c(\mathbf{x})}, \tag{4.164}$$

where $\delta_{ik}$ is the Kronecker delta. The geodesics of (4.163) are called rays, in accordance with the terminology already introduced above for the case of $m = 3$.

At the same time, together with the usual arc length $\mathbf{x} = \mathbf{x}(\sigma) = \int \sqrt{\sum_{i=1}^{3} (dx_i)^2}$, we define the Riemannian length $\int \frac{\sqrt{\sum_{i=1}^{3}(dx_i)^2}}{c(\mathbf{x})}$ that introduces the Riemannian structure with conformally Euclidean metric. The *Riemannian (conformally Euclidean) length* of the curve $\ell$, connecting points $M_o$ and $M$, i.e.,

$$\int_\ell \frac{ds}{c} = \int_{M_o}^{M} \frac{ds}{c},$$

has the physical meaning of the time that is needed to run along $\ell$ from $M_o$ to $M_1$ with speed $c(\mathbf{x})$.

It is not difficult to see that the theory of rays satisfying Fermat's principle (4.32) is mathematically equivalent to the Riemannian geometry with conformally Euclidean metric for $m = 3$. Then geodesics correspond to rays of the ray method, the Riemannian transversality reduces to the usual Euclidean orthogonality, and the semigeodesic coordinates correspond to ray coordinates.

Curiously, B. Riemann did not mention geometric optics of inhomogeneous media as a field of application of his new geometry in his celebrated lecture on the foundations of geometry (see Riemann 1948 [55]). Actually, the theory of wave propagation in inhomogeneous media had not been developed at Riemann's time.

### Equations of geodesics

We shortly outline derivation of equations describing geodesics. Of course, these equations are equivalent to the Euler equations for the Fermat functional (4.160).[11] Let

$$P := g_{ik}x'_i x'_k, \quad x'_j := \frac{dx_j}{d\sigma}.$$

The Euler equations for the functional (4.160) that are equivalent to (4.161) take the form

$$\frac{d}{d\sigma}\left(\frac{1}{\sqrt{P}}\frac{\partial P}{\partial x'_i}\right) - \frac{1}{\sqrt{P}}\frac{\partial P}{\partial x_i} = 0. \tag{4.165}$$

The parameter on the curve $\ell$ can be chosen arbitrarily. We put $\sigma = \tau$ in order that the relation

$$P = g_{ik}x'_i x'_k = g_{ik}\frac{dx_i}{d\tau}\frac{dx_k}{d\tau} = 1 \tag{4.166}$$

take place. This means that the parameter is the Riemannian length of $\ell$ counted from a certain point. Equations (4.165) take the form

$$\frac{d}{d\tau}\frac{\partial P}{\partial x'_i} - \frac{\partial P}{\partial x_i} = 0. \tag{4.167}$$

We note that (4.167) are the Euler equations for the functional

$$\int P d\tau = \int g_{ik}\frac{dx_i}{d\tau}\frac{dx_k}{d\tau}d\tau. \tag{4.168}$$

In more detail, (4.167) is

$$2\frac{\partial g_{ik}}{\partial x_j}\frac{dx_j}{d\tau}\frac{dx_k}{d\tau} + 2g_{ik}\frac{d^2x_k}{d\tau^2} - \frac{\partial g_{jk}}{\partial x_i}\frac{dx_j}{d\tau}\frac{dx_k}{d\tau} = 0. \tag{4.169}$$

Some manipulations (see, for example, Smirnov 1964b [59]) transform (4.169) to the classical form:

$$\frac{d^2x_i}{d\tau^2} + \Gamma^i_{js}\frac{dx_j}{d\tau}\frac{dx_s}{d\tau} = 0, \tag{4.170}$$

[11] These equations will be needed in Section 4.6.2 for the interpretation of the Rytov law for the polarization of waves $S$ in inhomogeneous media.

where

$$\Gamma^i_{js} := \sum_{p,j=1}^{m} \frac{1}{2} g^{ip} \left( \frac{\partial g_{pj}}{\partial x_s} + \frac{\partial g_{ps}}{\partial x_j} - \frac{\partial g_{js}}{\partial x_p} \right) \tag{4.171}$$

are the *Christoffel symbols*, and the $g^{ij}$ are the entries of the matrix $\|g^{ij}\|$ that is inverse to $\|g_{ij}\|$ (the inverse matrix exists owing to the positive definiteness of the metric tensor).

### 4.6.2 Parallel translation in Riemannian metric and the Rytov law

Now we describe the (well-known) connection between the Rytov law and the parallel translation in the corresponding Riemannian metric. In a domain $\Omega \subset \mathbb{R}^3$ we will consider a conformally-Euclidean metric (4.164) together with the standard Euclidean metric. Here, $c$ will have the meaning of the velocity $b$ of the wave $S$.

#### Parallel translation

Let us address now the vector structure. A real vector $\boldsymbol{\zeta}$ at a point $\mathbf{x} \in \Omega$ is an ordered segment starting at $\mathbf{x}$. Let its projections to the axes $x_1, x_2, x_3$ be $\zeta_1, \zeta_2, \zeta_3$, and $\boldsymbol{\eta} = (\eta_1, \eta_2, \eta_3)$ be a vector at $\mathbf{x}$ as well. A *Riemannian scalar product* of $\boldsymbol{\zeta}$ and $\boldsymbol{\eta}$ is

$$(\boldsymbol{\zeta}, \boldsymbol{\eta})_{\mathbf{g}(\mathbf{x})} := g_{ij}(\mathbf{x}) \zeta_i \eta_j. \tag{4.172}$$

It is obvious that this definition is consistent with the proper axioms (see Gelfand 1961 [26], for example). Let $x_i = x_i(\tau)$, $i = 1, 2, 3$, $-\infty < \tau_0 \leqslant \tau \leqslant \tau_1 < \infty$ be a curve $\ell$ parameterized by the Riemannian arc length: $d\tau = \frac{1}{c}\sqrt{\sum_{j=1}^{3} (dx_i)^2}$.

An important concept of *the parallel translation of vectors in the Riemannian metric* can be introduced as follows. Let a vector $\boldsymbol{\zeta}$ be set at a point $\mathbf{x}^o = \mathbf{x}(\tau^o)$ of the curve $\ell$. The parallel translation of a vector along a curve is the solution of a system of linear equations with the initial value $\zeta_j|_{\tau=\tau^o}$:

$$\frac{d\zeta_i}{d\tau} + \Gamma^i_{js} \zeta_j v_s = 0, \quad v_s := \frac{dx_s}{d\tau}, \tag{4.173}$$

where $\Gamma^i_{js}$ are the Christoffel symbols (4.171). Such a *parallel translated* vector changes its Euclidean length and direction during the translation.

#### Some calculations

Let us demonstrate that the Riemannian parallel translation of a vector orthogonal to a geodesic along this geodesic changes its direction in accordance with the Rytov law. We will see that in the case where the metric is assumed to be conformally-Euclidean, that is, $(\boldsymbol{\zeta}, \boldsymbol{\eta})_{\mathbf{g}(\mathbf{x})} = \frac{1}{c^2}(\boldsymbol{\zeta}, \boldsymbol{\eta})$, Riemannian orthogonality and Euclidean orthogonality coincide.

It is obvious that $g^{ik} = c^2\delta_{ik}$; consequently,

$$\Gamma^i_{js} = \frac{c^2}{2}\left(\frac{\partial g_{is}}{\partial x^j} + \frac{\partial g_{ij}}{\partial x^s} - \frac{\partial g_{js}}{\partial x^i}\right).$$

After some manipulations, equation (4.173) takes the form

$$\frac{d\boldsymbol{\zeta}}{d\tau} + \frac{c^2}{2}\left[\left(\boldsymbol{\zeta}, \nabla\frac{1}{c^2}\right)\boldsymbol{\upsilon} + \left(\boldsymbol{\upsilon}, \nabla\frac{1}{c^2}\right)\boldsymbol{\zeta} - (\boldsymbol{\zeta}, \boldsymbol{\upsilon})\nabla\frac{1}{c^2}\right] = 0, \tag{4.174}$$

where

$$\boldsymbol{\upsilon} = \frac{d\mathbf{x}}{d\tau}.$$

Further, we note that the equations describing a ray are the Euler equations (4.161) for the Fermat functional (4.170):

$$\frac{d\boldsymbol{\upsilon}}{d\tau} + \frac{c^2}{2}\left[2\left(\boldsymbol{\upsilon}, \nabla\frac{1}{c^2}\right)\boldsymbol{\upsilon} - \boldsymbol{\upsilon}^2\nabla\frac{1}{c^2}\right] = 0. \tag{4.175}$$

### Parallel translation of vectors either tangent or normal to a ray

We show that the Riemannian length of the vector $\boldsymbol{\upsilon} = \frac{d\mathbf{x}}{d\tau}$ tangent to a ray and having, from the Euclidean viewpoint, the physical meaning of the velocity of wave propagation, is constant and equals one. Indeed, by the definition (4.172),

$$(\boldsymbol{\upsilon}, \boldsymbol{\upsilon})_{\mathbf{g}(\mathbf{x})} = g_{ij}(\mathbf{x})\upsilon_i\upsilon_j = \frac{\delta_{ij}}{c^2(\mathbf{x})}\upsilon_i\upsilon_j = \frac{1}{c^2(\mathbf{x})}\sum_{j=1}^{3}(\upsilon^j)^2 = \frac{\delta_{ij}}{c^2(\mathbf{x})}\left(\frac{d\mathbf{x}}{d\tau}\right)^2 = 1.$$

Consider now the parallel translation of a vector $\boldsymbol{\zeta}$ that is normal to the ray at a certain point, i.e.,

$$(\boldsymbol{\upsilon}, \boldsymbol{\zeta})_{\mathbf{g}(\mathbf{x})} = 0. \tag{4.176}$$

As is known from the Riemannian geometry (see, e.g., Dubrovin, Fomenko, and Novikov 1984 [16]), the Riemannian scalar product (4.172) of a couple of vectors keeps its value during parallel translation along a curve. Obviously,

$$(\boldsymbol{\upsilon}, \boldsymbol{\zeta})_{\mathbf{g}(\mathbf{x})} = \frac{\delta_{ij}}{c^2(\mathbf{x})}\upsilon_i\zeta_j = \frac{1}{c^2(\mathbf{x})}(\boldsymbol{\upsilon}, \boldsymbol{\zeta}) = \text{const}.$$

Consequently, parallel translation preserves orthogonality in both Riemannian and Euclidean senses. It is clear that in the course of parallel translation of a vector, its Euclidean length is changing as const $c(\mathbf{x})$.

Consider how the Euclidean direction of a vector $\boldsymbol{\zeta}$ normal to a ray changes under its parallel translation along a ray. From (4.176) it follows that the system (4.174) reduces to

$$\frac{d\boldsymbol{\zeta}}{d\tau} + \frac{c^2}{2}\left[\left(\boldsymbol{\zeta}, \nabla\frac{1}{c^2}\right)\boldsymbol{\upsilon} + \left(\boldsymbol{\upsilon}, \nabla\frac{1}{c^2}\right)\boldsymbol{\zeta}\right] = 0. \tag{4.177}$$

We set

$$\boldsymbol{\zeta} = \zeta\boldsymbol{\zeta}^0, \quad \zeta = |\boldsymbol{\zeta}| = \sqrt{\sum_{m=1}^{3} \zeta_m^2}, \quad \boldsymbol{v} = c\mathbf{s}.$$

After some bulky calculation using (4.175), for the unit vector $\boldsymbol{\zeta}^0$ we get

$$\frac{d\boldsymbol{\zeta}^0}{d\tau} = -\mathbf{s}\left(\boldsymbol{\zeta}^0, \frac{d\mathbf{s}}{d\tau}\right) \Leftrightarrow \frac{d\boldsymbol{\zeta}^0}{ds} = -\mathbf{s}\left(\boldsymbol{\zeta}^0, \frac{d\mathbf{s}}{ds}\right). \tag{4.178}$$

Here, $ds$ is the Euclidean differential of the arc length of the ray $\ell$. Let us express $\boldsymbol{\zeta}^0$ through the principal normal $\mathbf{n}$ and the binormal $\boldsymbol{\nu}$ to the ray:

$$\boldsymbol{\zeta}^0 = \mathbf{n}\cos\varphi + \boldsymbol{\nu}\sin\varphi.$$

Using the Frenet formulas (4.111) in the notation of Section 4.4.2, from (4.178), we get

$$(-K\mathbf{s} - T\boldsymbol{\nu})\cos\varphi + T\mathbf{n}\sin\varphi + (-\mathbf{n}\sin\varphi + \boldsymbol{\nu}\cos\varphi)\frac{d\varphi}{ds} = -K\mathbf{s}\cos\varphi,$$

or

$$\boldsymbol{\nu}\cos\varphi\left(\frac{d\varphi}{ds} - T\right) - \mathbf{n}\sin\varphi\left(\frac{d\varphi}{ds} - T\right) = 0,$$

i.e.,

$$\frac{d\varphi}{ds} = T. \tag{4.179}$$

To summarize, we have established that the direction of the parallel translation of a vector normal to a ray satisfies the *Rytov law*, which we had already met in Section 4.4.2 when considering the polarization of the wave $S$. As we have seen, the length of the normal vector to the ray is proportional to the velocity of wave propagation.

---

## 4.7 Geometrical spreading in a homogeneous medium

The case of a central field of rays in a homogeneous medium was considered in Section 4.3.3. For a noncentral field of rays, the derivation of a formula for geometrical spreading requires addressing geometric terms such as lines of curvature and Gaussian curvature of a surface. The necessary geometric background is presented, e.g., in Smirnov 1964a [58].

Thus, let $\Sigma_{t_0}$ be a fixed position of the wavefront, and let the direction of its propagation be given. The eikonal $\tau$ is constant on $\Sigma_{t_0}$. To construct the position of the wavefront at a time instant $t > t_0$ (for not too distant future), we erect normals to the wavefront at each point $M_0 \in \Sigma_{t_0}$. Let us draw intervals $MM_0$ of length $c(t - t_0)$ in the direction of propagation. The set of points $M$ forms a surface $\Sigma_t$, which is the position of the wavefront at a time instant $t$. We see that the wavefronts are surfaces parallel to $\Sigma_{t_0}$. To find the geometrical spreading $J$, it is convenient to introduce coordinates on the surface $\Sigma_{t_0}$ in such a way that their coordinate lines are the lines of curvature.

### 4.7.1 Lines of curvature and Rodrigues' formula

Let us present necessary results from the differential geometry of surfaces (see any relevant textbook, for example, Smirnov 1964a [58]). Let $\mathring{g}_{\alpha\beta}$ and $\mathfrak{b}_{\alpha\beta}$, $\alpha, \beta = 1, 2$, be the coefficients of the first and second fundamental forms of the surface $\Sigma_{t_0}$ (see Section A.4). The curvature of a normal section is defined by the formula (A.21). Let $\frac{1}{\mathfrak{R}^0_{1,2}}$ be curvatures of normal sections of the surface $\Sigma_{t_0}$ at a fixed point $M_0 = M_0(q_0^1, q_0^2) \in \Sigma_{t_0}$. We will characterize directions of normal sections by the differentials $dq^1$ and $dq^2$. There are two possibilities. First, the curvatures of normal sections may coincide at $M_0$ (like at any point of a sphere); we will not consider such a case, known as the *umbilic case*. Second, they may be distinct, which we will assume in the sequel. Thus, there are the direction in which the curvature of a normal section is maximal and the direction in which it is minimal. The corresponding values of curvature, denoted by $\frac{1}{\mathfrak{R}^0_1}$ and $\frac{1}{\mathfrak{R}^0_2}$, are called the *principal curvatures*; the values $\mathfrak{R}^0_1$ and $\mathfrak{R}^0_2$ are called the *principal radii of curvature*. The corresponding directions which are called the *principal directions* are orthogonal to each other.

A *line of curvature* is a curve on a surface, that is always tangent to a principal direction. Let the coordinate lines of the orthogonal coordinate system $(q^1, q^2)$ be the curvature lines of the surface $\Sigma_{t_0}$. Essential is the fact that normals to the surface erected at the points of a fixed line of curvature have an envelop curve (see, e.g., Smirnov 1964a [58]). This envelop curve may be degenerate (for example, in the case where the surface is a sphere). Important is that the distance between a point $M_0$ on a curvature line and the envelope is equal to $|\mathfrak{R}^0_{1,2}|$, where $\frac{1}{\mathfrak{R}^0_{1,2}}$ stands for the corresponding principal curvatures.[12]

Let $\mathscr{L}_\alpha \in \Sigma_{t_0}$ be a line of curvature along a coordinate line $q^\alpha$ (where either $\alpha = 1$ or $\alpha = 2$) and $M_0$ be a point on $\mathscr{L}_\alpha$. Denote the corresponding envelope curve of normals to $\Sigma_{t_0}$ by $\mathscr{N}_\alpha$, a point of tangency of the normal erected at $M_0$ to $\mathscr{N}_\alpha$ by $Q_\alpha$, and the origin of Cartesian coordinates by $O$ (see Fig. 4.8 in which $\alpha = 1$). As obvious,

$$\overrightarrow{OQ_\alpha} + \overrightarrow{Q_\alpha M_0} = \overrightarrow{OM_0} \quad \Leftrightarrow \quad \mathbf{r}_\alpha + a^\alpha \boldsymbol{n} = \mathbf{r}_0, \tag{4.180}$$

where

$$\mathbf{r}_0 := \overrightarrow{OM_0}, \quad \mathbf{r}_\alpha := \overrightarrow{OQ_\alpha},$$

$\boldsymbol{n}$ is the unit normal to $\Sigma_0$, and $a^\alpha$ is a scalar such that $|a^\alpha| = |\overrightarrow{Q_\alpha M_0}|$. The

[12] In the case of an umbilical point, any two orthogonal directions can be reputed to be the principal curvature directions.

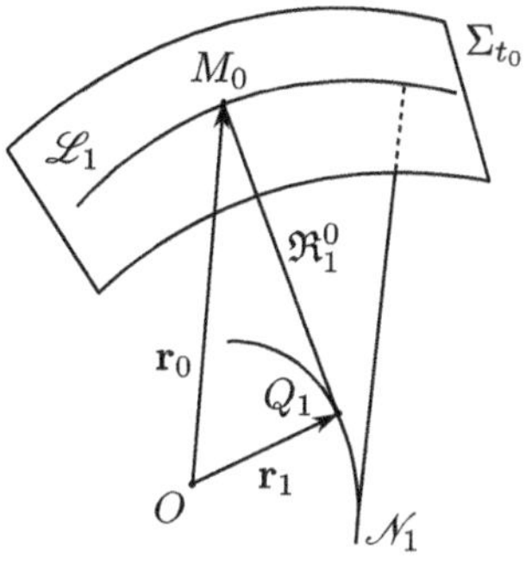


**Figure 4.8**: Geometric objects related to the line of curvature $q^1$

sign of $a^\alpha$ is defined by the relation $a^j \boldsymbol{n} = \overrightarrow{Q_\alpha M_0}$. For definiteness, we assume that $a^\alpha = \mathfrak{R}_\alpha > 0$, $\alpha = 1, 2$, which agrees with Fig. 4.8, where $\mathfrak{R}_1 > 0$.

Differentiating the relation (4.180) with respect to $q^\alpha$ gives

$$\frac{\partial \mathbf{r}_\alpha}{\partial q^\alpha} + \frac{\partial \mathfrak{R}_\alpha}{\partial q^\alpha}\boldsymbol{n} + \mathfrak{R}_\alpha \frac{\partial \boldsymbol{n}}{\partial q^\alpha} = \frac{\partial \mathbf{r}_0}{\partial q^\alpha}. \tag{4.181}$$

We do not assume summation over $\alpha$ here and henceforth in this section. Observe that

$$\frac{\partial \mathbf{r}_\alpha}{\partial q^\alpha} + \frac{\partial \mathfrak{R}_\alpha}{\partial q^\alpha}\boldsymbol{n} = 0. \tag{4.182}$$

Indeed, multiply the formula (4.181) by $\boldsymbol{n}$, and take notice of the obvious relations $(\boldsymbol{n}, \partial \boldsymbol{n}/\partial q^\alpha) = \frac{1}{2}\partial \boldsymbol{n}^2/\partial q^\alpha = 0$ and $(\boldsymbol{n}, \partial \mathbf{r}_0/\partial q^\alpha) = 0$ which imply $(\boldsymbol{n}, \partial \mathbf{r}/\partial q^\alpha + \boldsymbol{n}\partial \mathfrak{R}_\alpha/\partial q^\alpha) = 0$, whence the formula (4.182) follows. We thus arrive at the relation

$$\frac{\partial \mathbf{r}_0}{\partial q^\alpha} = \mathfrak{R}_\alpha \frac{\partial \boldsymbol{n}}{\partial q^\alpha}, \quad \alpha = 1, 2, \tag{4.183}$$

known as the *Rodrigues' formula* (see, e.g., the texbook Smirnov 1964a [58], which we closely follow here). In the relation (4.183), $\mathfrak{R}_\alpha^0 = \mathfrak{R}_\alpha(M_0)$ is taken at the point $M_0$.

### 4.7.2 Derivation of a formula for $J$

We assume that the wavefront moves in the same direction as the vectors $\overrightarrow{Q_\alpha M_0}$. The locus of points of intersection of normals to the wavefront $\Sigma|_{t_0}$ at its line of curvature with a wavefront $\Sigma|_t$ is a line of curvature on $\Sigma|_t$. At the points of $\Sigma|_t$, we have

$$\frac{\partial \mathbf{r}}{\partial q^\alpha} = (\mathfrak{R}_\alpha + \gamma)\frac{\partial \boldsymbol{n}}{\partial q^\alpha}, \quad \alpha = 1, 2, \tag{4.184}$$

where

$$\gamma := c(t - t_0), \tag{4.185}$$

$\gamma$ is the distance between the wavefronts $\Sigma_{t_0}$ and $\Sigma_t$ along the normal, and $\mathbf{r}$ is the radius vector of a point on $\Sigma_t$. The relations (4.183) and (4.185), together with the results of Section 4.3.3, yield

$$\frac{J(M_0)}{J(M_{t_0})} = \frac{\left|\left[\frac{\partial \mathbf{x}_0}{\partial q^1} \times \frac{\partial \mathbf{x}_0}{\partial q^2}\right]\right|}{\left|\left[\frac{\partial \mathbf{x}_{t_0}}{\partial q^1} \times \frac{\partial \mathbf{x}_{t_0}}{\partial q^2}\right]\right|} = \frac{(\mathfrak{R}_1^0 + \gamma)(\mathfrak{R}_2^0 + \gamma)}{\mathfrak{R}_1^0 \mathfrak{R}_2^0} = \frac{\mathfrak{R}_1 \mathfrak{R}_2}{\mathfrak{R}_1^0 \mathfrak{R}_2^0}, \tag{4.186}$$

where $\gamma$ is defined by (4.185), and $\frac{1}{\mathfrak{R}_1^0}$, $\frac{1}{\mathfrak{R}_2^0}$ are the principal curvatures at the intersection point of the ray and the wavefront $\Sigma_{t_0}$. The geometrical spreading at a time instant $t$ can be taken as

$$J = (\mathfrak{R}_1^0 + \gamma)(\mathfrak{R}_2^0 + \gamma) = \mathfrak{R}_1 \mathfrak{R}_2. \tag{4.187}$$

We can rewrite (4.187) as

$$J = \frac{1}{\mathfrak{H}}, \tag{4.188}$$

because the Gaussian curvature $\mathfrak{H}$ is equal to the product of the principal curvatures (see, e.g., a textbook by Smirnov 1964a [58]). The result (4.188) remains true in the presence of umbilical points as well.

### 4.7.3 On the vanishing of the geometrical spreading and caustics

Geometrical spreading is of utmost importance in the ray theory, because it occurs in all the expressions for the amplitudes $\boldsymbol{u}^m$ in the ansatz (4.1), as well as in all its analogs. In this chapter, we assume that $J \neq 0$, and similar assumptions will be accepted as concerning volume waves in anisotropic media and surface waves. Where the geometrical spreading vanishes, it is said that a *caustic* occurs, possibly a degenerate one. An example of the degenerate caustic is the center of a central field of rays. Wavefields in neighborhoods of caustics are described by formulas more complicated than those of the ray method.

---

## 4.8 ⋆ Geometrical spreading under reflection, transmission, and conversion in the planar case

### 4.8.1 Specific features of the planar case

Traditionally, by *planar* (or two-dimensional) problems of elastodynamics understood are the problems in which the displacement vector $\mathbf{u}$ is the same at any plane parallel to a selected one, and the component of $\mathbf{u}$ orthogonal to it

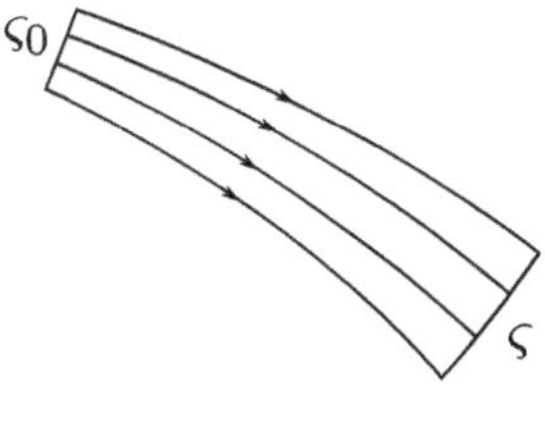


**Figure 4.9**: Ray strip

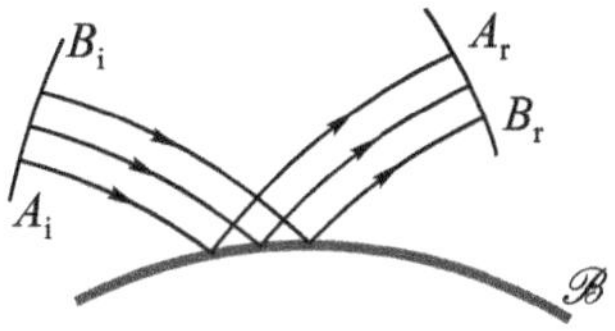


**Figure 4.10**: Rays of incident and reflected waves

identically vanishes[13] (see, e.g., Vekua and Kordzadze 1989 [66]). Assume that in the Cartesian coordinates $(x, y, z)$ the aforementioned plane is given by $z = 0$, and thus

$$\mathbf{u} = (u_1(x, y, t), u_2(x, y, t), 0). \tag{4.189}$$

Such wavefields exist, e.g., whenever the medium is isotropic, $\lambda = \lambda(x, y)$, $\mu = \mu(x, y)$, $\rho = \rho(x, y)$, and the volume forces are independent of $z$ and have no components along the $z$-axis. Motivation for considering solutions of the form (4.189) arises in the description of wavefields in plates (see, e.g., Muskhelishvili 1966 [49]).

The constructions of Sections 4.1–4.3.3 are applicable to the planar case mutatis mutandis. A point within an elastic medium will be described as $\mathbf{x} = (x, y, 0) := (x, y)$. In the ansatz (4.1), we take $\tau = \tau(x, y)$ and $\boldsymbol{u}^m = \boldsymbol{u}^m(x, y) = (u_1^m(x, y), u_2^m(x, y), 0)$, $m = 0, 1, \ldots$ Now a field of rays is a one-parameter family of curves $\mathbf{x} = \mathbf{x}(\eta, \tau)$, where $\tau$ is the eikonal, and the wave tube (see Fig. 4.5) is replaced by a *ray strip* pictured in Fig. 4.9. Here, the role of surfaces $\Sigma_0$ and $\Sigma$ is played by arcs $\varsigma_0$ and $\varsigma$ of smooth curves. Formulas (4.87) for the geometrical spreading are converted into

$$J = \left| \frac{\partial \mathbf{x}(\eta, \tau)}{\partial \eta} \right| \quad \text{and} \quad \widehat{J} = \left| \frac{\partial \mathbf{x}}{\partial \widehat{\eta}} \right|. \tag{4.190}$$

A given planar ray field corresponding to some eikonal $\tau$ falls on a surface $\mathscr{B}$ (see Fig. 4.10), and our task is to describe geometrical spreadings of reflected and transmitted rays. The geometrical spreading along a fixed ray is determined by the field of rays in its infinitesimal vicinity. Rays close to the

[13] In the sequel, the opposite can be assumed, that is, the only nonzero component is orthogonal to the selected plane.

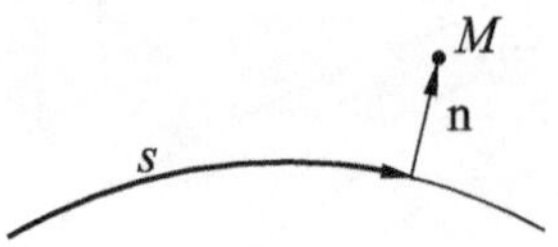


**Figure 4.11**: Coordinates $s$ and n for a ray of an incident or reflected wave

fixed ray are found in the linear approximation from the *Jacobi equation*, to which we proceed. Therefore, the task is reformulated as follows: knowing a solution of the Jacobi equation for a ray of an incident wave, find the initial data for the Jacobi equation of the corresponding reflected ray. We will solve this problem with the help of variational consideration.

## 4.8.2 Jacobi equation and geometrical spreading

### Derivation of the Jacobi equation within the framework of the calculus of variations

We start with the standard derivation of the Jacobi equation (see, e.g., Smirnov 1964b [59]). In the vicinity of a fixed ray $\ell$ of an incident (or reflected, or transmitted etc.) wave, we introduce the classical coordinate system "normal – arc length," i.e., points on the ray are characterized by the length $s$ of its arc measured from a fixed point on the ray, whereas the points outside $\ell$ are characterized by the distance n from $\ell$ and by the coordinate $s$ of the closest point on the ray; see Fig. 4.11. On the one side of $\ell$ we assume that n is positive, while on the other side it is negative. Near the ray, this system is regular and orthogonal.

Consider rays close to $\ell$; see Fig. 4.10. The variation of the Fermat functional $\int \frac{d\sigma}{c(s,\mathrm{n})}$, where the integral is taken along such a ray, vanishes. Here, $d\sigma = \sqrt{(1+K\mathrm{n})^2 + \left(\frac{d\mathrm{n}}{ds}\right)^2}\, ds$ is the differential of the arc length, and $K$ is the ray curvature. Assume that the ray is defined by the equation $\mathrm{n} = \mathrm{n}(s)$. Rays therefore satisfy the Euler equation for the functional

$$I = \int \frac{1}{c(s,\mathrm{n})} \sqrt{(1+K\mathrm{n})^2 + \left(\frac{d\mathrm{n}}{ds}\right)^2}\, ds. \tag{4.191}$$

For a given curve $\ell$ to be a ray, it is necessary and sufficient that the relation

$$K(s) - \frac{1}{c}\frac{\partial c}{\partial \mathrm{n}}\bigg|_{\mathrm{n}=0} = 0 \tag{4.192}$$

hold.[14] We note that the left-hand side of (4.192) is called the *effective cur-*

[14]Otherwise, the expansion of the integrand in (4.191) in powers of n and $d\mathrm{n}/ds$ would contain terms linear with respect to n. Then the curve $\ell$ would not be an extremal (see, e.g., Babič and Kirpičnikova 1979 [8], Babich and Buldyrev 2007 [6]).

*vature of a ray* (see Babič and Kirpičnikova 1979 [8], Babich and Buldyrev 2007 [6]).

Expand $c$ in powers of n. Let

$$c(s,\mathrm{n}) = c_0(s) + c_1(s)\mathrm{n} + c_2(s)\mathrm{n}^2 + \dots,$$

where

$$c_0(s) = c(s,\mathrm{n})|_{\mathrm{n}=0}, \quad c_1(s) = \left.\frac{\partial c(s,\mathrm{n})}{\partial \mathrm{n}}\right|_{\mathrm{n}=0}, \quad c_2(s) = \frac{1}{2}\left.\frac{\partial^2 c(s,\mathrm{n})}{\partial \mathrm{n}^2}\right|_{\mathrm{n}=0}. \tag{4.193}$$

Assuming that n and $d\mathrm{n}/ds$ are small, one can easily represent the Fermat functional (4.191), up to higher-order terms, in the form

$$\begin{aligned} I &= \int \frac{d\sigma}{c(s,\mathrm{n}(s))} \\ &\approx \int \left\{ \frac{1}{c_0(s)} - \frac{c_2(s)\mathrm{n}^2}{c_0^2(s)} + \frac{1}{2c_0(s)}\left(\frac{\partial \mathrm{n}}{\partial s}\right)^2 \right\} ds =: \int \Phi(\mathrm{n}, \frac{\partial \mathrm{n}}{\partial s}, s) ds. \end{aligned} \tag{4.194}$$

The *Jacobi equation* is the Euler equation for the functional (4.194):

$$\frac{d}{ds}\left(\frac{1}{c_0(s)}\frac{d\mathrm{n}}{ds}\right) + \frac{2c_2(s)}{c_0^2(s)}\mathrm{n} = 0. \tag{4.195}$$

Since the Jacobi equation is of the second order, its solution can be uniquely determined by two initial conditions at some point. These can be the values of n and

$$\mathrm{n}' := d\mathrm{n}/ds.$$

Where n and n′ are both small, a solution of the Jacobi equation approximately describes rays close to the ray $\ell$ at which $\mathrm{n}(s) \equiv 0$.

The Jacobi equation appeared above in describing the rays close to a fixed one. In the sequel, we will employ it for a different purpose, that is, for calculation of the geometrical spreading of a field of rays.

Now we come to considering a field of rays.

### Relationship between the Jacobi equation and the geometrical spreading

Analyzing the meaning of the geometrical spreading along a fixed ray $\ell$, we see that it is proportional to the distance between $\ell$ and an infinitesimally close ray of the same field of rays. We will show that the absolute value of a solution of the Jacobi equation has the same meaning. Thus, the corresponding solution of the Jacobi equation can be regarded as the geometrical spreading. Let us formally derive this assertion.

Let $\ell$ be a ray of a field of rays. Its rays close to $\ell$ are described in the related ray coordinates as $\mathrm{n} = \mathrm{n}(s,\eta)$, whereas $\ell$ corresponds to $\eta = \eta_0$. We find it convenient to equivalently parameterize the points of a ray by the arc length $s$ instead of the eikonal $\tau = \int^s ds/c$ and describe them as $\mathbf{x} = \mathbf{x}(s, \eta)$. Let $\ell$ itself be described by $\mathbf{x} = \mathbf{x}_0(\eta_0, s)$. Then a point on a ray close to $\ell$ is given by

$$\mathbf{x}(s,\eta) = \mathbf{x}_0(s) + \mathbf{n}\mathrm{n}(s,\eta), \tag{4.196}$$

where $\mathbf{x}_0(s)$ is a point on $\ell$, $\mathbf{n}$ is a unit normal vector to $\ell$, $|\mathbf{n}| = 1$, and $|n| \geqslant 0$ is the distance between $\mathbf{x}(s,\eta)$ and $\ell$ (see Fig. 4.11).

For the geometrical spreading along $\ell$, we have

$$\begin{aligned} J &= \left| \frac{\partial \mathbf{x}}{\partial \eta}\bigg|_{\eta=\eta_0} \right| = \left| \frac{\partial}{\partial \eta} \left[\mathbf{x}_0(s) + \mathbf{n}\mathrm{n}(s,\eta)\right]\bigg|_{\eta=\eta_0} \right| \\ &= \left| \mathbf{n}\frac{\partial \mathrm{n}(s,\eta)}{\partial \eta}\bigg|_{\eta=\eta_0} \right| = \left| \frac{\partial \mathrm{n}(s,\eta)}{\partial \eta}\bigg|_{\eta=\eta_0} \right| \end{aligned} \tag{4.197}$$

As is proved with little difficulty (see, e.g., Smirnov 1964b [59]), $\widehat{\mathrm{n}} := \frac{\partial \mathrm{n}(\eta,\tau)}{\partial \eta}|_{\eta=\eta_0}$ is a solution of the Jacobi equation. We thus established that the absolute value of a solution of the Jacobi equation coincides with the geometrical spreading.

It is worthy of note that in the vicinity of the selected ray $\ell$, the rays of the field of rays are approximately described by the relation

$$\mathbf{x} = \mathbf{x}_0(s) + (\eta - \eta_0)\mathbf{n}\widehat{\mathrm{n}}(s)|_{\eta=\eta_0}, \tag{4.198}$$

where $\mathbf{x}_0(s)$ is a point on $\ell$ and $\widehat{\mathrm{n}}(s)$ is a fixed solution of the Jacobi equation. In coordinates $(s, n)$, this reads

$$\mathrm{n} = \mathrm{n}(s,\eta) \approx \mathrm{n}(s,\eta) + (\eta - \eta_0)\frac{\partial \mathrm{n}}{\partial \eta}\bigg|_{\eta=\eta_0} = (\eta-\eta_0)\frac{\partial \mathrm{n}}{\partial \eta}\bigg|_{\eta=\eta_0}. \tag{4.199}$$

The approximate formula (4.199) links the two-dimensional geometrical spreading (see (4.197)) and a nonnegative solution n of the Jacobi equation.

### 4.8.3 Calculation of the initial data for the Jacobi equation in the case of monotype reflection

Let the curve $A_\mathrm{i}B_\mathrm{i}$ (see Fig. 4.10) be orthogonal to the rays of the incident wave, i.e., let it be its wavefront, and let $A_\mathrm{r}B_\mathrm{r}$ be the wavefront of the wave reflected from $\mathscr{B}$. The classical reflection law is assumed to hold, that is, the angle of incidence is equal to the angle of reflection. Consider a fixed broken polygonal line that consists of segments of the rays $M_\mathrm{i}D$ and $DM_\mathrm{r}$ of the incident and reflected waves, $D \in \mathscr{B}$; see Fig. 4.12. For such a broken line the variational relation

$$\delta I = \delta \int_{M_\mathrm{i}DM_\mathrm{r}} \frac{d\sigma}{c} = 0 \tag{4.200}$$

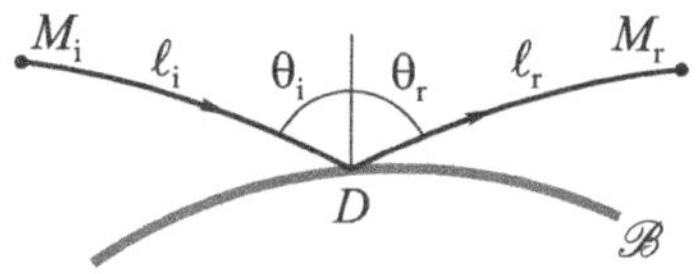


**Figure 4.12**: Broken line $M_i D M_r$

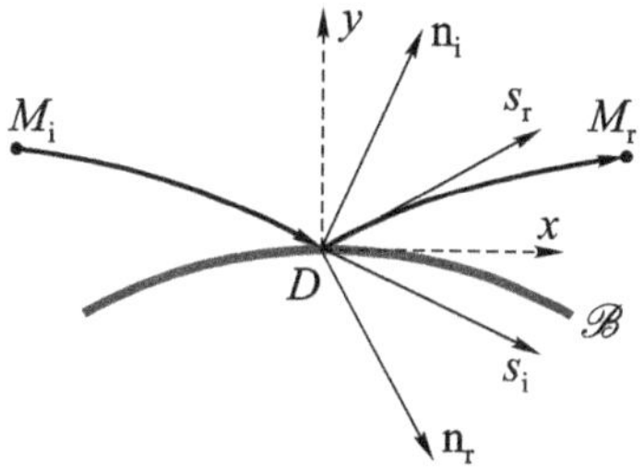


**Figure 4.13**: Geometry of incident and reflected waves

holds. In the sequel, we will always replace $\frac{d\sigma}{c(n,s)}$ by $\Phi(n, \frac{\partial n}{\partial s}, s)ds$ in the calculation of variations of the form (4.194). Comparison curves will be either the rays of the incident wave or those of the reflected wave (see Fig. 4.10).

Assuming that the solution of the Jacobi equation (4.195) for the incident wave is known, we find the initial data for the corresponding Jacobi equation of the reflected ray, $n_r|_{\mathscr{B}}$ and $\frac{dn_r}{ds}\big|_{\mathscr{B}}$. Let us introduce on the rays $M_i D$ and $DM_r$ the coordinate systems $(s_i, n_i)$ and $(s_r, n_r)$. We assume that at the point $D$, $s_i = 0$ and $s_r = 0$, while the direction in which $s$ grows and the normals to the rays are chosen in accordance with Fig. 4.13, i.e., the respective normals are positive on the side of the curve to which the arrow points. The consideration will be related to small neighborhoods of the rays $M_i D$ and $DM_r$, and the study of higher-order terms will be omitted.

Let us introduce Cartesian coordinates $(x, y)$ at the reflection point $D$, as shown in Fig. 4.13. The equation of the boundary $\mathscr{B}$, up to higher-order terms, can be written as

$$y = -\frac{\kappa}{2}x^2. \tag{4.201}$$

Here, $\kappa$ is the curvature of the boundary at the point $D$, not necessarily positive.

Figure 4.14 shows a close broken line with the reflection point $\widetilde{D}$. Let us calculate, in the principal approximation, the variation of the Fermat integral along the broken line $\widetilde{M_i}\widetilde{D}\widetilde{M_r}$, that is, $\delta\left(\int_{\widetilde{M_i}\widetilde{D}} + \int_{\widetilde{D}\widetilde{M_r}}\right) d\sigma/c$. We use the formula for the first variation (4.42). The integrals along the rays vanish (because the rays are extremals of the Fermat functional). Using (4.42), in the

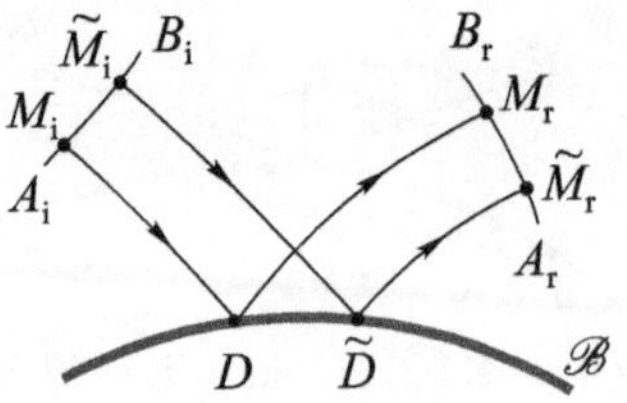


**Figure 4.14**: Reflection of two close rays

first approximation we obtain

$$0 = \delta\left(\int_{\widetilde{M}_i\widetilde{D}} + \int_{\widetilde{D}\widetilde{M}_r}\right)$$
$$= [\Phi_{n'}\delta n + (\Phi - n'\Phi_{n'})\delta s]\big|_{\widetilde{M}_i}^{\widetilde{D}} + [\Phi_{n'}\delta n + (\Phi - n'\Phi_{n'})\delta s]\big|_{\widetilde{D}}^{\widetilde{M}_r}. \qquad (4.202)$$

Here,

$$\Phi = \frac{1}{c_0(s)} - \frac{c_2(s)}{c_0^2(s)}n^2 + \frac{1}{2c_0(s)}(n')^2, \quad n' := \frac{dn}{ds}, \quad \Phi_{n'} := \frac{1}{c_0(s)}n'.$$

By the orthogonality of the rays of the incident wave to $A_iB_i$, and that of the reflected one to $A_rB_r$ at the points $M_i$ and $M_r$, the expressions $[\Phi_{n'}\delta n' + (\Phi - n'\Phi_{n'})\delta s]|_{M_{i,r}}$ can be replaced by zeros. Relation (4.202) is reduced to the following relation at $\widetilde{D}$:

$$\Phi_{n'}|_i\,\delta n_i - \Phi_{n'}|_r\delta n_r + (\Phi - n'\Phi_{n'})_i\delta s_i - (\Phi - n'\Phi_{n'})_r\delta s_r = 0. \qquad (4.203)$$

Expressions with the index i (respectively, r) relate to the incident (respectively, reflected) wave.

Let the coordinates of the boundary point $\widetilde{D}$ close to $D$ in the coordinate systems related to the rays beginning and ending at $\widetilde{D}$, see Fig. 4.13, (because, in small, the systems are approximately Cartesian) be $(s_i, n_i)$ and $(s_r, n_r)$, and let the point $\widetilde{D}$ have coordinates $(x, y)$ in the Cartesian system. If $\theta = \theta_i = \theta_r$ is the angle of incidence of the ray $M_iD$ (and, what is the same, the angle of reflection of the ray $DM_r$), then

$$\begin{aligned} s_i &= x\sin\theta - y\cos\theta, & s_r &= x\sin\theta + y\cos\theta, \\ n_i &= x\cos\theta + y\sin\theta, & n_r &= x\cos\theta - y\sin\theta. \end{aligned} \qquad (4.204)$$

The second line in the formulas (4.204) shows that in this approximation we can put $n_i = n_r = x\cos\theta$, because in consequence of (4.201) $y$ has a higher order of vanishing than $x$. We thus obtain one initial condition for the Jacobi equation:

$$n_r|_{\widetilde{D}} = n_i|_{\widetilde{D}}. \qquad (4.205)$$

It remains to find $\mathrm{n}'_\mathrm{r}|_{\widetilde{D}} := \frac{d\mathrm{n}_\mathrm{r}}{ds}|_{\widetilde{D}}$. The relation (4.203) makes it possible. We require that the equality hold at $\widetilde{D}$, assuming $x$, $y$, $n_\mathrm{i}$, and $n_\mathrm{r}$ to be small. When transforming the formula (4.203), we bear in mind that in the expression $[\Phi_{\mathrm{n}'}\delta\mathrm{n} + (\Phi - \mathrm{n}'\Phi_{\mathrm{n}'})\delta s]\,|_{\widetilde{M}_\mathrm{i}}^{\widetilde{D}}$ (respectively, $[\Phi_{\mathrm{n}'}\delta\mathrm{n} + (\Phi - \mathrm{n}'\Phi_{\mathrm{n}'})\delta s]\,|_{\widetilde{D}}^{\widetilde{M}_\mathrm{r}}$) $(s_\mathrm{i}, n_\mathrm{i})$ are the coordinates of the point $\widetilde{D}$ in the system associated with the ray $M_\mathrm{i}D$ (respectively, $(s_\mathrm{r}, n_\mathrm{r})$ are the coordinates of the point $\widetilde{D}$ in the system associated with the ray $DM_\mathrm{i}$). Further, $\delta s_\mathrm{i}, \delta n_\mathrm{i}$ and $\delta s_\mathrm{r}, \delta n_\mathrm{r}$ are the components of the same vector tangent to the boundary $\mathscr{B}$ at the point $\widetilde{D}$ in the coordinate systems $(s_\mathrm{i}, n_\mathrm{i})$ and $(s_\mathrm{r}, n_\mathrm{r})$, respectively. In the vicinity of the point $\widetilde{D}$, both systems can be regarded as Cartesian. The components of vectors in the passage from one Cartesian system to another are transformed by the same formulas as for coordinates, that is, in accordance with (4.204).

In the formula (4.203), the components of the vector tangent to $\mathscr{B}$ are written in different coordinate systems. Now it is convenient to write these formulas in the coordinates $x, y$, that is, to express $\delta s_i, \ldots$ via $\delta x$ and $\delta y$ which are the Cartesian components of the vector tangent to $\widetilde{D}$. Let us replace in the formulas (4.204) $s_\mathrm{i}, n_\mathrm{i}, s_\mathrm{r}, n_\mathrm{r}, x$, and $y$ by $\delta s_\mathrm{i}, \delta n_\mathrm{i}, \delta s_\mathrm{r}, \delta n_\mathrm{r}, \delta x$, and $\delta y$, respectively. Substituting the result into the formula (4.203) and discarding infinitesimal terms of higher order, we obtain

$$\frac{\delta x}{c_0}(\mathrm{n}'_\mathrm{i} - \mathrm{n}'_\mathrm{r})\Big|_{\widetilde{D}} \cos\theta - \frac{2\delta y}{c_0}\cos\theta = 0. \tag{4.206}$$

Taking into account (4.201),

$$\delta y = -\kappa x \delta x,$$

we find

$$(\mathrm{n}'_\mathrm{i} - \mathrm{n}'_\mathrm{r} + 2\kappa x)|_{\widetilde{D}} = 0.$$

The second line in (4.204) $\mathrm{n}_\mathrm{i} = \mathrm{n}_\mathrm{r} = x\cos\theta$ gives the second initial condition

$$\left(\mathrm{n}'_\mathrm{i} - \mathrm{n}'_\mathrm{r} + 2\frac{\kappa}{\cos\theta}\mathrm{n}_\mathrm{r}\right)\Big|_{\widetilde{D}} = 0. \tag{4.207}$$

Replacing in the formulas (4.205) and (4.207) $\widetilde{D}$ by $D$, which introduces an error of higher order of smallness, we eventually obtain

$$\mathrm{n}_\mathrm{r}|_D = \mathrm{n}_\mathrm{i}|_D, \tag{4.208}$$

$$\left(\mathrm{n}'_\mathrm{i} - \mathrm{n}'_\mathrm{r} + 2\frac{\kappa}{\cos\theta}\mathrm{n}_\mathrm{r}\right)\Big|_D = 0. \tag{4.209}$$

Here, $\mathrm{n}_\mathrm{i}|_D$ and $\mathrm{n}'_\mathrm{i}|_D$ (respectively, $\mathrm{n}_\mathrm{r}|_D$ and $\mathrm{n}'_\mathrm{r}|_D$) are the values of $\mathrm{n}_\mathrm{i}$ and $\mathrm{n}'_\mathrm{i}$ ($\mathrm{n}_\mathrm{r}$ and $\mathrm{n}'_\mathrm{r}$) for $s_\mathrm{i} = 0$ ($s_\mathrm{r} = 0$).

### 4.8.4 Calculation of the initial data for the Jacobi equation for the case of reflection with conversion

Now let the incident wave propagate with velocity $c_i$, and the reflected one travel with velocity $c_r$. We exclude the case of the total internal reflection. The previous constructions can easily be modified.

First, the formula (4.203) is kept as it is, whereas the formula (4.204) is changed: now $\theta$ in the formula for $s_i, n_i$ equals $\theta_i$, whereas in the formula for $s_r, n_r$ it equals $\theta_r$, respectively, where $\theta_i$ and $\theta_r$ are related by the Snell law:

$$\frac{\sin\theta_i}{\sin\theta_r} = \frac{c_i}{c_r}. \tag{4.210}$$

Now we have

$$\begin{aligned} s_i &= x\sin\theta_i - y\cos\theta_i, \quad & s_r &= x\sin\theta_r + y\cos\theta_r, \\ n_i &= x\cos\theta_i + y\sin\theta_i, \quad & n_r &= x\cos\theta_r - y\sin\theta_r. \end{aligned} \tag{4.211}$$

In this case, the counterpart of (4.208) is more complicated. From the second line in (4.211) it follows that $\frac{n_i}{\cos\theta_i}\Big|_{\widetilde{D}} = \frac{n_r}{\cos\theta_r}\Big|_{\widetilde{D}}$, which, up to terms of higher order, becomes

$$\left.\frac{n_i}{\cos\theta_i}\right|_D = \left.\frac{n_r}{\cos\theta_r}\right|_D. \tag{4.212}$$

The initial condition for $n_r'$ is obtained from the relation (4.203) becoming now approximately

$$\delta x \left.\frac{n_i'}{c_i}\right|_{\widetilde{D}} \cos\theta_i - \delta x \left.\frac{n_r'}{c_r}\right|_{\widetilde{D}} \cos\theta_i - \delta y \left.\left(\frac{\cos\theta_i}{c_i} + \frac{\cos\theta_r}{c_r}\right)\right|_{\widetilde{D}} = 0. \tag{4.213}$$

Again replacing, with an infinitesimal error, $\widetilde{D}$ by $D$, we obtain

$$\left.\frac{n_i'}{c_i}\right|_D \cos\theta_i - \left.\frac{n_r'}{c_r}\right|_D \cos\theta_r + \kappa \left.\left(\frac{n_i}{c_i} + \frac{n_r}{c_r}\right)\right|_D = 0, \tag{4.214}$$

which, together with (4.212), gives the initial data for the reflected ray of the converted wave.

### 4.8.5 Calculation of the initial data for the case of transmission

Let a wave transmit from a medium with a velocity $c_i$ to a medium with a velocity $c_t$ (no matter whether it is monotype or converted), in accordance with the Snell law

$$\frac{\sin\theta_i}{\sin\theta_t} = \frac{c_i}{c_t}, \tag{4.215}$$

and again let no total internal reflection occur.

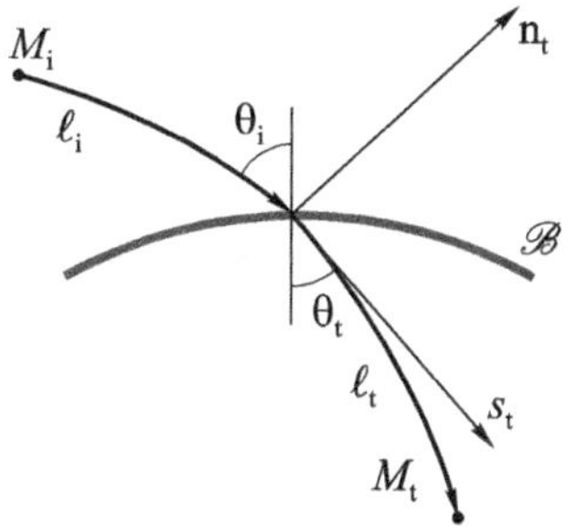


**Figure 4.15**: Geometry of a transmitted wave

The coordinates of the point $D$ in all three systems, those related to the incident and transmitted waves $(s_i, \mathrm{n}_i)$ and $(s_t, \mathrm{n}_t)$, and the Cartesian one $(x, y)$, see Fig. 4.15, are again zeros. Similarly to the above, in a higher order,

$$\begin{aligned} s_i &= x \sin\theta_i - y\cos\theta_i, & s_t &= x\sin\theta_t - y\cos\theta_t, \\ \mathrm{n}_i &= x\cos\theta_i + y\sin\theta_i, & \mathrm{n}_t &= x\cos\theta_t + y\sin\theta_t, \end{aligned} \tag{4.216}$$

and with (4.201) taken into account, replacing again $\widetilde{D}$ by $D$, we find

$$\left.\frac{\mathrm{n}_i}{\cos\theta_i}\right|_D = \left.\frac{\mathrm{n}_t}{\cos\theta_t}\right|_D. \tag{4.217}$$

The relation, which follows from variational calculus,

$$[\Phi_{\mathrm{n}'}\delta\mathrm{n}_i' + (\Phi - \mathrm{n}'\Phi_{\mathrm{n}'})\delta s_i]\Big|_{\widetilde{D}}^{\widetilde{M}_i} + [\Phi_{\mathrm{n}'}\delta\mathrm{n}_t' + (\Phi - \mathrm{n}'\Phi_{\mathrm{n}'})\delta s_t]\Big|_{\widetilde{M}_t}^{\widetilde{D}} = 0$$

takes in the principal approximation the form

$$\left.\frac{\mathrm{n}_i'}{c_i}\cos\theta_i\right|_{\widetilde{D}}\delta x - \left.\frac{\mathrm{n}_t'}{c_t}\cos\theta_t\right|_{\widetilde{D}}\delta x - \left.\frac{\mathrm{n}_i}{c_i}\cos\theta_i\right|_{\widetilde{D}}\delta y + \left.\frac{\mathrm{n}_t}{c_t}\cos\theta_t\right|_{\widetilde{D}}\delta y = 0, \tag{4.218}$$

whence, passing with a small relative error to the point $D$, we obtain

$$\left.\left[\frac{\mathrm{n}_i'}{c_i}\cos\theta_i - \frac{\mathrm{n}_t'}{c_t}\cos\theta_t + \kappa\left(\frac{\mathrm{n}_i}{c_i} - \frac{\mathrm{n}_t}{c_t}\right)\right]\right|_D = 0. \tag{4.219}$$

The relations (4.212) and (4.219) define the sought-for values of $\mathrm{n}_t$ and $\mathrm{n}_t'$ for $s_t = 0$.

### 4.8.6 Case of constant velocities

As an important example, we consider the case of constant velocities. We keep in mind now that $n_i$, $n_r$, and $n_t$ describe geometrical spreadings and are not necessarily small.

If the velocity is constant, then the general solution of the Jacobi equation (4.195) before and after reflection or transmission obviously has the form

$$\mathrm{n} = As + B \tag{4.220}$$

with different constants $A = A_\mathrm{j}$ and $B = B_\mathrm{j}$, $\mathrm{j} = \mathrm{i}, \mathrm{r}, \mathrm{t}$ and $s = s_\mathrm{j}$. Consider an incident wave from a point source. Let it be spaced from the reflection point by a distance $\ell_\mathrm{i}$, see Fig. 4.12. We normalize the solution for the incident wave by the condition $A_\mathrm{i} = 1$; then to satisfy the relation $s_\mathrm{i} = 0$ at the reflection point $D$, we put

$$\mathrm{n}_\mathrm{i} = s_\mathrm{i} + \ell_\mathrm{i}. \tag{4.221}$$

Here, $s_\mathrm{i}$ is the distance from the point $D$. For the reflected wave,

$$\mathrm{n}_\mathrm{r} = A_\mathrm{r} s_\mathrm{r} + B_\mathrm{r}. \tag{4.222}$$

The condition (4.208) gives $B_\mathrm{r} = B_\mathrm{i} = \ell_\mathrm{i}$, that is,

$$\mathrm{n}_\mathrm{r} = A_\mathrm{r} s_\mathrm{r} + \ell_\mathrm{i}. \tag{4.223}$$

In the case of monotype reflection, from (4.209) it follows that $A_\mathrm{r} = 1 + \frac{2\kappa\ell_\mathrm{i}}{\cos\theta}$, where $\kappa$ is the curvature of the boundary at the reflection point, whence

$$\mathrm{n}_\mathrm{r} = A_\mathrm{r} s + \ell_\mathrm{i} = \left(1 + \frac{2\kappa\ell_\mathrm{i}}{\cos\theta}\right) s_\mathrm{r} + \ell_\mathrm{i}. \tag{4.224}$$

Here, $\theta := \theta_\mathrm{i} = \theta_\mathrm{r}$. Putting $s_\mathrm{r} = \ell_\mathrm{r}$ (see Fig. 4.12), for the geometrical spreading we obtain the expression

$$\mathrm{n}_\mathrm{r} = \ell_\mathrm{i} + \ell_\mathrm{r} + \frac{2\kappa\ell_\mathrm{i}\ell_\mathrm{r}}{\cos\theta}. \tag{4.225}$$

In the case of reflection with conversion, (4.214) gives

$$A_\mathrm{r} = \frac{c_\mathrm{r}}{\cos\theta_\mathrm{r}} \left[\frac{\cos\theta_\mathrm{i}}{c_\mathrm{i}} + \kappa\ell_\mathrm{i}\left(\frac{1}{c_\mathrm{i}} + \frac{1}{c_\mathrm{r}}\right)\right],$$

$$\mathrm{n}_\mathrm{r} = A_\mathrm{r} s + \ell_\mathrm{i} = \ell_\mathrm{i} + \frac{c_\mathrm{r}\cos\theta_\mathrm{i}}{c_\mathrm{i}\cos\theta_\mathrm{r}}\ell_\mathrm{r} + \frac{\kappa\ell_\mathrm{i}\ell_\mathrm{r}}{\cos\theta_\mathrm{r}}\left(1 + \frac{c_\mathrm{r}}{c_\mathrm{i}}\right), \tag{4.226}$$

which naturally coincides with (4.224) for $c_\mathrm{r} = c_\mathrm{i}$ and $\theta_\mathrm{i} = \theta_\mathrm{r} = \theta$.

For transmission, denoting by $\ell_\mathrm{t}$ the length of the transmitted ray, see Fig. 4.15, from (4.217) and (4.219) we obtain

$$B_\mathrm{t} = \ell_\mathrm{t}, \quad A_\mathrm{t} = \frac{c_\mathrm{t}}{\cos\theta_\mathrm{t}} \left[\frac{\cos\theta_\mathrm{i}}{c_\mathrm{i}} + \kappa\ell_\mathrm{i}\left(\frac{1}{c_\mathrm{t}} + \frac{1}{c_\mathrm{t}}\right)\right],$$

whence

$$\mathrm{n}_\mathrm{t} = A_\mathrm{t} s + \ell_\mathrm{i} = \ell_\mathrm{i} + \frac{c_\mathrm{r}\cos\theta_\mathrm{i}}{c_\mathrm{i}\cos\theta_\mathrm{t}}\ell_\mathrm{r} + \frac{\kappa\ell_\mathrm{i}\ell_\mathrm{t}}{\cos\theta_\mathrm{t}}\left(-1 + \frac{c_\mathrm{r}}{c_\mathrm{i}}\right). \tag{4.227}$$

### 4.8.7 Focusing under reflection

Formula (4.224) in the case where $\kappa < 0$ gives a simple example of focusing rays, which may result in occurrence of a caustic; see Section 4.7.3. For a source located in a homogeneous medium on the side to which the reflecting surface is concave, the geometrical spreading may vanish on a line, namely a caustic. In a nondegenerate case, it is an envelope of rays. When approaching a caustic, the wavefield becomes dissimilar to a plane wave. Anomalous polarization near the caustic becomes more pronounced (see, e.g., Kiselev 1987 [34], Kiselev and Roslov 1991 [35], Fradkin and Kiselev 1997 [22], Kiselev, Yarovoy, and Vsemirnova 2005 [38]).

---

## 4.9 ⋆ Nonstationary versions of the ray method

Ray method is multifaceted, that is, it has several distinct versions. In this section, we will touch on several such versions, based on ansatzses different from (4.1).

### 4.9.1 High-frequency asymptotics and asymptotics with respect to smoothness

The ansatz (4.1), which is studied in Chapters 4 and 5, in view of the footnote to the formula (4.1) can be written in the form

$$\mathbf{u}(\mathbf{x}, t) = \boldsymbol{u}^0(\mathbf{x}) f^{(0)}(\theta) + \boldsymbol{u}^1(\mathbf{x}) f^{(1)}(\theta) + \boldsymbol{u}^2(\mathbf{x}) f^{(2)}(\theta) + \dots, \tag{4.228}$$

where

$$\theta = t - \tau. \tag{4.229}$$

Here, the functions of one variable $f^{(j)}(\theta)$ have the form

$$f^{(j)}(\theta) = \frac{e^{-i\omega\theta}}{(-i\omega)^{j+\xi}}, \tag{4.230}$$

and the following relationship holds:

$$\frac{df^{(j)}(\theta)}{d\theta} = f^{(j-1)}(\theta), \quad f^{(j)}(\theta) = O\left(\frac{1}{\omega^{j+\operatorname{Re}\xi}}\right). \tag{4.231}$$

One may consider an expansion of the form (4.228), (4.229), (4.231) as the ansatz taking instead of (4.230) a sequence of functions that have singularities on the wavefront $\theta = 0$. For example, the leading term may have a delta-function-type singularity, the next may be of the Heaviside-type, etc.

As $f^{(j)}$, the classical homogeneous generalized functions (see, e.g., Gelfand and Shilov 1964 [28])

$$f^{(j)}(\theta) = \frac{\theta_+^{\alpha+j}}{\Gamma(\alpha+j+1)} \tag{4.232}$$

can be taken. Here, $\Gamma$ is the Euler gamma function, and the function $\theta_+^\beta$ is defined for $\operatorname{Re}\beta > -1$ by

$$\theta_+^\beta = \begin{cases} \theta^\beta & \text{for } \theta > 0, \\ 0 & \text{for } \theta \leqslant 0. \end{cases}$$

The function $\theta_+^\beta/\Gamma(\beta+1)$ is generalized to complex values of $\beta$ lying in the half-plane $\operatorname{Re}\beta \leqslant -1$, by analytic continuation. For negative integer $\beta = -n$, $n = 1, 2, \ldots$, as is known,

$$\left.\frac{\theta_+^\beta}{\Gamma(\beta+1)}\right|_{\beta=-n} = \frac{d^{n-1}}{d\theta^{n-1}}\delta(\theta), \tag{4.233}$$

where $\delta$ is the delta function; see Section 3.1. For $\beta = 0$ we have $\theta_+^\beta/\Gamma(\beta+1) = H(\theta)$, where $H$ is the Heaviside function (3.10). Generalizations and analogs of the functions (4.232) are considered in Gelfand and Shilov 1964 [28].

The statement that a "singularity on the wavefront $\tau(\mathbf{x}) = t$ is described by the expansion (4.228), (4.232)" means that in a neighborhood of the wavefront,

$$\begin{aligned} &\mathbf{u}(\mathbf{x}, t) - \sum_{j=0}^{N} \boldsymbol{u}^j(\mathbf{x}) f^{(j)}(\theta) \\ &\qquad = \mathrm{u}(\mathbf{x}, t) - \sum_{j=0}^{N} \boldsymbol{u}^j(\mathbf{x}) \frac{\theta_+^{\alpha+j}}{\Gamma(\alpha+j+1)} \in C^{[\mathrm{N}-\operatorname{Re}\alpha-1]}(\mathbb{R}^4) \end{aligned} \tag{4.234}$$

for $\mathrm{N} - \operatorname{Re}\alpha - 1 \geqslant 0$, with [P] standing for the integer part of a real number P and $C^{\mathrm{M}}$ being the space of infinite functions having M continuous derivatives. The expansion (4.228), (4.232) is called in the Russian literature (see, e.g., Vainberg 1989 [65]) *asymptotic with respect to smoothness*. Functions $f^{(j)}$ allowing such an asymptotic and different from (4.232) are considered, e.g., by Babich 1989 [5], Gelfand and Shilov 1964 [28], Courant and Hilbert 1962 [14].

The duality of high-frequency asymptotic and smoothness asymptotic is well known: they are connected via the Fourier transform. In the particular case of (4.232), this reads as

$$\int_{-\infty}^{\infty} e^{i\omega t} \frac{(t-\tau)_+^\beta}{\Gamma(\beta+1)} dt = \frac{e^{i\omega\tau}}{(-i\omega)^{\beta+1}} = \frac{e^{i\omega\tau}}{e^{-i\frac{\pi}{2}(\beta+1)}\omega^{\beta+1}}, \quad \omega > 0, \tag{4.235}$$

(see, e.g., Gelfand and Shilov 1963 [28]). This duality has long been known (see, e.g., Klein 1955 [39]).

### 4.9.2 Other nonstationary versions

Consider now the ansatz

$$\mathbf{u}(\mathbf{x},t) = e^{ip\theta(\mathbf{x},t)}\left\{\boldsymbol{u}^0(\mathbf{x},t) + \frac{\boldsymbol{u}^1(\mathbf{x},t)}{-ip} + \dots\right\}, \quad p\to\infty, \tag{4.236}$$

which is another generalization of the ray expansion (4.1). The presence of a large parameter $p$ secures a high-frequency nature of the wave process. The expansion (4.236) is reduced to (4.1) when $p=\omega$ and the phase function $\theta = \theta(\mathbf{x},t)$ has the special form of (4.229). The theory based on the ansatz (4.236) and not requiring that (4.229) hold, is known as the *space–time ray method, the STRM*. Its very detailed presentation is given by Babich, Buldyrev, and Molotkov 1985 [7] (with an account of early history of STRM). The STRM expansion describes waves having both frequency and amplitude modulation. The STRM was generalized to elastic waves by Babich 1979 [5].

A further generalization of the above constructions is the ray method for general modulated waves, based on the ansatz

$$\boldsymbol{u} = \boldsymbol{u}^0(\mathbf{x},t)f^{(0)}(p\theta) + \boldsymbol{u}^1(\mathbf{x},t)\frac{f^{(1)}(p\theta)}{p} + \boldsymbol{u}^2(\mathbf{x},t)\frac{f^{(2)}(p\theta)}{p^2} + \dots, \quad p\to\infty, \tag{4.237}$$

where the prescribed functions $f^{(j)}(p\theta)$ are successive integrals

$$f^{(j)}(s) = \int^s f^{(j-1)}(s')ds', \quad j=1,2,\dots \tag{4.238}$$

It is assumed that $f^{(j)} = O\left(p^{-1}f^{(j-1)}\right)$. Such a theory was considered in the scalar case by Buldyrev 1958 [11], and for elastodynamics, under the condition (4.229), by Kiselev 1982, 1984 [32, 33]. In the latter case, $\tau$ satisfies the eikonal equation (either (4.24) or (4.27)); expressions for the amplitudes $\boldsymbol{u}^{(j)}$ show much similarity to those for harmonic waves $P$ and $S$; the corresponding diffraction coefficients $\psi^j = \psi^j(t+\tau,\eta_1,\eta_2)$ and $\boldsymbol{\chi}^j = \boldsymbol{\chi}^j(t+\tau,\eta_1,\eta_2)$ are functions of three variables.

---

## 4.10 ⋆ Comments to Chapter 4

As is typical of a truly fundamental theory, the ray method (its widely used synonyms are *geometrical acoustics* and *geometrical approximation*) originated from several sources. Among these, we mention geometrical optics, traceable to the 18th century theory of plane waves, and the WKB method (which should be rather named after Green and Liouville) in the theory of ODEs. Also, deserving mention are Umov's ideas on energy flow in elastic bodies and those by Hamilton, Reynolds, and Rayleigh on the group velocity and its interpretation as the velocity of propagation of energy. It is pertinent to speak well of the research by celebrated French mathematician Hadamard 1923 [29] concerning the Cauchy problem for second-order hyperbolic equations. Hadamard presented asymptotics with respect to the order of singularity for waves generated by point sources. His monograph (Hadamard 1923 [29]) much impressed pure mathematicians, though physicists and applied mathematicians did not pay it due attention. Early history of the ray method is described in detail in the monograph by Babich, Buldyrev, and Molotkov 1985 [7].

To our best knowledge, a scalar version of the transport equation for the zeroth-order term, as well as its solution, were first encountered in Umov's thesis (Umov 1974, 1950 [63, 64]). This result was later rediscovered. Solutions of transport equations for transient waves in the scalar case were presented by Hadamard 1923 [29]. For the wave equation with a variable velocity, transport equations related to the Cauchy problem were solved by Sobolev 1930 [61], whose approach is presented in Smirnov 1964 [59]. The basic identity in this approach equivalent to (4.78) for $\mathcal{A} \equiv 1$, is sometimes called the "Smirnov lemma[15]" (see, e.g., Dahlen and Tromp 1998 [15]).

Of fundamental importance in solving vector transport equations is the Umov equation; see (4.67). Some of Umov's statements (Umov 1874, 1950 [63, 64]) lack clarity. In the 1950s, the patriarch of Leningrad mathematics V. I. Smirnov expressed to one of the authors of the present book his certitude of a possibility of a transparent derivation of the Umov equation. Indeed, such a derivation is given in Section 4.3.1, where $\bar{\bar{\mathcal{E}}}$ is treated as a reduced averaged density of the energy flow of a local plane wave. The connection of the ray theory with plane waves was mentioned by several researchers; see, e.g., Babich, Buldyrev, and Molotkov 1985 [7]. The notion of local plane wave is first treated as a substantive self-sufficient object in the present book.

The history of the formula (4.80), which can be written as

$$\bar{\bar{\mathcal{E}}}\mathscr{D}|_{\mathbf{x}} = \bar{\bar{\mathcal{E}}}\mathscr{D}|_{\mathbf{y}}, \tag{4.239}$$

is traceable back to the 19th century. It yields a version of the classical continuity equation in Lagrangian coordinates, which are counterparts of the coordinates $(\eta_1, \eta_2, \tau)$ (see, e.g., Kochin, Kibel, and Roze 1964 [40]).

Aimed at applications (mainly to acoustics and electromagnetics), versions of the ray theory emerged in the 1930s and 1940s were due to several authors: the paper by Rytov 1938 [56] and books by Luneburg 1944, 1964 [46, 47] and Friedlander 1958 [23]. In the 1950s, reinforcing each other, investigations by Babich 1956 [4], Filippov 1957 [18], Levin and Rytov 1957 [45] were published, where the main formulas for inhomogeneous isotropic elastic media were derived; see also Karal and Keller 1959 [31]. A review of early research was given

[15]This agrees with the *Arnold principle* stating that whenever a statement or notion is named after someone, the name typically is not that of the pioneer discoverer.

by Alekseev, Babich, and Gel'chinsky 1961 [1], who pioneered in finding formulas for higher-order terms for the isotropic case. The Rytov law which had been first established in electrodynamics by Rytov 1938 [56], was extended to elastodynamics by Levin and Rytov 1957 [45]. There, the zeroth-order amplitude was assumed real.

Clear mathematical definitions related to formal asymptotic expansions go back to Poincaré and can be found in Olver 1974 [51]. The concept of asymptotic expansion of great generality was presented by Bourbaki 2001 [10]. Buslaev and Skriganov 1974 [12] introduced a convenient definition of a *series having an asymptotic nature*, which is inherent to a series whether or not it is an expansion of any function. The definition is as follows. A series $\sum_{j=0}^{\infty} \mathcal{U}_j(p)$, where the functions $\mathcal{U}_j(p)$ are defined for $p \geqslant \text{const}$, has an *asymptotic nature* as $p \to \infty$ if the estimates $|\mathcal{U}_j(p)| \leqslant C_j |p|^{-\beta_j}$ hold, with $\beta_j \nearrow \infty$ as $j \to \infty$.

A detailed exposition of application of the ray theory of volume waves to numerical simulation of seismic wavefields was given by Červený 2001 [13]; see also Kucher and Kashtan 1999 [42], Slawinski 2003 [57], Dahlen and Tromp 1998 [15], Popov 2002 [53]. Such simulations with a few exclusions are restricted to the employment of the zeroth-order term. Interest in the anomalous polarization (see, e.g., Hudson 1980 [30], Kiselev 1984 [33], Gavrilov and Kiselev 1986 [24], Yanovskaya and Roslov [67], Kiselev and Roslov 1987 [35], Eisner and Pšenčík 1996 [17], Kucher and Kashtan 1999 [42]) was motivated by implementation of polarization observations in seismic exploration (see, e.g., Galperin 1984 [25]). Kiselev and Tsvankin 1989a, 1989b [36, 37], and Fradkin and Kiselev 1997 [22] proposed to describe the waves $P$ and $S$ by expressions of the form

$$\widehat{\boldsymbol{u}} = e^{i\omega\tau} \left\{ \boldsymbol{u}^0 + \frac{\boldsymbol{u}^{anom}}{-i\omega} \right\}, \tag{4.240}$$

where $\boldsymbol{u}^{anom}$ stands for the corresponding additional component (which equals $\boldsymbol{u}^{P1\perp}$ for the wave $P$ and $\boldsymbol{u}^{S1\parallel}$ for the wave $S$, respectively[16]). Expressions of the form (4.240), called *double-term representations* of the wavefield by Kiselev and Tsvankin 1989a, 1989b [36, 37] (and *two-component representations* by Fradkin and Kiselev 1997 [22]), allow easy numerical finding of the anomalous component via the zeroth-order term of the asymptotics. Formula (4.240) shows that for a wave the zeroth term of which is linearly polarized, the elliptic polarization is typical, the effect which is described within the first-order approximation.

The reflection of a wave defined by its ray expansion from a boundary has been addressed by numerous authors. As an example we mention a paper by Alekseev and Tsepelev 1956 [2]. The formula (4.187) for the geometrical spreading in homogeneous media has long been known, and its various derivations can be found, e.g., in the monographs by Friedlander 1958 [23] and

[16] In Kiselev and Tsvankin 1989a, 1989b [36, 37] nonstationary counterparts of (4.240) were considered.

Babich and Buldyrev 2007 [6]. In relation to the calculation of the geometrical spreading in 3D, including that for reflected waves, see, e.g., Fock 1965 [20], Popov 2002 [53], Červený 2001 [13]. In Section 4.8, we follow the approach due to M. M. Popov who described it in his 9th chapter of the monograph by Babich and Buldyrev 2007 [6], see also his own monograph Popov 2002 [53]. The calculation in the case of three dimensions is essentially much alike that of Section 4.8 but looks more bulky (see Popov 2002). We note that the important formula for the geometrical spreading in a vertically inhomogeneous medium with the velocity given by $c = c(z) = Az + B$ can be found in Popov 2002 [53]. Popov and Camerlynck 1996 [54] and Fradkin 1989 [21] studied the applicability of ray formulas at large distances. High-frequency wavefields in the vicinity of caustics have been explored since the beginning of the 19th century (see, e.g., Landau and Lifshitz 2000 [44]). In general, a *caustic* is defined as a locus of points where the geometrical spreading vanishes. Modern methods of calculation of wavefields in the presence of caustics are described, e.g., in monographs by Babich and Buldyrev 2007 [6], Babič and Kirpičnikova 1979 [8], Maslov and Fedoryuk 1981 [48], Kravtsov and Orlov 1990 [41], Popov 2002 [53], among others.

---

## References to Chapter 4

[1] Alekseev, A. S., Babich, V. M. and Gel'chinsky, B. Ya. 1961. Ray method of calculation of intensity of wave fronts, in *Voprosy Dynamicheskoi Teorii Rasprostraneniya Seismicheskikh Voln [Problems in the Dynamic Theory of Propagation of Seismic Waves]*, Vol. 5, ed. G. I. Petrashen'. Leningrad: Leningrad Univ. Press, Leningrad, 3–24 [in Russian]. *Алексеев А. С., Бабич В. М., Гельчинский Б. Я.* Лучевой метод вычисления интенсивности волновых фронтов. Вопр. динам. теории распростр. сейсм. волн, 1961. Т. 5. С. 3–24.

[2] Alekseev, A. S. and Tsepelev, N. V. 1956. Intensity of reflected waves in inhomogeneous layered media. *Izv. AN SSSR. Ser. Geophys.* 9:1022–35. *Алексеев А. С., Цепелев Н. В.* Интенсивность отраженных волн в слоисто-неоднородной упругой среде. Изв. АН СССР. Сер. геофиз., 1956. № 9. С. 1022–1035.

[3] Arnol'd, V. I. 1998. On teaching mathematics. *Russ. Math. Surveys.* 53:229–36. *Арнольд В. И.* О преподавании математики. Успехи матем. наук, 1998. Т. 53:1(319). С. 229–234.

[4] Babich, V. M. 1956. Ray method of calculation of intensity of wave fronts. *Dokl. Akad. Nauk SSSR.* 110:355–7 [in Russian]. *Бабич В. М.* Луче-

вой метод вычисления интенсивности волновых фронтов. Докл. АН СССР, 1956. Т. 110(3). С. 355–357.

[5] Babich, V. M., 1979. The space-time ray method in elastic wave theory. *Izv. Acad. Sci. USSR. Phys. Solid Earth.* 15(3):87–93. *Бабич В. М.* О пространственно-временном лучевом методе в теории упругих волн. Изв. АН СССР. Физика Земли, 1979. № 3. С. 3–13.

[6] Babich, V. M. and Buldyrev, V. S. 2007. *Asymptotic methods in short-wavelength diffraction theory.* Oxford: Alpha Science. *Бабич В. М., Булдырев В. С.* Асимптотические методы в задачах дифракции коротких волн. Метод эталонных задач. М.: Наука, 1972.

[7] Babich, V. M., Buldyrev, V. S. and Molotkov, I. A. 1985. *Space–time ray method. Linear and nonlinear waves.* Leningrad: Leningrad University Press [in Russian]. *Бабич В. М., Булдырев В. С., Молотков И. А.* Пространственно-временной лучевой метод. Линейные и нелинейные волны. Л.: Изд-во ЛГУ, 1985.

[8] Babič, V. M. and Kirpičnikova, N. Ya. 1979. *The boundary-layer method in diffraction problems.* Berlin: Springer-Verlag. *Бабич В. М., Кирпичникова Н. Я.* Метод пограничного слоя в задачах дифракции. Л.: Изд-во ЛГУ, 1975.

[9] Babich, V. M. and Kiselev, A. P. 1989. Nongeometrical waves — are there any? An asymptotic description of some “nongeometrical” phenomena in seismic wave propagation. *Geophys. J. Intern.*, 99:415–20.

[10] Bourbaki, N. 2001. *Functions of a real variable.* Berlin: Springer. *Бурбаки Н.* Функции действительного переменного. М.: Наука, 1965.

[11] Buldyrev, V. S. 1958. Propagation of modulated oscillations. *Vestn. Leningr. Univ. Ser. Phys. Chem.* 1(4): 45–52. *Булдырев В. С.* Распространение модулированных колебаний. Вестн. ЛГУ. Физика. Химия, 1958. № 1(4). С. 45–52.

[12] Buslaev, V. S. and Skriganov, M. M. 1974. Coordinate asymptotic behavior of the solution of the scattering problem for the Schrödinger equation. *Theor. Math. Phys.* 19:465–76. *Буслаев В. С., Скриганов М. М.* Координатная асимптотика решения задачи рассеяния для уравнения Шредингера. Теор. матем. физ., 1974. Т. 19(2). С. 217–232.

[13] Červený, V. 2001. *Seismic ray theory.* Cambridge: Cambridge University Press.

[14] Courant, R. and Hilbert, D. 1962. *Methods of mathematical physics. Vol. 2.* New York: Interscience. *Курант Р.* Уравнения с частными производными. М.: Мир, 1964.

[15] Dahlen, F. A. and Tromp, J. 1998. *Theoretical global seismology.* Princeton University Press.

[16] Dubrovin, B. A, Fomenko, A. T. and Novikov, S. P. 1984. *Modern geometry — methods and applications. Part I. The geometry of surfaces, transformation groups, and fields.* New York: Springer-Verlag. *Дубровин Б. А., Новиков С. П., Фоменко, А. Т.* Современная геометрия. Методы и приложения. Т. 1. М.: Наука, 1998.

[17] Eisner, L. and Pšenčík, I. 1996. Computation of additional components of the first-order ray approximation in isotropic media. *Pure Appl. Geophys.* 148:227–53.

[18] Filippov, A. F. 1957. On approximate evaluation of reflected waves. *Izv. AN SSSR. Ser. Geophys.* 7:841–57 [in Russian]. *Филиппов А. Ф.* О приближенном вычислении отраженных волн. Изв. АН СССР. Сер. геофиз., 1957. № 7. С. 841–857.

[19] Fine, B. and Rosenberger, G. 1991. *The fundamental theorem of algebra.* New York: Springer-Verlag.

[20] Fock, V. A. 1965. *Electromagnetic diffraction and propagation problems.* Oxford: Pergamon. *Фок В. А.* Проблемы дифракции и распространения электромагнитных волн. М.: Советское радио, 1970.

[21] Fradkin, L. Ju. 1989. Limits of validity of geometrical optics in weakly irregular media. *J. Opt. Soc. Amer. A.* 6:1315–9.

[22] Fradkin, L. Ju. and Kiselev, A. P. 1997. The two-component representations of time-harmonic elastic body waves in the high- and intermediate frequency regimes *J. Acoust. Soc. Amer.* 101:52–65.

[23] Friedlander, F. G. 1958. *Sound pulses.* Cambridge: Cambridge University Press.

[24] Gavrilov, A. V. and Kiselev, A. P. 1986. Influence of media inhomogeneity and the source pattern on the polarization of elastic $P$-waves *Izv. Acad. Sci. USSR. Phys. Solid Earth.* 22:494–6. Гаврилов А. В., Киселев А. П. Влияние неоднородности среды и направленности источника на поляризацию упругих $P$-волн. Изв. АН СССР. Физика Земли, 1986. № 6. С. 84–87.

[25] Galperin, E. I. 1984. *Polarization method of seismic investigations.* Dordrecht: Reidel. *Гальперин Е. И.* Поляризационный метод сейсмических исследований. М.: Недра, 1977.

[26] Gelfand, I. M. 1961. *Lectures on linear algebra.* New York: Dover. *Гельфанд И. М.* Лекции по линейной алгебре. М.: МЦНМО, 2007.

[27] Gelfand, I. M. and Fomin, S. V. 1963. *Calculus of variations.* New York: Prentice-Hall, Englewood Cliffs. *Гельфанд И. М., Фомин С. В.* Вариационное исчисление. М.: Физматгиз, 1961.

[28] Gelfand, I. M. and Shilov, G. E. 1964. *Generalized functions. Vol. I: Properties and operations.* New York: Academic Press. *Гельфанд И. М., Шилов Г. Е.* Обобщенные функции и действия над ними (Обобщенные функции вып. 1). М.: Добросвет, 2000.

[29] Hadamard, J. 1923. *Lectures on Cauchy's problem in linear partial differential equations.* New Haven: Yale University Press. *Адамар Ж.* Задача Коши для уравнений в частных производных гиперболического типа. М.: Наука, 1978.

[30] Hudson, J. A. 1980. *Excitation and propagation of elastic waves.* Cambridge: Cambridge University Press.

[31] Karal, F. G. and Keller, J. B. 1959. Elastic wave propagation in homogeneous and inhomogeneous media. *J. Acoust. Soc. Amer.* 31:694–705.

[32] Kiselev, A. P. 1982. Excitation of modulated oscillations in inhomogeneous media. *J. Sov. Math.* 20: 1818–25. *Киселев А. П.* Возбуждение модулированных колебаний в неоднородных средах. Зап. научн. семин. ЛОМИ АН СССР, 1981. Т. 104. С. 111–122.

[33] Kiselev, A. P. 1984. Extrinsic components of elastic waves. *Izv. Acad. Sci. USSR, Phys. Solid. Earth,* 19:707–10. *Киселев А. П.* Примесные компоненты упругих волн. Изв. АН СССР. Физика Земли, 1983. № 9. С. 51–56.

[34] Kiselev, A. P. 1987. Depolarization in diffraction. *Sov. Phys. – Tech. Phys.* 32:695–6. *Киселев А. П.* Дифракционная деполяризация. Ж. техн. физ., 1987. Т. 57(6). С. 1184–1185.

[35] Kiselev, A. P. and Roslov, Yu. V. 1991. Use of additional components for numerical modelling of polarization anomalies of elastic body waves, *Sov. Geol. Geophys.* 32:105–14. *Киселев А. П., Рослов Ю. В.* Использование примесных компонент при численном моделировании поляризационных аномалий объемных упругих волн. Геология и геофизика, 1991. № 4. С. 121–131.

[36] Kiselev, A. P. and Tsvankin, I. D. 1989a. Comparison of numerical and asymptotic estimates of elastic wave field. *Doklady. Earth Sci. Sections.* 304:4–7. *Киселев А. П., Цванкин И. Д.* О сопоставлении асимптотических и численных расчетов упругих волновых полей. Докл. АН СССР, 1989. Т. 304(1). С. 61–65.

[37] Kiselev, A. P. and Tsvankin, I. D. 1989b. A method of comparison of exact and asymptotic wave field computations. *Geophys. J. Intern.* 96:253–8.

[38] Kiselev, A. P., Yarovoy, V. O. and Vsemirnova, E. A. 2005. Diffraction, interference, and depolarization of elastic waves. Caustic and penumbra. *J. Math. Sci.* 127:2413–23. *Киселев А. П., Яровой В. О., Всемирнова Е. А.* Аномалии поляризации упругих волн. Каустика и полутень. Зап. научн. семин. ЛОМИ РАН, 2003. Т. 297. С. 136–153.

[39] Kline, M. 1955. Asymptotic solutions of Maxwell's equations involving fractional powers of the frequency. *Comm. Pure Appl. Math.* 8:595–614.

[40] Kochin, N. E., Kibel, I. A. and Roze, N. V. 1964 Theoretical hydrodynamics. New York: John Wiley and Sons. *Кочин Н. Е., Кибель Н. А., Розе Н. В.* Теоретическая гидромеханика. Т. 1. М.: Физматгиз, 1963.

[41] Kravtsov, Yu. A. and Orlov, Yu. I. 1990. *Geometrical optics of inhomogeneous media.* Berlin: Springer Verlag.

[42] Kucher, V. I. and Kashtan, B. M. 1999. *Ray method for homogeneous isotropic media.* St. Petersburg: St. Petersburg Univerity Press [in Russian]. *Кучер В. И., Каштан Б. М.* Лучевой метод для изотропной неоднородной упругой среды. СПб: Изд. СПбГУ, 1999.

[43] Landau, L. D. and Lifshitz, E. M. 1987. *Fluid dynamics.* Oxford: Pergamon. *Ландау Л. Д., Лифшиц Е. М.* Гидродинамика. М.: Наука, 2006.

[44] Landau, L. D. and Lifshitz, E. M. 2000. *The classical theory of fields.* Oxford: Pergamon. *Ландау Л. Д., Лифшиц Е. М.* Теория поля. М.: Физматлит, 2012.

[45] Levin, M. L. and Rytov, S. M. 1957. On the transition to the geometrical approximation in the theory of elasticity. *Sov. Phys. Acoust.* 2:179–84. *Левин М. Л., Рытов С. М.* О переходе к геометрическому приближению в теории упругости. Акуст. журн., 1956. Т. 2(2). С. 173–176.

[46] Luneburg, R. K. 1944. *Mathematical theory of optics. Lecture notes.* Providence, Rhode Island: Brown University.

[47] Luneburg, R. K. 1964. *Mathematical theory of optics.* Berkeley: University of California Press.

[48] Maslov, V. P. and Fedoryuk, M. V. 1981. *Semi-classical approximation in quantum mechanics.* Boston: Springer. *Маслов В. П., Федорюк М. В.* Квазиклассическое приближение для уравнений квантовой механики. М.: Наука, 1976.

[49] Muskhelishvili, N. I. 1966. *Some basic problems of the mathematical theory of elasticity.* Cambridge: Cambridge University Press.

[50] Novikov, S. P. and Taimanov, I. A. 2006. *Modern geometric structures and fields.* AMS. *Новиков С. П., Тайманов И. А.* Современные геометрические структуры и поля. М.: МЦНМО, 2005.

[51] Olver, F. W. J. 1974. *Asymptotics and special functions.* San Diego: Academic Press. *Олвер Ф.* Асимптотика и специальные функции. М.: Наука, 1990.

[52] Popov, M. M. 1977. On calculation of geometrical spreading in inhomogeneous media with interfaces. *Preprint P-3-77. Leningrad Branch of Steklov Math. Institute* [in Russian]. *Попов М. М.* О вычислении геометрического расхождения в неоднородной среде с границами раздела. Препринт ЛОМИ P-3-77. Л., 1977.

[53] Popov, M. M. 2002. *Ray theory and Gaussian beams for geophysicists.* Salvador–Bahia: EDUFBA.

[54] Popov, M. M. and Camerlynck, C. 1996. Second term of the ray series and validity of the ray theory. *J. Geophys. Res.: Solid Earth.* 101(B1):817–26.

[55] Riemann, B. 1948. On the hypotheses which lie at the bases of geometry, in *Selected works.* Moscow: OGIZ [in Russian]. *Риман Б.* Сочинения. М.–Л., 1948.

[56] Rytov, S. M. 1938. On the transition from wave to geometrical optics. *Doklady AN SSSR.* 18:263–6 [in Russian]. *Рытов С. М.* О переходе от волновой к геометрической оптике. Докл. АН СССР, 1938. Т. 18(4-5). С. 263–266.

[57] Slawinski, M. A. 2003. *Seismic waves and rays in elastic media.* Amsterdam: Pergamon.

[58] Smirnov, V. I. 1964a. *A course of higher mathematics, Volume 2.* Oxford: Pergamon. *Смирнов В. И.* Курс высшей математики. Т. 2. М.: Наука, 1981.

[59] Smirnov, V. I. 1964b. *A course of higher mathematics, Volume 4.* Reading MA: Addison-Wesley. *Смирнов В. И.* Курс высшей математики. Т. 4. Ч. 1. М.: Наука, 1981.

[60] Smirnov, V. I. 1964c. *A course of higher mathematics, Volume 3, Part 1.* (Oxford: Pergamon Press, 1964). *Смирнов В. И.* Курс высшей математики. Т. 3. Ч. 1. М.: Наука, 1974.

[61] Sobolev, S. L. 1930. Wave equation for inhomogeneous medium. *Trudy Seismolog. Institute.* 6:1–57 [in Russian]. *Соболев С. Л.* Волновое уравнение для неоднородной среды. Труды Сейсмологического института, 1930. № 6. С. 1–57.

[62] Sommerfeld, A. 1950. *Mechanics of deformable bodies.* New York: Academic Press. *Зоммерфельд А.* Механика деформируемых сред. М.: ИЛ, 1954.

[63] Umov, N. A. 1874. *Equations of the movement of energy in bodies.* Odessa: Ulrich & Schulze Printing House [in Russian]. *Умовъ Н. А.* Уравненія движенія энергіи въ тѣлахъ. Одесса: Въ типографіи Ульриха и Шульце, 1874.

[64] Umov, N. A. 1950. *Selected papers.* Moscow: GTTI [in Russian]. *Умов Н. А.* Избранные сочинения. М.: ГТТИ, 1950.

[65] Vainberg, B. R. 1989. *Asymptotic methods in equations of mathematical physics.* New York: Gordon and Breach. *Вайнберг Б. Р.* Асимптотические методы в уравнениях математической физики. М.: Изд-во МГУ, 1982.

[66] Vekua, I. N. and Kordzadze, R. A., 1989. Elasticity theory, planar problem of. In: *Encyclopedia of Mathematics*, ed. Kluwer.

[67] Yanovskaya, T. V. and Roslov, Yu. V. 1987. A contribution of first order ray approximation to the wavefield reflected from a free surface of inhomogeneous halfspace. *Vestn. Leningr. Univ. Ser. Fiz. Khim.* 2:66–72. *Яновская Т. Б., Рослов Ю. В.* Вклад первого лучевого приближения в поле волн, отраженных от свободной границы однородного полупространства. Вестник ЛГУ. Физика. Химия, 1987. № 2. С. 66–72.

# Chapter 5

## Ray Method for Volume Waves in Anisotropic Media

In this chapter, we construct asymptotic solutions of elastodynamics equations for homogeneous anisotropic media in the form of ray ansatz (4.1). The same reasoning as given in Section 4.1.1 shows that in this case the leading term of the ray ansatz is still a perturbed local plane wave, with its amplitude and phase smoothly varying from point to point. The formulas in this chapter look relatively simple, because we do not specify the structure of the tensor of elastic stiffnesses.

## 5.1 Recurrent system and eikonal equation

### 5.1.1 Recurrent system

Similarly to Chapter 4, we substitute the ray ansatz (4.1) into equation (4.15) where the operator $\mathfrak{L}$ is defined now by (1.37). Equating to zero the coefficients of powers of $\frac{1}{-i\omega}$, we obtain equations of the form (4.20)–(4.23). Now (compare with (2.83)), $\mathfrak{N}$ is a matrix defined by

$$\mathfrak{N}_{im} = c_{ijkm}p_j p_k - \rho\delta_{im}, \tag{5.1}$$

where $p_j = \partial_j\tau$ are the components of the slowness vector $\boldsymbol{p} = \operatorname{grad}\tau$ (see (4.14)), and $\mathfrak{M}$ and $\mathfrak{L}$ are homogeneous matrix differential operators of the first and second orders, respectively, given by

$$(\mathfrak{M}\boldsymbol{u})_i := \mathfrak{M}_{im}u_m = -\partial_j(c_{ijkm}p_k u_m) - c_{ijkm}p_j\partial_k u_m, \tag{5.2}$$

$$\mathfrak{L}_{im} = -\partial_j c_{ijkm}\partial_k, \quad \partial_j := \partial/\partial x_j. \tag{5.3}$$

As in the case of isotropic media, the system (4.20)–(4.23) involves solely operators with real coefficients. We assume that the vector functions $\boldsymbol{u}^0$, $\boldsymbol{u}^1$, ... are also real, which is sufficient for many applications. Consideration of complex amplitudes complicates the theory and makes it devoid of elegance.

### 5.1.2 Eikonal equation

The equation (4.20), with the relation (5.1) taken into account, can be written in the form

$$\frac{1}{\rho(\mathbf{x})}\mathbf{\Gamma}(\boldsymbol{p};\mathbf{x})\boldsymbol{u}^0 = \boldsymbol{u}^0. \tag{5.4}$$

As distinct from Section 2.4.1, the matrix $\frac{1}{\rho}\mathbf{\Gamma}$ depends now on the parameter $\mathbf{x}$. Let $H^2(\boldsymbol{p};\mathbf{x})$ be any of its eigenvalues. As follows from (5.4), the corresponding slowness vector satisfies the relation (cf. (2.22))

$$H^2(\boldsymbol{p};\mathbf{x}) = 1. \tag{5.5}$$

The equation (5.5) will prove to be an analog of the eikonal equation.

Now we make an important assumption that the eigenvalue $H^2(\boldsymbol{p};\mathbf{x})$ is nondegenerate, i.e., the corresponding eigenspace is one-dimensional. Let $\boldsymbol{h} = \boldsymbol{h}(\mathbf{x})$ be the related eigenvector, for which we do not specify any normalization. Then

$$\boldsymbol{u}^0 = \phi^0\boldsymbol{h} \tag{5.6}$$

with $\phi^0 = \phi^0(\mathbf{x})$, that is, describing the vector $\boldsymbol{u}^0$ is reduced to finding a scalar $\phi^0$ (which depends on the normalization of $\boldsymbol{h}$). Thus, the local plane wave at $\mathbf{x}$ is characterized by the vector (5.6) and the wave vector $\boldsymbol{k} = \omega\boldsymbol{p} = \omega\,\mathrm{grad}\,\tau$.

Taking into account the fact that $\boldsymbol{p} = \mathrm{grad}\,\tau$, we see that the nonlinear partial differential equation (5.5) for the function $\tau(\mathbf{x})$ is an analog of the eikonal equation $\tau(\mathbf{x})c^2(\mathrm{grad}\,\tau)^2 = 1$ (4.30), with which we dealt in the case of isotropic media. We will call the equation (5.5) also the eikonal equation for the case of anisotropic media. The theory of the equation (5.5) is that of rays, wavefronts, and the Fermat principle. The next section is devoted to the theory of equations a bit more general than (5.5). The results will be helpful also in Chapter 8 when considering the propagation of Rayleigh waves along a curved surface.

The ray theory requires solving first-order nonlinear partial differential equations (e.g., the classical eikonal equation in the case of isotropic media). A general approach to such equations can be based on the method of characteristics, which reduces the problem to specialized systems of ordinary differential equations. For the case of the classical eikonal equation, the method of characteristics is equivalent to the approach based on the Fermat principle described in Chapter 4. This equivalence is a consequence of the absence of dispersion.

## 5.2 Rays and wavefronts

### 5.2.1 Cauchy problem for a nonlinear equation

The eikonal equation (5.5) is a particular case of a nonlinear first-order partial differential equation with three variables, $\mathbf{x} = (x_1, x_2, x_3)$, of the form

$$P(\boldsymbol{p}; \mathbf{x}) = 0, \quad \boldsymbol{p} = (p_1, p_2, p_3), \quad p_j := \frac{\partial \tau}{\partial x_j}, \quad j = 1, 2, 3. \tag{5.7}$$

Here, $\tau = \tau(\mathbf{x})$ is the unknown function, and $P$ is a given smooth function, on which some additional restrictions will be imposed later. In Chapter 8, we will need to consider an equation of the form (5.7) for the case $\boldsymbol{p} = (p_1, p_2)$, $\boldsymbol{x} = (x_1, x_2)$. The whole of the theory presented below, with trivial changes, is carried over to this "two-dimensional" equation $P(p_1, p_2; x_1, x_2) = 0$.

Let the unknown function $\tau$ be given on a smooth surface $\Sigma$,

$$\tau^o := \tau|_\Sigma, \tag{5.8}$$

and

$$p_j|_\Sigma =: p_j^o, \quad j = 1, 2, 3, \tag{5.9}$$

be also given. The surface $\Sigma$ will be described in parametric form

$$x_i = x_i^o(\eta_1, \eta_2), \quad i = 1, 2, 3, \tag{5.10}$$

where $\eta_1, \eta_2$ define a regular coordinate system on $\Sigma$. Naturally, $\tau^o$ and $p_j^o$ are given functions of the coordinates on $\Sigma$: $\tau^o = \tau^o(\eta_1, \eta_2)$ and $p_i^o = p_i^o(\eta_1, \eta_2)$. We assume that they are smooth and real.[1] Of course, the initial data must satisfy the *"consistency conditions"* (see Courant and Hilbert 1962 [7])

$$P(\boldsymbol{p}^o; \mathbf{x}^o)|_{\mathbf{x}^o \in \Sigma} = 0 \tag{5.11}$$

and

$$d\tau^o = p_j^o dx_j^o \quad \Leftrightarrow \quad \frac{\partial \tau^o}{\partial \eta_l} = p_j^o \frac{\partial x_j^o}{\partial \eta_l}, \quad l = 1, 2, 3. \tag{5.12}$$

The Cauchy problem for the equation (5.7) consists in finding such a $\tau$ that the equation (5.7) is satisfied for given Cauchy data (5.8), (5.9) and conditions (5.11), (5.12). Its solvability requires a certain additional condition of nondegeneracy, which will be discussed in the next section. In the sequel, the study of the Cauchy problem is based on the method of characteristics proposed by Cauchy himself. In the important case where (5.7) is the eikonal

[1]The theory presented here is essentially real. Consideration of the complex case would require a different mathematical apparatus, similar to that used in Section 4.2.5 and in Chapter 7.

equation, the theory of characteristics is essentially the theory of rays. Alternatively, the mathematical theory of rays can be constructed on the basis of the Fermat principle, which we followed in Chapter 4. One could proceed this way in the case of anisotropic media as well. However, here we develop the theory of rays, going "along the way of Cauchy," which will lead to the goal a bit faster. In our case, the latter approach based on the characteristic system of equations reduces to the so-called canonical system of equations (see, e.g., Smirnov 1964 [14]).

### 5.2.2 Characteristic system

The *characteristic system* for the equation (5.7) is the system of ordinary differential equations

$$\frac{dx_i}{d\sigma} = \frac{\partial P}{\partial p_i}, \tag{5.13}$$

$$\frac{dp_i}{d\sigma} = -\frac{\partial P}{\partial x_i}, \tag{5.14}$$

$$\frac{d\tau}{d\sigma} = p_i \frac{\partial P}{\partial p_i}, \tag{5.15}$$

$i = 1, 2, 3$. Here, $\sigma$ is a parameter along the ray. The solution of the Cauchy problem for (5.13)–(5.15) with initial data

$$x_i|_\Sigma = x_i^o, \quad p_i|_\Sigma = p_i^o, \quad \tau|_\Sigma = \tau^o, \quad i = 1, 2, 3\,, \tag{5.16}$$

exists, at least near $\Sigma$. The solution of the equations (5.13)–(5.15) can be regarded as a parametrically defined curve $x_i = x_i(\sigma)$, $i = 1, 2, 3$, in $\mathbb{R}^3$, at each point of which $\tau$ and $p_j(\sigma)$, $j = 1, 2, 3$, are given. These curves are called the *characteristic curves* or simply the *characteristics* of the equation (5.7).

In the theory of propagation of waves, $\frac{dx_j}{d\sigma}$ typically has the meaning of a component of the group velocity. The values of $\frac{dx_j}{d\sigma}$ (see (5.13)) define the absolute value and the direction of the group velocity. Characteristic curves have the meaning of rays. The group velocity is tangent to a ray and defines its direction that coincides with the direction of the averaged energy flow.

We also make the assumption of nondegeneracy: let at $\Sigma$

$$\begin{vmatrix} \frac{\partial x_1}{\partial \sigma} & \frac{\partial x_2}{\partial \sigma} & \frac{\partial x_3}{\partial \sigma} \\ \frac{\partial x_1}{\partial \eta_1} & \frac{\partial x_2}{\partial \eta_1} & \frac{\partial x_3}{\partial \eta_1} \\ \frac{\partial x_1}{\partial \eta_2} & \frac{\partial x_2}{\partial \eta_2} & \frac{\partial x_3}{\partial \eta_2} \end{vmatrix} \neq 0. \tag{5.17}$$

The inequality (5.17) means that characteristic curves are nontangent to $\Sigma$. The smoothness of $x_i = x_i(\sigma, \eta_1, \eta_2)$, together with the condition (5.17), implies that in a neighborhood of $\Sigma$, $(\sigma, \eta_1, \eta_2)$ can be regarded as a regular system of (generally speaking, curvilinear) coordinates. Passing from them to the Cartesian coordinates $(x_1, x_2, x_3)$, we arrive at $\tau = \tau(x_1, x_2, x_3)$ and $p_j = p_j(x_1, x_2, x_3)$. It turns out that the functions $\tau$ and $\boldsymbol{p} = \operatorname{grad} \tau$ solve the Cauchy problem for the equation (5.7) with data on $\Sigma$. The proof can be found, e.g., in books by Courant and Hilbert 1962 [7] and Smirnov 1964 [14].

Another and also important for applications solution of the equation (5.7), with the help of characteristic curves, can be gained if we consider all the characteristic curves $x_i(\sigma)$, passing at $\sigma = 0$ through the fixed point $\mathbf{x}^{\mathrm{O}} = (x_1^{\mathrm{O}}, x_2^{\mathrm{O}}, x_3^{\mathrm{O}})$ (with all possible $p_j|_{\sigma=0} = p_j^{\mathrm{o}}$ such that $P(\boldsymbol{p}^{\mathrm{o}}; \mathbf{x}^{\mathrm{O}}) = 0$).

Characterizing (in a smooth manner) the direction of a unit vector tangent to the characteristic curve at $\sigma = 0$ by a couple of parameters $\eta_1, \eta_2$, e.g., by putting $\tau|_{\sigma=0} = \tau(x_1^{\mathrm{o}}, x_2^{\mathrm{o}}, x_3^{\mathrm{o}}) = 0$,[2] we arrive at the solutions of the equation (5.7) corresponding to a central field of rays. As usual, the direction on a ray is chosen to where $\tau$ grows.

### 5.2.3 Special case: eikonal equation

Of much importance is the case where the equation $P = 0$ can be reduced to the form

$$H(\boldsymbol{p}; \mathbf{x}) = 1, \tag{5.18}$$

where $H$ is a first-order homogeneous function with respect to $\boldsymbol{p}$

$$H(\mathrm{C}\boldsymbol{p}; \mathbf{x}) = \mathrm{C}H(\boldsymbol{p}; \mathbf{x}), \quad \mathrm{C} > 0. \tag{5.19}$$

In such a form, one can write the eikonal equation for the isotropic case (4.30) with $H(\boldsymbol{p}; \mathbf{x}) = c(\mathbf{x})|\nabla\tau| = c(\mathbf{x})|\boldsymbol{p}| = 1$, and also the equation (5.5).

Let us discuss the relationship between the equation (5.5), the characteristic system for this equation, and the theory of local plane waves. In the case of (5.18), the equations (5.13)–(5.15) take the form

$$\frac{dx_i}{d\sigma} = \frac{\partial H}{\partial p_i}, \quad \frac{dp_i}{d\sigma} = -\frac{\partial H}{\partial x_i}, \tag{5.20}$$

$$\frac{d\tau}{d\sigma} = p_i \frac{\partial P}{\partial p_i} = H \equiv 1. \tag{5.21}$$

One can recognize the classical *canonical system* of Hamiltonian mechanics (see, e.g., Arnold 1978 [1], Goldstein 1950 [10], Dubrovin, Fomenko, and Novikov 1984 [8]) in the equations (5.20), where $H$ plays the role of the Hamilton function. If $x_i$ and $p_i$ are solutions of the system (5.20), then $\tau$ can be immediately found from (5.21) by integration.

Equations (5.20)–(5.21) allow us to understand the kinematic meaning of differentiation with respect to $\sigma$. Return to the ray ansatz (4.1). We have already observed that the wavefronts, i.e., the surfaces on which the phase factor $e^{-i\omega(t-\tau)}$ is constant, are moving surfaces $\tau(\mathbf{x}) = t + \text{const}$. From the

---

[2] Any other real constant can be chosen instead of 0.

formula (5.18) with $\boldsymbol{p} = \operatorname{grad}\tau$, it follows that

$$c(\mathbf{x};\mathbf{s}) = \frac{1}{|\operatorname{grad}\tau|} = \frac{1}{|\boldsymbol{p}|} = \frac{H(\boldsymbol{p};\mathbf{x})}{|\boldsymbol{p}|} = H(\mathbf{s};\mathbf{x}). \tag{5.22}$$

The function $c(\mathbf{x};\mathbf{s})$ is seen to be the velocity of the wavefront $\tau(\mathbf{x}) = t+\text{const}$. The equation (5.21) means that at the points of intersection of the curve $x_i = x_i(\sigma)$ (which is a characteristic curve for the equation (5.18)) and the surface $\tau(\mathbf{x}) = t + \text{const}$, the parameter $\sigma$ may differ from the value of $\tau(\mathbf{x})$ only by an additive constant. We put

$$\sigma = \tau = t + \text{const} \quad \Leftrightarrow \quad \frac{d}{d\sigma} = \frac{d}{dt}.$$

The derivatives $\frac{dx_i}{d\sigma}$ prove to be the components of the group velocity $\frac{d\mathbf{x}}{dt}$. As follows from (5.20)–(5.21), they equal $\frac{\partial H}{\partial p_i}$. For the respective plane wave, in consequence of the homogeneity of $H(\boldsymbol{k})$ with respect to $\boldsymbol{k}$, we have $\frac{\partial H(\boldsymbol{p})}{\partial p_i} = \frac{\partial H(\boldsymbol{k})}{\partial k_i} = \frac{\partial \omega}{\partial k_i}$, and, according to (2.20), $\boldsymbol{p} = \frac{\boldsymbol{k}}{\omega}$. Thus, we obtained the formula (4.68) for the group velocity.

The characteristic curve $\mathbf{x} = \mathbf{x}(\tau)$ will be called a ray. We choose the traditional direction on the ray, namely, that in which $\tau$ grows, or, what is the same, the direction of the group velocity. At each point of the ray, the relation (4.68) holds, the vector $\boldsymbol{p} = \boldsymbol{p}(\tau)$ is uniquely defined, and the characteristic equations (5.20)–(5.21) are satisfied.

---

## 5.3 ⋆ Fermat principle and Finsler geometry

### 5.3.1 Rays as extremals of a certain functional of the calculus of variations

In analytical mechanics (see, e.g., Arnold 1978 [1], Goldstein 1950 [10], Dubrovin, Fomenko, and Novikov 1984 [8]), it is common to use the Legendre transformation for transition from the equations of motion in Lagrangian form to the canonical Hamiltonian equations (and further to the Hamilton–Jacobi PDE). In a sense, we reverse the path and come from the canonical equations to rays and the Fermat principle.

We start by demonstrating that rays are extremals of a certain functional of the calculus of variations $\int \frac{\mathscr{L}^2(\mathbf{x}';\mathbf{x})}{2} d\sigma$, where $\mathbf{x}'$ plays the role of the velocity. For the velocity, the notation $\mathbf{x}'$ is employed instead of $\boldsymbol{v}$ in order to respect the traditions of the calculus of variations. The subsequent analysis based on transition by means of the Legendre transformation from the function $\frac{H^2(\boldsymbol{p};\mathbf{x})}{2}$ to the function $\frac{\mathscr{L}^2(\mathbf{x}';\mathbf{x})}{2}$ will have much in common with the consideration given in Section 2.4.4.

Rewrite the equation (5.5) in the form $\frac{H^2}{2} = \frac{1}{2}$. Considering $\frac{H^2}{2}$ as a function of the variables $\boldsymbol{p} = (p_1, p_2, p_3)$, transform it by Legendre to the variables $\mathbf{x}' = (x'_1, x'_2, x'_3)$,[3]

$$x'_j = \frac{\partial}{\partial p_j}\frac{H^2}{2}, \tag{5.23}$$

and let

$$\frac{1}{2}\mathscr{L}^2 = x'_j p_j - \frac{H^2}{2}. \tag{5.24}$$

Assuming that the equation (5.23) can be solved with respect to $p_j$ and substituting $p_j = p_j(\mathbf{x}';\mathbf{x})$ into (5.24), we obtain the function $\mathscr{L} = \mathscr{L}(\mathbf{x}';\mathbf{x})$. By Euler's homogeneity relation,

$$\frac{1}{2}\mathscr{L}^2 = x'_j p_j - \frac{H^2}{2} = p_j \frac{\partial \frac{H^2}{2}}{\partial p_j} - \frac{H^2}{2} = H^2 - \frac{H^2}{2} = \frac{H^2}{2}. \tag{5.25}$$

Acting in the traditional manner (see, e.g., Dubrovin, Fomenko, and Novikov 1984 [8]), consider the differential of the function $\frac{1}{2}\mathscr{L}^2 = \frac{1}{2}\mathscr{L}^2(\boldsymbol{p};\mathbf{x})$, $\boldsymbol{p} = \boldsymbol{p}(\mathbf{x}';\mathbf{x})$. We get

$$\begin{aligned} d\left(\frac{\mathscr{L}^2}{2}\right) &= \frac{\partial(\frac{\mathscr{L}^2}{2})}{\partial x_j}dx_j + \frac{\partial(\frac{\mathscr{L}^2}{2})}{\partial x'_j}dx'_j = d\left(x'_j p_j - \frac{H^2}{2}\right) \\ &= x'_j dp_j + p_j dx'_j - \frac{\partial \frac{H^2}{2}}{\partial x_j}dx_j - \frac{\partial \frac{H^2}{2}}{\partial p_j}dp_j. \end{aligned}$$

By the definition (5.23), cancelation occurs, and we arrive at the relation

$$\frac{\partial\left(\frac{\mathscr{L}^2}{2}\right)}{\partial x_j}dx_j + \frac{\partial\left(\frac{\mathscr{L}^2}{2}\right)}{\partial x'_j}dx'_j = p_j dx'_j - \frac{\partial\left(\frac{H^2}{2}\right)}{\partial x_j}dx_j. \tag{5.26}$$

Equating the coefficients of differentials of the independent variables, we find

$$\frac{\partial}{\partial x_j}\left(\frac{\mathscr{L}^2}{2}\right)\bigg|_{\mathbf{x}'=\text{const}} = -\frac{\partial}{\partial x_j}\left(\frac{H^2}{2}\right)\bigg|_{\boldsymbol{p}=\text{const}}, \quad \frac{\partial}{\partial x'_j}\left(\frac{\mathscr{L}^2}{2}\right) = p_j. \tag{5.27}$$

From the characteristic system (5.20)–(5.21), it follows that

$$\frac{dp_i}{d\sigma} = -\frac{\partial H}{\partial x_i} = -H\frac{\partial H}{\partial x_i} = \frac{\partial}{\partial x_j}\frac{\mathscr{L}^2}{2},$$

and, taking into account (5.27), we obtain $\frac{dp_i}{d\sigma} = -\frac{\partial\left(\frac{H^2}{2}\right)}{\partial x_i}$, which is equivalent to the equations

$$\frac{d}{d\sigma}\frac{\partial\left(\frac{\mathscr{L}^2}{2}\right)}{\partial x'_j} - \frac{\partial\left(\frac{\mathscr{L}^2}{2}\right)}{\partial x_j} = 0. \tag{5.28}$$

[3]The situation here is more complicated than in Section 2.4.4, because we must take into account the dependence of $H^2$ on $\mathbf{x}$.

Note that the equations (5.28) are the Euler equations for the functional $\int \frac{\mathscr{L}^2}{2} d\sigma = \int \frac{\mathscr{L}^2}{2}(\mathbf{x}';\mathbf{x}) d\sigma$, where $\mathbf{x}' = \frac{d\mathbf{x}}{d\sigma}$. In view of the fact that $\frac{\mathscr{L}^2}{2}$ does not directly depend on $\sigma$, the equation has the classical first integral (see, e.g., Smirnov 1964 [14], Gelfand and Fomin 1963 [9])

$$x'_j \frac{\partial \left(\frac{\mathscr{L}^2}{2}\right)}{\partial x'_j} - \frac{\mathscr{L}^2}{2} \equiv \frac{\mathscr{L}^2}{2} = \text{const.} \tag{5.29}$$

Since $\frac{\mathscr{L}^2}{2}$ is second-degree homogeneous with respect to $x'_j$ (see (5.24)), we come to the conclusion that the function $\mathscr{L}$ is first-degree homogeneous with respect to $x'_j$.

By virtue of the relations (5.5) and (5.25), we put const $= \frac{1}{2}$ in (5.29), so that (5.29) takes the form $\frac{\mathscr{L}^2}{2} = \frac{1}{2}$. Assuming that $\mathscr{L} > 0$, we find that $\mathscr{L} = 1$ on the curve $x_i = x_i(\sigma)$, which is a solution of the equations (5.28). From (5.28) it follows that the equations

$$\frac{d}{d\sigma}\frac{\partial \mathscr{L}}{\partial x'_j} - \frac{\partial \mathscr{L}}{\partial x_i} = 0 \tag{5.30}$$

hold, which are the Euler equations for the functional $\int \mathscr{L} d\sigma$.

### 5.3.2 Finsler metric

Define the distance between points $M_0$ and $M$ along a curve connecting them by the formula

$$\int_{M_0}^{M} \mathscr{L} d\sigma = \int_{M_0}^{M} \mathscr{L}(\mathbf{x}';\mathbf{x})\, d\sigma, \quad \mathbf{x}' = \frac{d\mathbf{x}'}{d\sigma}, \tag{5.31}$$

and assume certain conditions of nondegeneracy for $\mathscr{L}$. Then the expression (5.31) defines the *Finsler metric* (see, e.g., Rund 1959 [13]). By virtue of the first-degree homogeneity of $\mathscr{L}(\mathbf{x}';\mathbf{x})$ with respect to $\mathbf{x}'$, the value of the integral is independent of the parameterization of the curve. We choose the parameter as $\sigma = \tau = t + \text{const}$, whence the relations (5.5), (4.68), (5.24), and (5.31) hold. Therefore, the Finsler length of the segment $M_0M$ is the time required for running along $M_0M$ with group velocity (see also the next section).

The Finsler metric is a generalization of the Riemannian metric, where $\mathscr{L}^2(\mathbf{x}';\mathbf{x})$ is a quadratic form with respect to $\mathbf{x}' = \frac{d\mathbf{x}}{d\sigma}$. A domain in which a distance is defined by the expression (5.31) is called the Finsler space; a theory of such spaces is known as Finsler geometry.

The rays (see Section 5.2.3) are extremals of the functional (5.31), i.e., *geodesics of the Finsler metric.* As is easy to see, the inverse also holds, that is, geodesics of the Finsler metric are rays. We omit a simple proof of this fact.

### 5.3.3 Fermat principle

Since the definition of ray is introduced, we can expect that a ray satisfies the Fermat principle stating that the related travel time is extremal. On the one hand, we know that rays are extremals of the integral $\int_{M_0}^{M} \mathscr{L} d\tilde{\sigma}$, where $\tilde{\sigma}$ is a parameter along the curve (the value of the integral is independent of the parameterization, because $\mathscr{L}$ is homogeneous with respect to $\mathbf{x}'$). On the other hand, we know that the components of the group velocity, by definition, equal

$$v_j^{gr} = \frac{dx_j}{d\sigma} = \frac{\partial H}{\partial p_j} \tag{5.32}$$

(see Section 5.2.3). Let $\tilde{\sigma}$ be an arbitrary parameter on the curve, then, as follows from the homogeneity,

$$\mathscr{L} d\tilde{\sigma} = \mathscr{L}\left(\frac{d\mathbf{x}}{d\tilde{\sigma}}; \mathbf{x}\right) d\tilde{\sigma} = \mathscr{L}\left(\frac{\frac{d\mathbf{x}}{d\tilde{\sigma}}}{\left|\frac{d\mathbf{x}}{d\tilde{\sigma}}\right|}; \mathbf{x}\right) \left|\frac{d\mathbf{x}}{d\tilde{\sigma}}\right| d\tilde{\sigma} = \mathscr{L}(\mathbf{s}; \mathbf{x})\, ds. \tag{5.33}$$

Here $\mathbf{s}$, $|\mathbf{s}| = 1$ is a unit vector tangent to the curve, and $ds$ is the differential of its arc length. If we draw a ray at a point $\mathbf{x}$ along $\mathbf{s}$, then $\mathscr{L}\left(\frac{d\mathbf{x}}{d\sigma}; \mathbf{x}\right) = 1$, where $\frac{d}{d\sigma} = \frac{d}{d\tau}$ is differentiation with respect to the eikonal $\tau$. Further (see (5.25)),

$$1 = \mathscr{L}\left(\frac{d\mathbf{x}}{d\sigma}; \mathbf{x}\right) = \mathscr{L}\left(\frac{\frac{d\mathbf{x}}{d\sigma}}{\left|\frac{d\mathbf{x}}{d\sigma}\right|}; \mathbf{x}\right) \left|\frac{d\mathbf{x}}{d\sigma}\right| = \mathscr{L}(\mathbf{s}; \mathbf{x})\, |\boldsymbol{v}^{gr}|.$$

Therefore,

$$\mathscr{L}(\mathbf{s}; \mathbf{x}) = \frac{1}{v^{gr}}, \qquad v^{gr} := |\boldsymbol{v}^{gr}|, \tag{5.34}$$

and the variational relation $\delta \int_{M_0}^{M} \mathscr{L} d\sigma = 0$ defining a ray and equivalent to (5.30) can be written as

$$\delta \int_{M_0}^{M} \mathscr{L} d\sigma = \delta \int_{M_0}^{M} \mathscr{L}(\mathbf{s}; \mathbf{x}) ds = \delta \int_{M_0}^{M} \frac{ds}{v^{gr}} = 0, \tag{5.35}$$

which is a form of the Fermat principle.

In (5.35), comparison curves are smooth curves connecting $M_0$ and $M$, and the parameter on the curve is the Euclidean arc length. For a given unit tangent vector $\mathbf{s}$, the group velocity in its direction is defined as $\boldsymbol{v}^{gr} = v^{gr}\mathbf{s}$. A ray proves to be a curve along which the relation (5.35) holds or, equivalently, the system of equations (5.30) holds.

### 5.3.4 Concluding remarks

Thus, we provide a recipe for solving the Cauchy problem for the eikonal equation (5.5): it is sufficient to solve a characteristic system with proper initial data. Also, in Section 5.2.1 we have constructed the coordinate system $(\eta_1, \eta_2, \sigma)$, where $\sigma$ can be replaced by $\tau$ (which is a consequence of the relation $\frac{d\tau}{d\sigma} = 1$ and of the fact that the parameter $\sigma$ is defined up to an arbitrary additive constant).

---

## 5.4 Solution of transport equation for $\mathbf{u}^0$

### 5.4.1 Consistency condition and the Umov equation

Let us turn to the transport equations (4.20) and (4.21). The equation (4.20) implies the formula (5.6). If $\boldsymbol{h}$ in (5.6) is somehow fixed, the problem is reduced to finding the scalar factor $\phi^0$. Similarly to Chapter 4, we address the consistency condition for the equation (4.21), which has the form

$$(\mathfrak{M}\boldsymbol{u}^0, \boldsymbol{u}^0) = 0. \tag{5.36}$$

We will use energy considerations close to those employed in Section 4.3.

The relation (5.36) can easily be written in divergent form:

$$\partial_j \left(c_{jmqr} u^0_m u^0_q p_r\right) = 0. \tag{5.37}$$

Using the equality (2.94) for the local plane wave, we find

$$\frac{\rho\omega^2}{2} u^0_i u^0_i v^{gr}_j = \frac{\omega^2}{2} c_{jmqr} u^0_m u^0_q p_r. \tag{5.38}$$

We again arrived at the Umov equation, that is, the continuity equation (4.67) for the energy fluid. Cancelling by $\frac{\omega^2}{2}$, we turn to reduced quantities, as in Section 4.3.1.

It is convenient to rewrite the equation (5.37) in the form (4.67):

$$\operatorname{div}(\overline{\overline{\mathcal{E}}}{}^0 \boldsymbol{v}^{gr}) = \frac{\partial}{\partial x_j}\left(\rho u^0_i u^0_i v^{gr}_j\right) = 0, \quad \overline{\overline{\mathcal{E}}}{}^0 := \rho u^0_i u^0_i. \tag{5.39}$$

Just as in Section 4.3.2, the equation (5.39) implies the formula (4.80) taking now the form

$$\rho|\boldsymbol{u}^0|^2 \mathscr{D} = \rho|\phi^0|^2 \boldsymbol{h}^2 \mathscr{D} = \Psi(\eta_1, \eta_2), \tag{5.40}$$

where $\mathscr{D}$ is defined in (4.76), and $\Psi(\eta_1, \eta_2)$ is an arbitrary nonnegative smooth function. Let $\mathbf{x}$ and $\mathbf{x}^{\text{o}}$ be points on the same ray. From (5.40), the relation (cf. (4.83)) easily follows:

$$\left|\boldsymbol{u}^0(\mathbf{x})\right| = \left|\boldsymbol{u}^P(\mathbf{x}^{\text{o}})\right| \frac{\sqrt{\rho(\mathbf{x}^{\text{o}})\mathscr{D}(\mathbf{x}^{\text{o}})}}{\sqrt{\rho(\mathbf{x})\mathscr{D}(\mathbf{x})}}. \tag{5.41}$$

Also, (5.40) implies that

$$\phi^0 = \frac{\psi^0(\eta_1, \eta_2)}{\sqrt{\rho\mathscr{D}}|\boldsymbol{h}|} \tag{5.42}$$

and

$$\boldsymbol{u}^0 = \frac{\psi^0(\eta_1, \eta_2)}{\sqrt{\rho\mathscr{D}}} \frac{\boldsymbol{h}}{|\boldsymbol{h}|}. \tag{5.43}$$

Here, $\psi^0(\eta_1, \eta_2)$ is the *diffraction coefficient*, that is, an arbitrary smooth function, and $\boldsymbol{h}$ is a real eigenvector satisfying the equation (4.20).

## 5.5 Higher-order terms

Having thus ensured the consistency of the equation (4.22), we seek $\boldsymbol{u}^1$, similarly to the isotropic case, as a sum of terms parallel and perpendicular to the direction of the vector of polarization of the local plane wave $\boldsymbol{h}$:

$$\boldsymbol{u}^1 = \boldsymbol{u}^{1\|} + \boldsymbol{u}^{1\perp}, \quad \boldsymbol{u}^{1\|} = \phi^1 \boldsymbol{h}, \quad (\boldsymbol{u}^{1\perp}, \boldsymbol{h}) = 0. \tag{5.44}$$

The vector function $\boldsymbol{u}^{1\perp}$ describes the *anomalous polarization* of the wave.

### Anomalous polarization

The matrix $\mathfrak{M}$ is degenerate on a one-dimensional subspace spanned on the vector $\boldsymbol{h}$, and is invertible on its orthogonal complement. Let

$$\mathfrak{M}\boldsymbol{U} = \boldsymbol{F}, \quad (\boldsymbol{F}, \boldsymbol{h}) = 0, \tag{5.45}$$

and let $\boldsymbol{U} = \mathfrak{M}^{-1}\boldsymbol{F}$ be a solution of the equation (5.45), satisfying the orthogonality condition

$$(\boldsymbol{U}, \boldsymbol{h}) = 0. \tag{5.46}$$

Denote two eigenvalues of the matrix $\frac{1}{\rho}\boldsymbol{\Gamma}(\mathbf{s})$, distinct from $c^2$, (i.e., the squared related phase velocities) by $c_{\mathrm{I}}^2$ and $c_{\mathrm{II}}^2$ and the corresponding eigenvectors by $\boldsymbol{h}^{\mathrm{I}}$ and $\boldsymbol{h}^{\mathrm{II}}$:

$$\boldsymbol{\Gamma}(\mathbf{s})\boldsymbol{h}^{\mathrm{I}} = \rho c_{\mathrm{I}}^2 \boldsymbol{h}^{\mathrm{I}}, \quad \boldsymbol{\Gamma}(\mathbf{s})\boldsymbol{h}^{\mathrm{II}} = \rho c_{\mathrm{II}}^2 \boldsymbol{h}^{\mathrm{II}}. \tag{5.47}$$

The vectors $\boldsymbol{h}^{\mathrm{I}}$ and $\boldsymbol{h}^{\mathrm{II}}$ are taken real and unit, $(\boldsymbol{h}^{\mathrm{I}}, \boldsymbol{h}^{\mathrm{I}}) = (\boldsymbol{h}^{\mathrm{II}}, \boldsymbol{h}^{\mathrm{II}}) = 1$. In the case where $c_{\mathrm{I}}^2 \neq c_{\mathrm{II}}^2$, $\boldsymbol{h}^{\mathrm{I}}$ and $\boldsymbol{h}^{\mathrm{II}}$ are orthogonal, and in the case where $c_{\mathrm{I}}^2 = c_{\mathrm{II}}^2$, they can be chosen orthogonal, and we assume them to be orthogonal. The condition (5.45) implies the relation

$$\boldsymbol{U} = \gamma_{\mathrm{I}} \boldsymbol{h}^{\mathrm{I}} + \gamma_{\mathrm{II}} \boldsymbol{h}^{\mathrm{II}} \tag{5.48}$$

with certain constants $\gamma_{\mathrm{I,II}} = \gamma_{\mathrm{I,II}}(\boldsymbol{F})$. Substituting the expression (5.48) into the left-hand side of the equation (5.45) and using (5.47), we obtain

$$\mathfrak{M}\boldsymbol{U} = \rho \left\{ \gamma_{\mathrm{I}} \left(c_{\mathrm{I}}^2 - c^2\right) \boldsymbol{h}^{\mathrm{I}} + \gamma_{\mathrm{II}} \left(c_{\mathrm{II}}^2 - c^2\right) \boldsymbol{h}^{\mathrm{II}} \right\}.$$

Since $\boldsymbol{F} = (\boldsymbol{F}, \boldsymbol{h}^{\mathrm{I}})\boldsymbol{h}^{\mathrm{I}} + (\boldsymbol{F}, \boldsymbol{h}^{\mathrm{II}})\boldsymbol{h}^{\mathrm{II}}$, we have $\gamma_{\mathrm{I,II}} = \frac{(\boldsymbol{F},\boldsymbol{h}^{\mathrm{I,II}})}{\rho(c^2_{\mathrm{I,II}}-c^2)}$, and

$$\mathfrak{N}^{-1}\boldsymbol{F} = \frac{1}{\rho}\left\{\frac{(\boldsymbol{F}, \boldsymbol{h}^{\mathrm{I}})\boldsymbol{h}^{\mathrm{I}}}{c^2_{\mathrm{I}} - c^2} + \frac{(\boldsymbol{F}, \boldsymbol{h}^{\mathrm{I}})\boldsymbol{h}^{\mathrm{II}}}{c^2_{\mathrm{II}} - c^2}\right\}, \tag{5.49}$$

so that $\boldsymbol{u}^{1\perp} = \mathfrak{N}^{-1}\mathfrak{M}\boldsymbol{u}^0$ is found uniquely and given by the expression (5.49) with $\boldsymbol{F} = \mathfrak{M}\boldsymbol{u}^0$. As in the isotropic case, the term $\boldsymbol{u}^{P1\perp}$ is called the additional, or anomalously polarized, component.

### "Normal" polarization

We call polarization corresponding to the local plane wave with the amplitude vector $\boldsymbol{h}$ the "normal" polarization, in contrast to the anomalous one. Similarly to the isotropic case, we put

$$\boldsymbol{u}^{1\|} = \phi^1\boldsymbol{h}. \tag{5.50}$$

Now $\phi^1$ can be found from the consistency condition for the equation for $\boldsymbol{u}^2$. Recall that all quantities are assumed to be real. Equating to zero the scalar product of the right-hand side of the relation (4.22) and $\boldsymbol{h}$, we write the consistency condition for the equation (4.22) in the form

$$(\mathfrak{M}\boldsymbol{u}^1, \boldsymbol{h}) - (\mathfrak{L}\boldsymbol{u}^0, \boldsymbol{h}) = 0.$$

With (5.44), we arrive at the equation

$$(\mathfrak{M}\boldsymbol{u}^{1\|}, \boldsymbol{h}) = (\mathfrak{L}\boldsymbol{u}^0 - \mathfrak{M}\boldsymbol{u}^{1\perp}, \boldsymbol{h}). \tag{5.51}$$

Rearrangement of the left-hand side of the equation (5.51), with (4.78) and (5.50) taken into account, gives:

$$\begin{aligned}(\mathfrak{M}\boldsymbol{u}^{1\|}, \boldsymbol{h}) &= \frac{1}{\phi^1}(\mathfrak{M}\boldsymbol{u}^{1\|}, \boldsymbol{u}^1) = \frac{1}{\phi^1}\operatorname{div}\left(\rho(\boldsymbol{u}^{1\|})^2\boldsymbol{v}^{gr}\right)\\ &= \frac{1}{\phi^1}\frac{1}{\mathscr{D}}\frac{\partial(\rho\mathscr{D}(\boldsymbol{u}^{1\|})^2)}{\partial\tau} = \frac{1}{\phi^1}\frac{1}{\mathscr{D}}\frac{\partial(\rho\mathscr{D}(\phi^1)^2\boldsymbol{h}^2)}{\partial\tau}.\end{aligned}$$

The relation (5.51) takes the form

$$\frac{1}{\phi^1}\frac{1}{\mathscr{D}}\frac{\partial}{\partial\tau}\left\{\rho\mathscr{D}(\boldsymbol{u}^{1\|})^2\right\} = \mathfrak{F}, \quad \mathfrak{F} := (\mathfrak{L}\boldsymbol{u}^0 - \mathfrak{M}\boldsymbol{u}^{1\perp}, \boldsymbol{h}). \tag{5.52}$$

This is a linear ordinary differential equation along the ray

$$\frac{1}{\phi^1}\frac{1}{\mathscr{D}}\frac{\partial\left(\rho\mathscr{D}(\phi^1)^2\boldsymbol{h}^2\right)}{\partial\tau} = 2\rho\boldsymbol{h}^2\frac{\partial\phi^1}{\partial\tau} + \phi^1\frac{1}{\mathscr{D}}\frac{\partial(\rho\mathscr{D}\boldsymbol{h}^2)}{\partial\tau} = \mathfrak{F},$$

from which, with little difficulty, we obtain

$$\phi^1 = \frac{1}{\sqrt{\rho\boldsymbol{h}^2\mathscr{D}}}\left\{\psi^1 + \int^{\tau}\frac{\mathscr{D}\left(\mathfrak{L}\boldsymbol{u}^0 - \mathfrak{M}\boldsymbol{u}^{1\perp}, \boldsymbol{h}\right)}{\sqrt{\rho\boldsymbol{h}^2\mathscr{D}}}d\tau\right\}, \tag{5.53}$$

where $\psi^1 = \psi^1(\eta_1, \eta_2)$ is the diffraction coefficient.

**Higher-order terms**

For next approximations

$$\boldsymbol{u}^j = \boldsymbol{u}^{j\|} + \boldsymbol{u}^{j\perp}, \quad \boldsymbol{u}^{j\|} = \phi^j \boldsymbol{h}, \quad (\boldsymbol{u}^{j\perp}, \boldsymbol{h}) = 0, \quad j = 2, 3, \ldots, \tag{5.54}$$

we likewise get

$$\boldsymbol{u}^{j\perp} = \mathfrak{N}^{-1}\left[\mathfrak{M}\boldsymbol{u}^{j-1} - \mathfrak{L}\boldsymbol{u}^{j-2}\right], \tag{5.55}$$

$$\phi^j = \frac{1}{\sqrt{\rho h^2 \mathscr{D}}}\left\{\psi^j + \int^{\tau} \frac{\mathscr{D}\left(\mathfrak{L}\boldsymbol{u}^{j-1} - \mathfrak{M}\boldsymbol{u}^{j\perp}, \boldsymbol{h}\right)}{\sqrt{\rho h^2 \mathscr{D}}} d\tau\right\}. \tag{5.56}$$

Here, $\psi^j = \psi^j(\eta_1, \eta_2)$ is the corresponding diffraction coefficient.

In conclusion, we note that the formulas in Chapter 4 for the wave $P$ are specializations of those found in the current section, while the formulas for the wave $S$ are not. The reason is that we essentially used here the nondegeneracy of the eigenvalue $H^2$, which holds in the first case and breaks down in the second.

---

## 5.6 ⋆ Comments to Chapter 5

The ray method for anisotropic inhomogeneous media was originally developed by Babich 1961 [2] (for English translation see Babich 1994 [4]). Babich succeeded in integration of the corresponding transport equations via development of the techniques used by Hadamard 1923 [11] in the case of a second-order scalar equation. Unsuccessful attempts to generalize Hadamard's theory to hyperbolic equations of order larger than 2 (see Théodoresco 1938, 1940 [15, 16]) were not completely fruitless: an important relationship between the theory of characteristics for such equations and Finsler geometry was observed.

The theory of characteristics is described in detail, e.g., in Courant and Hilbert 1962 [7] and Smirnov 1964 [14]. A very detailed exposition of the theory of characteristics with respect to propagation of waves was presented in Babich, Buldyrev, and Molotkov 1985 [5]. A space-time version of the ray method for the anisotropic case was given by Babich 1979 [3]. The above description of the relationship between rays arising in the theory characteristics and the Fermat principle given in Section 5.3.3 is based on the paper by Bóna and Slawinski 2003 [6]. Simple exact solutions for a general anisotropic homogeneous medium, which have nonzero additional components, were presented by Kiselev 2008 [12].

## References to Chapter 5

[1] Arnold, V. I. 1978. *Mathematical methods of classical mechanics.* New York: Springer. *Арнольд В. И.* Математические методы классической механики. М.: Едиториал УРСС, 2003.

[2] Babich, V. M. 1961. Ray method of calculation of intensity of wave fronts in the case of anisotropic medium, in *Voprosy Dynamicheskoi Teorii Rasprostraneniya Seismicheskikh Voln [Problems in the Dynamic Theory of Propagation of Seismic Waves]*, Vol. 5, ed. G. I. Petrashen'. Leningrad: Leningrad Univ. Press, Leningrad, 36–46 [in Russian]. *Бабич В. М.* Лучевой метод вычисления интенсивности волновых фронтов в случае неоднородной анизотропной среды. Вопр. динам. теории распростр. сейсм. волн, 1961. Т. 5. С. 36–46.

[3] Babich, V. M. 1979. The space–time ray method in elastic wave theory. *Izv. Acad. Sci. USSR. Phys. Solid Earth.* 15(3):87–93. *Бабич В. М.* Пространственно-временной лучевой метод в теории упругих волн. Изв. АН СССР. Физика Земли, 1979. № 3. С. 3–13.

[4] Babich, V. M. 1994. Ray method of calculating the intensity of wavefronts in the case of a heterogeneous, anisotropic, elastic medium. *Geophys. J. Int.* 118:379–83.

[5] Babich, V. M., Buldyrev, V. S. and Molotkov, I. A. 1985. *Space–time ray method. Linear and nonlinear waves.* Leningrad: Leningrad University Press [in Russian]. *Бабич В. М., Булдырев В. С., Молотков И. А.* Пространственно-временной лучевой метод. Линейные и нелинейные волны. Л.: Изд-во ЛГУ, 1985.

[6] Bóna, A. and Slawinski, M. A. 2003. Fermat's principle for seismic rays in elastic media. *J. Appl. Geophys.* 54:445–51.

[7] Courant, R. and Hilbert, D. 1962. *Methods of mathematical physics. Vol. 2.* New York: Interscience. *Курант Р.* Уравнения с частными производными. М.: Наука, 1964.

[8] Dubrovin, B. A, Fomenko, A. T. and Novikov, S. P. 1984. *Modern geometry – methods and applications. Part I. The geometry of surfaces, transformation groups, and fields.* New York: Springer-Verlag. *Дубровин Б. А., Новиков С. П., Фоменко А. Т.* Современная геометрия. Методы и приложения. Т. 1. М.: Наука, 1998.

[9] Gelfand, I. M. and Fomin, S. V. 1963. *Calculus of variations.* New York: Prentice-Hall, Englewood Cliffs. *Гельфанд И. М., Фомин С. В.* Вариационное исчисление. М.: Физматгиз, 1961.

[10] Goldstein, H. 1950. *Classical mechanics.* Mass.: Addison-Wesley. *Голдстейн Г.* Классическая механика. М.: Гостехиздат, 1957.

[11] Hadamard, J. 1923. *Lectures on Cauchy's problem in linear partial differential equations.* New Haven: Yale University Press. *Адамар Ж.* Задача Коши для уравнений в частных производных гиперболического типа. М.: Наука, 1978.

[12] Kiselev, A. P. 2008. Plane waves with a transverse structure in arbitrarily anisotropic elastic medium. *Doklady Physics.* 53:48–50. *Киселев А. П.* Плоские волны с поперечной структурой в произвольно анизотропной упругой среде. Доклады Академии Наук, 2008. Т. 418(3). С. 336–338.

[13] Rund, H. 1959. *The differential geometry of Finsler spaces.* Berlin: Springer. *Рунд Х.* Дифференциальная геометрия финслеровых пространств. М.: Наука, 1981.

[14] Smirnov, V. I. 1964. *A course of higher mathematics, Volume 4.* Reading MA: Addison–Wesley. *Смирнов В. И.* Курс высшей математики. Т. 4. Ч. 2. М.: Наука, 1981.

[15] Théodoresco, N. 1938. Recherches sur équationes aux dérivées partielles, linéares, d'ordre quelquenque. Les solutions élémentaires. *Ann. Sci. Univ. Jassy, Sect. I, Math.* 24:263–321.

[16] Théodoresco, N. 1940. Géométrie Finslerienne et propagation des ondes. *Bull. Sect. Sci. Acad. Roum.* 23:138–44.

[17] Umov, N. A. 1950. *Selected papers.* Moscow: GTTI [in Russian]. *Умов Н. А.* Избранные сочинения. М.: ГТТИ, 1950.

# Chapter 6

## Point Sources in Inhomogeneous Isotropic Media. Wave $S$ from a Center of Expansion. Wave $P$ from a Center of Rotation

The ray method allows a high-frequency asymptotic description of a wavefield, provided that the diffraction coefficients that define the amplitude distribution along the wavefront are known. The problem of finding diffraction coefficients is beyond the scope of the ray method.

Consider the wavefield generated by a high-frequency point source placed at a point $\mathbf{x}^\circ$ of a slowly-varying inhomogeneous medium. In the current chapter, the medium is assumed to be isotropic. Outside a small neighborhood of the source point $\mathbf{x}^\circ$, the wavefield is described by a sum of two ray expansions corresponding to central ray fields centered at $\mathbf{x}^\circ$.[1] How can one find the diffraction coefficients? The leading one for each of the waves can easily be found via an elementary *locality approach*, which is based on the replacement of the inhomogeneous medium by the related homogeneous one in the vicinity of the source point $\mathbf{x}^\circ$. Such an approach is applicable to any point source. For the wave $S$ from a center of expansion and the wave $P$ from a center of rotation, the leading diffraction coefficients prove to be zeros, because in a homogeneous medium such waves are not generated. Finding the amplitudes of these waves requires finding the diffraction coefficients of next order. This is a far more challenging task.

In the current chapter, we develop a method for finding diffraction coefficients for the waves $P$ and $S$ generated in an inhomogeneous medium by an arbitrary time-harmonic point source. We find in explicit form the leading nonzero diffraction coefficients for the wave $S$ from a center of expansion and for the wave $P$ from a center of rotation — the waves that do not arise in homogeneous media. For finding higher-order diffraction coefficients, near the source point $\mathbf{x}^\circ$ we use the boundary layer method based on the introduction of stretched coordinates, the construction of an expression for the wavefield in a small neighbourhood of $\mathbf{x}^\circ$, and matching the asymptotic expansions in an intermediate zone. This is not an easy task.

[1] We are not concerned with the description of the wavefield near caustics that may arise, in general, at some distance from $\mathbf{x}^\circ$.

## 6.1 Statement of the problem and elementary consideration

### 6.1.1 Statement of the problem

A time-harmonic point source $\boldsymbol{F}(\mathbf{x})$ is acting at a point $\mathbf{x}^{\circ}$ within an inhomogeneous isotropic elastic medium with $\lambda = \lambda(\mathbf{x})$, $\mu = \mu(\mathbf{x})$, and $\rho = \rho(\mathbf{x})$ being smooth functions. The wavefield is described by the equation (1.82), where the operator $\boldsymbol{l}$ is given by the expressions (1.83), (1.60), and the frequency $\omega$ is a large parameter as was assumed throughout Chapters 4 and 5. The unique solution is determined by the limiting absorption principle.

We expect that outside a small (and diminishing as $\omega \to \infty$) vicinity of the source point $\mathbf{x}^{\circ}$, the wavefield has a ray structure, i.e., it is described by the sum of ray expansions of the waves $P$ and $S$:

$$\boldsymbol{u} = \boldsymbol{u}^a + \boldsymbol{u}^b, \quad \omega \to \infty, \tag{6.1}$$

$$\begin{aligned} \boldsymbol{u}^a &= (-i\omega)^{\xi} e^{i\omega\tau^a} \left\{ \boldsymbol{u}^{a0} + \frac{\boldsymbol{u}^{a1}}{-i\omega} + \ldots \right\}, \\ \boldsymbol{u}^b &= (-i\omega)^{\xi} e^{i\omega\tau^b} \left\{ \boldsymbol{u}^{b0} + \frac{\boldsymbol{u}^{b1}}{-i\omega} + \ldots \right\}. \end{aligned} \tag{6.2}$$

Each of the expansions (6.2) is related to the corresponding velocity of propagation, and both fields of rays are central with center $\mathbf{x}^{\circ}$. In order to deal with the waves issuing out of the source point (and without incoming to it), we impose a condition that in some finite neighborhood of the source point, the vectors $\operatorname{grad} \tau^a$ and $\operatorname{grad} \tau^b$ are directed from $\mathbf{x}^{\circ}$. The constants $\xi$ in expansions (6.2) are chosen equal.

Also, we require that at least one of the expansions (6.2) start with a nonzero amplitude with number 0, that is,

$$|\boldsymbol{u}^{a0}| + |\boldsymbol{u}^{b0}| \not\equiv 0. \tag{6.3}$$

The problem is to find (for a given source $\boldsymbol{F}$ in (3.47)):

a) the constant $\xi$;

b) the diffraction coefficients for the wave $P$, which are smooth scalar functions $\psi^0$, $\psi^1$, ... of two variables defined on the unit sphere;

c) the diffraction coefficients for the wave $S$, which are smooth vector functions $\boldsymbol{\chi}^0$, $\boldsymbol{\chi}^1$, ... also defined on the unit sphere.

Of particular interest in application are the leading nonzero terms for each of the waves.

### 6.1.2 Non-applicability of ray formulas near the source point

As we know, ray expressions are perturbed plane waves. They were derived under the assumption that the wavefronts can be locally approximated by planes. Approaching the source point, the curvatures of the fronts unrestrictedly grow, and the ray theory fails.

We are going to estimate the distance $r$ at which the ray series cease to have an asymptotic nature. Near the source point, the rays are close to straight half-lines drawn from $\mathbf{x}^{\mathrm{o}}$; their geometrical spreadings are of order $O(r^2)$, $r = |\mathbf{x} - \mathbf{x}^{\mathrm{o}}|$, and if $\boldsymbol{u}^0 \not\equiv 0$, then $|\boldsymbol{u}^0| = O(\frac{1}{r})$. Next, as is clear from the formulas of Chapter 4, for each of the waves $P$ and $S$, the vector $\boldsymbol{u}^1$ involves the terms resulting from single differentiation of the vector $\boldsymbol{u}^0$ (as well as those obtained by double differentiation and subsequent integration with respect to $\tau \approx cr$). Therefore, $|\boldsymbol{u}^1| = O\left(\frac{1}{r^2}\right)$. Proceeding in a similar manner, we derive that $|\boldsymbol{u}^j| = O\left(\frac{1}{r^{1+j}}\right)$ and

$$\left|\frac{\boldsymbol{u}^m}{(-i\omega)^j}\right| = O\left(\frac{1}{r(\omega r)^j}\right), \quad r \to 0.$$

We have thus established that the ray series have an asymptotic nature under the condition[2]

$$r \gg \omega^{-1}. \tag{6.4}$$

The condition (6.4) formally coincides with that of the far-zone for a homogeneous medium. In the case of a homogeneous medium, the radiation pattern of a spherical wave can thus be identified with the corresponding leading-order diffraction coefficient.

### 6.1.3 Elementary locality approach

The zero-order terms of the asymptotic (6.1) can be found by means of a simple procedure, which we are going to describe.

It is obvious that in a very small vicinity of the source point, the wavefield of the point source is close to that of the same source in a medium with "frozen" constant coefficients. For the eikonals we have (see (4.90))

$$\tau^a(\mathbf{x}) \approx \frac{r}{a(\mathbf{x}^{\mathrm{o}})}, \quad \tau^b(\mathbf{x}) \approx \frac{r}{b(\mathbf{x}^{\mathrm{o}})}, \quad r = |\mathbf{x} - \mathbf{x}^{\mathrm{o}}|. \tag{6.5}$$

Accordingly, for a unit vector tangent to the ray we have

$$\frac{\operatorname{grad}\tau}{|\operatorname{grad}\tau|} \approx \mathbf{s} := \operatorname{grad} r. \tag{6.6}$$

[2] In this chapter, we are still not concerned with the expansion of wavefields in powers of a dimensionless parameter, though this can be done at the cost of a certain complication of formulas.

Diffraction coefficients are functions of coordinates parameterizing the rays. In the case under consideration, rays can by parameterized by unit vectors $\mathbf{s}$ tangent to them at the source point. As was mentioned in Section 4.3.3, by choosing coordinates on the sphere in a special way (see (4.94)), we get $J = r^2$. As a consequence of this assertion, the leading term as $\omega \to \infty$ takes the form

$$\boldsymbol{u} \approx (-i\omega)^{\xi} \left\{ \frac{e^{i\omega\alpha r}}{r} \frac{1}{\sqrt{a\rho}}\bigg|_{\mathbf{x}=\mathbf{x}^{\mathrm{o}}} \psi^0(\mathbf{s})\mathbf{s} + \frac{e^{i\omega\beta r}}{r} \frac{1}{\sqrt{\rho b}}\bigg|_{\mathbf{x}=\mathbf{x}^{\mathrm{o}}} \boldsymbol{\chi}^0(\mathbf{s}) \right\}, \tag{6.7}$$

with the notation

$$\alpha := \frac{1}{a(\mathbf{x}^{\mathrm{o}})}, \quad \beta := \frac{1}{b(\mathbf{x}^{\mathrm{o}})}. \tag{6.8}$$

Now consider the specific sources that we have met in Chapter 3.

**Concentrated force (3.72)**

Compare (6.7) with the solution for a homogeneous medium. The condition $\omega r \to \infty$, which is required for the applicability of the ray theory, coincides with the condition of the far-field. Near the point $\mathbf{x}^{\mathrm{o}}$ (but under the assumption that $\omega r \to \infty$), we replace the exact solution by its far-field asymptotics (3.87) and, making use of (6.5), we obtain

$$\boldsymbol{u} \approx \frac{e^{i\omega\alpha r}}{r} \frac{\mathbf{s}}{\rho a^2} + \frac{e^{i\omega\beta r}}{r} \frac{\boldsymbol{m} - (\boldsymbol{m}, \mathbf{s})\mathbf{s}}{\rho b^2}. \tag{6.9}$$

Compare the coefficients of rapidly oscillating exponents $e^{i\omega\alpha r}$ and $e^{i\omega\beta r}$ in (6.7) and (6.9) separately. We get

$$\xi = 0, \tag{6.10}$$

$$\psi^0(\mathbf{s}) = \frac{(\boldsymbol{m}, \mathbf{s})}{\sqrt{\rho(\mathbf{x}^{\mathrm{o}}) a^3(\mathbf{x}^{\mathrm{o}})}}, \quad \boldsymbol{\chi}^0(\mathbf{s}) = \frac{\boldsymbol{m} - (\boldsymbol{m}, \mathbf{s})\mathbf{s}}{\sqrt{\rho(\mathbf{x}^{\mathrm{o}}) b^3(\mathbf{x}^{\mathrm{o}})}}. \tag{6.11}$$

**Center of expansion (3.49)**

In this case, (3.84) implies that no wavefield describing the wave $S$ is present in the leading term, and

$$\boldsymbol{u} \approx i\omega \frac{e^{i\omega\alpha r}}{r} \frac{\mathbf{s}}{\rho a^2(\mathbf{x}^{\mathrm{o}})}. \tag{6.12}$$

Comparison with (6.7) provides

$$\xi = 1, \tag{6.13}$$

$$\psi^0 = -\frac{1}{\sqrt{\rho(\mathbf{x}^{\mathrm{o}}) a^5(\mathbf{x}^{\mathrm{o}})}}, \quad \boldsymbol{\chi}^0 = 0. \tag{6.14}$$

**Center of rotation (3.54)**

Dealing with a center of rotation (3.87) along this pathway, we obtain the same constant $\xi = 1$ (6.13), and

$$\psi^0(\mathbf{s}) = 0, \quad \boldsymbol{\chi}^0(\mathbf{s}) = -\frac{[\mathbf{s} \times \boldsymbol{m}]}{\sqrt{\rho(\mathbf{x}^\circ)b^5(\mathbf{x}^\circ)}}. \tag{6.15}$$

**Agenda**

In what follows we find an explicit expression for the leading nonzero diffraction coefficient $\boldsymbol{\chi}^1$ of the wave $S$ in the case of a center of expansion, which is of interest in explosive seismic exploration. Similarly, we find $\psi^1$ for a center of rotation. It is natural to expect that these functions are linear with respect to the gradients of $\lambda$, $\mu$, and $\rho$ (or, equivalently, of $a$, $b$, and $\rho$) at the source point. The calculation is not very difficult in its essence, but is a bit long.

## 6.2 Structure of the wavefield near the source point in more detail

We proceed with the construction of an asymptotic representation of the wavefield near the point $\mathbf{x}^\circ$, considering inhomogeneity of the medium as a perturbation. The choice of a characteristic scale is prompted by the size of the domain where the ray formulas fail.

### 6.2.1 Recurrent system

We introduce the *stretched coordinates*

$$\mathbf{X} = \omega(\mathbf{x} - \mathbf{x}^\circ), \tag{6.16}$$

which have the order of $O(1)$ as $|\mathbf{x} - \mathbf{x}^\circ| \sim \frac{1}{\omega}$ and $\omega \to \infty$.[3] Expand the Lamé coefficients and the volume density in powers of $\mathbf{X}$:

$$\left.\begin{aligned}
\lambda(\mathbf{x}) &= \lambda^0 + \lambda^1(\mathbf{x}) + \ldots = \lambda^0 + \frac{\lambda^1(\mathbf{X})}{\omega} + \ldots, \\
\mu(\mathbf{x}) &= \mu^0 + \mu^1(\mathbf{x}) + \ldots = \mu^0 + \frac{\mu^1(\mathbf{X})}{\omega} + \ldots, \\
\rho(\mathbf{x}) &= \rho^0 + \rho^1(\mathbf{x}) + \ldots = \rho^0 + \frac{\rho^1(\mathbf{X})}{\omega} + \ldots
\end{aligned}\right\} \tag{6.17}$$

[3] At this stage, one may pass to dimensionless coordinates by defining the *characteristic scale* of variation of properties of the medium near $\mathbf{x}^\circ$ as $\mathscr{M} = \left(\max\left(\frac{|\nabla a|}{a}, \frac{|\nabla b|}{b}, \frac{|\nabla \rho|}{\rho}\right)\right)^{-1}$ and putting $\mathbf{X} = \omega(\mathbf{x} - \mathbf{x}^\circ)/\mathscr{M} b(\mathbf{x}^\circ)$. Needless to say that resulting expressions for diffraction coefficients will be the same while intermediate formulas complicate.

Here, $\lambda^m$, $\mu^m$, and $\rho^m$ are homogeneous polynomials of degree $m$ with respect to the components $\mathrm{X}_1$, $\mathrm{X}_2$, $\mathrm{X}_3$ of the vector $\mathbf{X}$. In particular,

$$\lambda^1(\mathbf{X}) = (\nabla_{\mathbf{x}}\lambda(\mathbf{x})|_{\mathbf{x}=\mathbf{x}^\circ}, \mathbf{X}), \quad \mu^1(\mathbf{X}) = (\nabla_{\mathbf{x}}\mu(\mathbf{x})|_{\mathbf{x}=\mathbf{x}^\circ}, \mathbf{X}), \\ \rho^1(\mathbf{X}) = (\nabla_{\mathbf{x}}\rho(\mathbf{x})|_{\mathbf{x}=\mathbf{x}^\circ}, \mathbf{X}). \tag{6.18}$$

Expansion of the left-hand side of the equation (1.82), (1.83) in powers of $\omega$ for the case of (1.60) yields

$$\boldsymbol{l}\boldsymbol{u} = \omega^2 \left\{ \text{Л}\boldsymbol{u} + \frac{\text{Л}^1\boldsymbol{u}}{\omega} + \frac{\text{Л}^2\boldsymbol{u}}{\omega^2} + \dots \right\}, \tag{6.19}$$

where Л is the operator with constant coefficients,

$$\text{Л} = (\lambda^\circ + 2\mu^\circ)\,\mathrm{grad}_{\mathbf{X}}\,\mathrm{div}_{\mathbf{X}} - \mu^\circ\,\mathrm{rot}_{\mathbf{X}}\,\mathrm{rot}_{\mathbf{X}} + \rho^\circ\mathbf{I}, \tag{6.20}$$

corresponding to the homogeneous medium with "frozen" parameters

$$\lambda^\circ = \lambda(\mathbf{x}^\circ), \quad \mu^\circ = \mu(\mathbf{x}^\circ), \quad \rho^\circ = \rho(\mathbf{x}^\circ). \tag{6.21}$$

Here $\mathbf{I}$ is the $3 \times 3$ identity matrix, and $\text{Л}^1$, $\text{Л}^2$, ... are not the powers of Л but certain operators with polynomial coefficients, which take into account the inhomogeneity of the medium near the source point,

$$(\text{Л}^m\boldsymbol{u})_p = \mathrm{D}_j\left\{\mu^m(\mathbf{X})[\mathrm{D}_p u_j + \mathrm{D}_j u_p]\right\} + \mathrm{D}_p\left\{\lambda^m(\mathbf{X})\,\mathrm{div}_{\mathbf{X}}(\boldsymbol{u})\right\} \\ + \rho^m(\mathbf{X})u_p, \quad m = 1, 2, \dots, \tag{6.22}$$

$$\mathrm{D}_l := \frac{\partial}{\partial X_l}, \quad l = 1, 2, 3. \tag{6.23}$$

As usual, the summation over the repeated lower indices is assumed.

Seeking the desired wavefield in the form of a *local expansion*

$$\mathbf{V} := \omega^\sigma \left\{ \mathbf{V}^0(\mathbf{X}) + \frac{\mathbf{V}^1(\mathbf{X})}{\omega} + \frac{\mathbf{V}^2(\mathbf{X})}{\omega^2} + \dots \right\}, \quad \sigma = \text{const}, \tag{6.24}$$

we arrive at

$$\text{Л}\mathbf{V}^0 = -\boldsymbol{F}(\mathbf{X}), \tag{6.25}$$

$$\text{Л}\mathbf{V}^1 = -\text{Л}^1\mathbf{V}^0, \tag{6.26}$$

$$\text{Л}\mathbf{V}^m = -\sum_{l=1}^{m} \text{Л}^l\mathbf{V}^{m-l}, \quad m \geqslant 1. \tag{6.27}$$

Unlike the ray series, no rapidly oscillating factor is separated out here, because at a distance from the source point comparable to the wavelength, the wavefield has not yet managed to get a ray structure.

Each of the equations (6.25)–(6.27) is a nonhomogeneous equation in the whole space for a system with constant coefficients. For uniqueness, we require that the limiting absorption principle hold. It means that the solutions of nonhomogeneous problems for the differential operator

$$\mathbf{J\!I} = \rho^{\circ}\left\{\frac{1}{\alpha^2}\operatorname{grad}_{\mathbf{X}}\operatorname{div}_{\mathbf{X}} - \frac{1}{\beta^2}\operatorname{rot}_{\mathbf{X}}\operatorname{rot}_{\mathbf{X}} + \mathbf{I}\right\} \tag{6.28}$$

tend to zero in a proper manner (see Section 3.5.1) as $\mathbf{X} \to \infty$ if $\operatorname{Im}\alpha > 0$, $\operatorname{Im}\beta > 0$. This will be discussed in Section 6.2.2 in more detail. The existence of the solutions will be established via their direct explicit construction.

Since inhomogeneity of the medium is regarded as a perturbation, we have a reason to expect that "in small," the wavefield will show similarity to (6.7), that is, it will be approximately a sum of two spherical waves.

### 6.2.2 ⋆ On solving equations (6.25)–(6.27)

The right-hand side of equation (6.25) is a generalized vector function localized at $\mathbf{x}^{\circ}$, and the vectors $\mathbf{V}^m$ should also be regarded as generalized vector functions. The solutions of the elliptic equations (the equation $\mathbf{J\!I}\mathbf{V} = \mathbf{\Phi}$ relates to this type) are smooth where their right-hand sides are smooth. We are not interested in all solutions of the equations (6.25)–(6.27), but solely in those which can be matched with the ray expansions in a certain intermediate zone surrounding the source point (which will be discussed in Section 6.2.3). Such are the vector functions $\mathbf{V}^m$ having the asymptotics of the form

$$\frac{\mathbf{Q}^a(\mathrm{R},\mathbf{s})}{\mathrm{R}}e^{i\alpha\mathrm{R}} + \frac{\mathbf{Q}^b(\mathrm{R},\mathbf{s})}{\mathrm{R}}e^{i\beta\mathrm{R}} + o\left(\frac{1}{\mathrm{R}}\right), \quad \mathrm{R} \to \infty, \tag{6.29}$$

where $\mathbf{Q}^{a,b}(\mathrm{R},\mathbf{s})$ stand for polynomials in $\mathrm{R} = |\mathbf{X}|$ with the coefficients smoothly dependent on $\mathbf{s} = \frac{\mathbf{X}}{|\mathbf{X}|}$. These solutions can be constructed as a result of a simple but long calculation, for the idea of which see Section 6.4. The construction gives explicit expressions for $\mathbf{V}^m$ in the form

$$\mathbf{V}^m = e^{i\alpha\mathrm{R}}\mathbf{W}^{am} + e^{i\beta\mathrm{R}}\mathbf{W}^{bm}, \tag{6.30}$$

where $\mathbf{W}^{am}$ and $\mathbf{W}^{bm}$ are polynomials in $\mathrm{X}_1, \mathrm{X}_2, \mathrm{X}_3, \mathrm{R}$, and $\frac{1}{\mathrm{R}}$. Obviously, the functions (6.30) are smooth everywhere, except for the origin of coordinates. The orders of their singularities at $\mathbf{X} = 0$ decrease with increasing $m$. For the case of, say, a concentrated force, $\mathbf{V}^m$ are continuous as $m > 1$, $\mathbf{V}^m \in C^{m-1}$. It can also be shown that $\mathbf{V}^m = O(\mathrm{R}^{2m-1})$ as $\mathrm{R} \to \infty$ (see Kiselev 1975 [12]).

The solutions of the form (6.30) satisfy the limiting absorption principle. The introduction of small attenuation requires a certain comment, because the operator $\mathbf{J\!I}$, unlike elliptic ones considered in Chapter 3, explicitly involves neither $k$ nor $\omega$. The attenuation can be introduced under the assumption that $\rho^{\circ}$ has a small positive imaginary part, and, consequently, so do $\alpha = \sqrt{\rho^{\circ}}/\sqrt{\lambda^{\circ} + 2\mu^{\circ}}$ and $\beta = \sqrt{\rho^{\circ}}/\sqrt{\mu^{\circ}}$.

Now, owing to the factors $e^{i\alpha\mathrm{R}}$ and $e^{i\beta\mathrm{R}}$, the solution (6.30) exponentially decays as $\mathrm{R} \to \infty$, together with its first derivatives. By the theorem in

Section 3.5.1, such a solution is unique. Turning $\operatorname{Im}\rho^{\circ}$ to zero, we arrive at solutions having asymptotic behavior (6.29) as $\mathrm{R} \to \infty$.

### 6.2.3 Intermediate zone

Describing the elementary locality approach in Section 6.1.3, we just required that inequality (6.4) hold and did not describe in detail the area where the ray expansion was matched with spherical waves. Now we deduce an extra condition, which was implicitly used in Section 6.1.3. It was the condition which enables us to approximately replace $e^{i\omega\tau^{a,b}}$ by $e^{i\omega\alpha r}$ and $e^{i\omega\beta r}$, respectively. The local solution is a sum of two exponents, each multiplied by an algebraic function, and the unique solution needed for matching is determined by the limiting absorption principle.

The formula (6.5) can be refined. As is well known (see, e.g., Babič and Kirpičnikova 1979 [7]), the eikonal corresponding to the central field of rays in inhomogeneous media with smooth velocity $a$ can be expanded near the center $\mathbf{x}^{\circ}$ as follows

$$\tau^{a}(\mathbf{x}) = \alpha r + r^2 \widetilde{\tau(\mathbf{x})} + O(r^3), \quad \widetilde{\tau(\mathbf{x})} = O(1), \quad r \to 0, \tag{6.31}$$

where $\alpha = 1/a(\mathbf{x}^{\circ})$, and $\widetilde{\tau(\mathbf{x})}$ is a smooth function of $\mathbf{s} = \frac{\mathbf{x}-\mathbf{x}^{\circ}}{|\mathbf{x}-\mathbf{x}^{\circ}|}$. Thus,

$$e^{i\omega\tau^{a}} = e^{i\omega\alpha r} e^{i\omega O(r^2)} = e^{i\omega\alpha r}\left(1 + O(\omega r^2)\right). \tag{6.32}$$

An analogous formula is true for velocity $b$. Thus, for matching the ray and local expansions, the condition

$$\omega^{-1} \ll r \ll \omega^{-\frac{1}{2}} \tag{6.33}$$

must be satisfied. A spherical annulus around the source point, described by (6.33), will be called the *intermediate zone.* It can be shown that the condition $r \ll \omega^{-\frac{1}{2}}$ guarantees that the local expansion has an asymptotic nature (see Kiselev 1975 [12]).

## 6.3 Preliminary notes on calculating diffraction coefficients $\chi^1$ for a center of expansion and $\psi^1$ for a center of rotation

The calculation we aim at will not require thorough consideration of higher-order approximation. What is needed is an elementary analysis of the ray formulas, and a rather cumbersome, though based on simple analytic background, analysis of the first correction term $\mathbf{V}^1$. Specifically, the vector $\mathbf{V}^1$ in each of the two problems is a sum of a finite number of terms, but only few of them are required in the matching procedure. To clarify this statement, we briefly discuss now the case of a center of expansion.

### 6.3.1 Wavefield in the homogeneous-medium approximation

Reconsider the calculation in Section 6.1.3 in a more formalized manner.

We focus on the case where the right-hand side of the equation (6.25) has the form (3.49)

$$\boldsymbol{F} = 4\pi \operatorname{grad}_{\mathbf{x}} \delta(\mathbf{x} - \mathbf{x}^{\mathrm{o}}) = 4\pi\omega^4 \operatorname{grad}_{\mathbf{X}} \delta(\mathbf{X}). \tag{6.34}$$

Substitute the ansatz (6.24) into the equation (1.82) and take into account the expansion (6.19):

$$\boldsymbol{l}\boldsymbol{u} = \omega^{2+\sigma} \left\{ \text{Л}\mathbf{V}^0 + \frac{\text{Л}^1\mathbf{V}^1}{\omega} + \ldots \right\} = -4\pi\omega^4 \operatorname{grad}_{\mathbf{X}} \delta(\mathbf{X}). \tag{6.35}$$

The requirement that the relation (6.25) which arises in equating the leading-order term on the left-hand side of (6.35) to the right-hand side not explicitly involve $\omega$ implies that

$$\sigma = 2. \tag{6.36}$$

The wavefield in the homogeneous-medium approximation $\mathbf{V}^0$, which solves the equation (6.25), in the case of (6.34) has the form

$$\mathbf{V}^0 = \mathbf{V}^0(\mathbf{X}) = \operatorname{grad}_{\mathbf{X}} \Phi, \quad \Phi = \frac{\alpha^2}{\rho^{\mathrm{o}}} \mathrm{A}, \tag{6.37}$$

$$\mathrm{A} := \frac{e^{i\alpha \mathrm{R}}}{\mathrm{R}}, \quad \mathrm{R} := |\mathbf{X}|.$$

As we have already seen in (6.24), $\sigma$ must be taken in accordance with (6.36), then

$$\boldsymbol{u} \approx \frac{e^{i\omega\alpha r}}{r} i\omega \frac{\alpha^3 \mathbf{s}}{\rho^{\mathrm{o}}} \approx \frac{\alpha^2}{\rho^{\mathrm{o}}} \operatorname{grad}_{\mathbf{x}} \frac{e^{i\omega\alpha r}}{r}. \tag{6.38}$$

Passing to the stretched coordinates $\mathbf{X}$ (see (6.16)), we find

$$\boldsymbol{u} \approx \omega^2 \frac{\alpha^2}{\rho^{\mathrm{o}}} \operatorname{grad}_{\mathbf{X}} \frac{e^{i\alpha \mathrm{R}}}{\mathrm{R}} = \omega^{\sigma} \mathbf{V}^0(\mathbf{X}), \quad \mathrm{R} = |\mathbf{X}| = \omega r. \tag{6.39}$$

### 6.3.2 How to find $\chi^1$, or discussion of the matching procedure

The correction term which takes into account the inhomogeneity of the medium near the source point, that is, the solution of the equation (6.26), proves to be expressible via a finite linear combination of elementary functions. It will be seen to have the form

$$\mathbf{V}^1 = e^{i\alpha \mathrm{R}}(\ldots) + e^{i\beta \mathrm{R}} \mathbf{W}^b \tag{6.40}$$

with

$$\mathbf{W}^b = \frac{\mathbf{Z}_1(\mathrm{s})}{\mathrm{R}} + \frac{\mathbf{Z}_2(\mathrm{s})}{\mathrm{R}^2} + \frac{\mathbf{Z}_3(\mathrm{s})}{\mathrm{R}^3}, \quad \mathrm{s} = \mathrm{grad}_{\mathbf{X}}\, \mathrm{R}, \tag{6.41}$$

where the $\mathbf{Z}_j$ denote certain functions smooth on a unit sphere, and the dots (...) denote a function analogous to the right-hand side of (6.41), but consisting of a larger number of homogeneous terms. The first term on the right-hand side of the relation (6.41) dominates in the intermediate zone (6.33). Obviously, it matches with the ray expansion of the wave $P$, and we denote the coefficient of $e^{i\alpha\mathrm{R}}$ by dots, because we are not interested in it.

The second term in (6.40) matches with the ray expansion of the wave $S$ in (6.7) in the same manner as in Section 6.1.3,

$$\frac{1}{\sqrt{\rho b}}\bigg|_{\mathbf{x}=\mathbf{x}^{\mathrm{o}}} \chi^1(\mathbf{s}) = \mathbf{Z}_1(\mathbf{s}),$$

whence

$$\chi^1(\mathbf{s}) = \sqrt{\rho b}\Big|_{\mathbf{x}=\mathbf{x}^{\mathrm{o}}} \mathbf{Z}_1(\mathbf{s}). \tag{6.42}$$

We observe that the vector $e^{i\alpha\mathrm{R}}\mathbf{Z}_1(\mathrm{s})/\mathrm{R}$ is the far-field asymptotics of the second term in (6.40), and $\mathbf{Z}_1(\mathbf{s})$ is the corresponding radiation pattern of the wave $S$. Important is that it turns out that $(\mathbf{Z}_1(\mathbf{s}), \mathbf{s}) = 0$.[4]

### 6.3.3 The case of a center of rotation

For the center of rotation (3.54), where in (6.25) we have

$$\boldsymbol{F} = 4\pi\, \mathrm{rot}_{\mathbf{x}}(\boldsymbol{m}\delta(\mathbf{x} - \mathbf{x}^{\mathrm{o}})) = 4\pi\omega^4\, \mathrm{rot}_{\mathbf{X}}(\boldsymbol{m}\delta(\mathbf{X})), \tag{6.43}$$

very similar reasoning is applicable. We obtain the same value for the constant $\sigma$ given by (6.36). The wavefield in the homogeneous-medium approximation is

$$\mathbf{V}^0 = \mathbf{V}^0(\mathbf{X}) = \mathrm{rot}_{\mathbf{X}}\boldsymbol{\psi}, \quad \boldsymbol{\psi} = \frac{\beta^2}{\rho^{\mathrm{o}}}\boldsymbol{m}\mathrm{B}, \tag{6.44}$$

where $\mathrm{B} := \frac{e^{i\beta\mathrm{R}}}{\mathrm{R}}$. The leading nonzero diffraction coefficient of the wave $P$ is expressed via the vector $\mathbf{V}^1$ (which is representable in terms of elementary functions) by:

$$\psi^1(\mathbf{s})\mathbf{s} = \sqrt{\rho a}\big|_{\mathbf{x}=\mathbf{x}^{\mathrm{o}}} \mathbf{Z}_1(\mathbf{s}),$$

whence

$$\psi^1(\mathbf{s}) = \sqrt{\rho a}\big|_{\mathbf{x}=\mathbf{x}^{\mathrm{o}}} (\mathbf{Z}_1(\mathbf{s}), \mathbf{s}), \tag{6.45}$$

where the vector $\mathbf{Z}_1(\mathbf{s})$ is obtained by replacement $b$ by $a$ and $\beta$ by $\alpha$ in the formulas (6.40), (6.41). Important is that $\mathbf{Z}_1(\mathbf{s}) \parallel \mathbf{s}$.

Similarly to the previous case, $(\mathbf{Z}_1(\mathbf{s}), \mathbf{s})$ proves to be the radiation pattern of the corresponding spherical wave $P$.

[4]This can be shown independently of a specific calculation (see Kiselev 1974 [11]).

## 6.4 ⋆ Operator background for constructing solutions of equations (6.25), (6.26), ...

Let us concisely describe the basic idea of explicitly finding the vector functions $\mathbf{V}^0, \mathbf{V}^1, \ldots$

### Reducing $\boldsymbol{Л}\boldsymbol{U} = \boldsymbol{\varphi}$ to scalar equations

In order to solve the vector equation

$$\boldsymbol{Л}\boldsymbol{U} = \boldsymbol{\varphi}, \tag{6.46}$$

first consider the scalar equation

$$Л_\gamma U = \varphi, \tag{6.47}$$

where $Л_\gamma$ is the Helmholtz operator:

$$Л_\gamma U := (\nabla^2_{\mathbf{x}} + \gamma^2)U. \tag{6.48}$$

We are interested only in the solutions of the equations (6.46) and (6.47) that satisfy the respective limiting absorption principles (see Section 3.5.1). Their uniqueness follows from the results of Section 3.5.1; their existence can be established in each case via explicit constructions.

The formal scheme is based on the reduction of the vector equation (6.46) to scalar equations for the operators $Л_\alpha$ and $Л_\beta$ with the help of the formula

$$\boldsymbol{Л}^{-1} = \frac{1}{\rho^{\mathrm{o}}}\left\{\beta^2\, Л_\beta^{-1}\mathbf{I} + \left(Л_\beta^{-1} - Л_\alpha^{-1}\right)\operatorname{grad}_{\mathbf{x}}\operatorname{div}_{\mathbf{x}}\right\}, \tag{6.49}$$

which can easily be verified, e.g., via the Fourier transform. However, the basic formula (6.49) will not be directly used in Sections 6.5–6.7.

### Commutativity of the operators $\boldsymbol{Л}$, $Л_\alpha^{-1}$, and $Л_\beta^{-1}$

In what follows, it is important that the operators $\boldsymbol{Л}$, $Л_\alpha^{-1}$, and $Л_\beta^{-1}$ (and their inverse) commute. Operators $P$ and $Q$ *commute* if $PQ = QP$ (and they are defined in the same domain).

The required commutativity can be established using the following considerations. Under the assumption of the existence of $P^{-1}$ and $Q^{-1}$, the commutativity of $P$ and $Q$ implies the commutativity of $P$ and $Q^{-1}$, $P^{-1}$ and $Q$, $P^{-1}$ and $Q^{-1}$. This follows from the chain of relations

$$\begin{aligned}
&PQ = QP \Rightarrow P = QPQ^{-1} \Rightarrow Q^{-1}P = QP^{-1} \Rightarrow Q^{-1} = PQ^{-1}P^{-1} \\
&\Rightarrow P^{-1}Q^{-1} = Q^{-1}P^{-1} \Rightarrow QP^{-1}Q^{-1} = P^{-1} \Rightarrow QP^{-1} = P^{-1}Q.
\end{aligned}$$

Scalar operators with constant coefficients obviously commute. Thus, $Л_\alpha$, $Л_\beta$,

$\text{Л}_\alpha^{-1}$, and $\text{Л}_\beta^{-1}$ commute with each other and with the operators of differentiation with respect to $X_1$, $X_2$, and $X_3$.

Clear is that the vector operator $\textbf{Л}$ commutes with the operators of differentiation with respect to coordinates and therefore with any scalar differential operators with constant coefficients, in particular, with $\text{Л}_\alpha$ and $\text{Л}_\beta$. Thus, $\textbf{Л}$ commutes with $\text{Л}_\alpha^{-1}$ and $\text{Л}_\beta^{-1}$.

---

## 6.5 Auxiliary formulas

We split the subsequent bulky calculation into several steps.

In this section, we derive two formulas which will be useful in explicitly solving the equation (6.26) for the sources (6.34) and (6.43). Recall the notation

$$D_j = \frac{\partial}{\partial X_j}, \quad D^2_{jp} = \frac{\partial^2}{\partial X_j \partial X_p};$$

$$\mathrm{A} = \frac{e^{i\alpha \mathrm{R}}}{\mathrm{R}}, \quad \mathrm{B} = \frac{e^{i\beta \mathrm{R}}}{\mathrm{R}}, \quad \mathrm{R} = |\mathbf{X}|, \quad \mathrm{s} = \frac{\mathbf{X}}{\mathrm{R}}.$$

We aim at solving the equations

$$\textbf{Л}\mathbf{U} = \mathbf{m}\mathrm{A} \tag{6.50}$$

and

$$\textbf{Л}\mathbf{W} = \mathbf{m}\mathrm{B} \tag{6.51}$$

with arbitrary constant vectors $\mathbf{m}$. Such equations occur in finding the vectors $\mathbf{V}^1$. In passing, we will solve the scalar analogs of these equations:

$$(\nabla^2_{\mathbf{X}} + \alpha^2)U = \mathrm{A}, \quad (\nabla^2_{\mathbf{X}} + \beta^2)V = \mathrm{B}. \tag{6.52}$$

### 6.5.1 Equations with Helmholtz operators

We denote solutions of the equations

$$\text{Л}_\alpha \mathrm{v} = (\nabla^2_{\mathbf{X}} + \alpha^2)\mathrm{v} = \widetilde{\mathrm{v}} \quad \text{and} \quad \text{Л}_\beta \mathrm{w} = (\nabla^2_{\mathbf{X}} + \beta^2)\mathrm{w} = \widetilde{\mathrm{w}}$$

satisfying the limiting absorption principle by

$$\mathrm{v} = (\nabla^2_{\mathbf{X}} + \alpha^2)^{-1}\widetilde{\mathrm{v}} = \text{Л}_\alpha^{-1}\widetilde{\mathrm{v}} \quad \text{and} \quad \mathrm{w} = (\nabla^2_{\mathbf{X}} + \beta^2)^{-1}\widetilde{\mathrm{w}} = \text{Л}_\beta^{-1}\widetilde{\mathrm{w}},$$

respectively. We assume that $\widetilde{\mathrm{v}}$ and $\widetilde{\mathrm{w}}$ either have the same form as the components of the vectors (6.30), or are generalized functions localized at the origin.

This guarantees the existence of solutions exponentially decaying at infinity as $\operatorname{Im}\alpha > 0$ and $\operatorname{Im}\beta > 0$.

Write the identity

$$(\nabla^2_{\mathbf{X}} + \alpha^2)\mathrm{A} = (\nabla^2_{\mathbf{X}} + \beta^2)\mathrm{B} = -4\pi\delta(\mathbf{X}) \tag{6.53}$$

known from Chapter 3 in the form

$$\begin{aligned}(\nabla^2_{\mathbf{X}} + \alpha^2)^{-1}(-4\pi\delta(\mathbf{X})) &= \mathrm{A},\\ (\nabla^2_{\mathbf{X}} + \beta^2)^{-1}(-4\pi\delta(\mathbf{X})) &= \mathrm{B}.\end{aligned} \tag{6.54}$$

Note that

$$(\nabla^2_{\mathbf{X}} + \alpha^2)(\mathrm{A} - \mathrm{B}) = (\nabla^2_{\mathbf{X}} + \alpha^2)\mathrm{A} - (\nabla^2_{\mathbf{X}} + \beta^2)\mathrm{B} - (\alpha^2 - \beta^2)\mathrm{B} = -(\alpha^2 - \beta^2)\mathrm{B}.$$

We employed the relations (6.53). Similarly,

$$(\nabla^2_{\mathbf{X}} + \beta^2)(\mathrm{A} - \mathrm{B}) = -(\alpha^2 - \beta^2)\mathrm{A},$$

whence

$$(\nabla^2_{\mathbf{X}} + \alpha^2)^{-1}\mathrm{B} = (\nabla^2_{\mathbf{X}} + \beta^2)^{-1}\mathrm{A} = -\frac{\mathrm{A} - \mathrm{B}}{\alpha^2 - \beta^2}. \tag{6.55}$$

In the same way, we get the identities $(\nabla^2_{\mathbf{X}} + \alpha^2)e^{i\alpha\mathrm{R}} = 2i\alpha\mathrm{A}$, $(\nabla^2_{\mathbf{X}} + \beta^2)e^{i\beta\mathrm{R}} = 2i\beta\mathrm{B}$, which we write in the form

$$(\nabla^2_{\mathbf{X}} + \alpha^2)^{-1}\mathrm{A} = \frac{e^{i\alpha\mathrm{R}}}{2i\alpha}, \quad (\nabla^2_{\mathbf{X}} + \beta^2)^{-1}\mathrm{B} = \frac{e^{i\beta\mathrm{R}}}{2i\beta}. \tag{6.56}$$

## 6.5.2 Equations (6.50) and (6.51)

It is convenient to solve the equations (6.50) and (6.51) by applying the operators $(\nabla^2_{\mathbf{X}} + \alpha^2)^{-1}$ and $(\nabla^2_{\mathbf{X}} + \beta^2)^{-1}$ to the solution for a concentrated force given in Chapter 3.

Let the equation

$$\mathbf{J\!I}\mathbf{V} = \widetilde{\mathbf{V}}$$

have a solution $\mathbf{V}$ satisfying the limiting absorption principle. It will be denoted by

$$\mathbf{V} = \mathbf{J\!I}^{-1}\widetilde{\mathbf{V}}.$$

The formula (3.79) for the wavefield of a concentrated force can thus be written as follows

$$\mathbf{J\!I}^{-1}(-4\pi\mathbf{m}\delta(\mathbf{X})) = \frac{1}{\rho^{\circ}}\left\{\beta^2\mathbf{m}\mathrm{B} + m_j\mathrm{D}_j \operatorname{grad}_{\mathbf{X}}(\mathrm{B} - \mathrm{A})\right\}, \tag{6.57}$$

$\mathrm{D}_j := \frac{\partial}{\partial \mathrm{X}_j}$. Applying the operator $(\nabla^2_{\mathbf{X}} + \alpha^2)^{-1}$ to both sides of the relation

(6.57) and using the formulas (6.54), (6.55), and (6.56), we find the solution of the equation (6.50):

$$\begin{aligned}\mathbf{U} &= \text{Л}^{-1}(\mathbf{m}\mathrm{A}) \\ &= -\frac{1}{\rho^{\circ}}\left\{\left(\beta^2\mathbf{m} + m_j \mathrm{D}_j \operatorname{grad}_{\mathbf{X}}\right)\frac{\mathrm{A}-\mathrm{B}}{\alpha^2-\beta^2} + m_j \mathrm{D}_j \operatorname{grad}_{\mathbf{X}} \frac{e^{i\alpha \mathrm{R}}}{2i\alpha}\right\}. \qquad (6.58)\end{aligned}$$

Similarly, applying the operator $\text{Л}_{\beta}^{-1} = (\nabla^2 + \beta^2)^{-1}$, we find the solution of the equation (6.50):

$$\begin{aligned}\mathbf{W} &= \text{Л}^{-1}(\mathbf{m}\mathrm{B}) \\ &= \frac{1}{\rho^{\circ}}\left\{m_j \mathrm{D}_j \operatorname{grad}_{\mathbf{X}} \frac{\mathrm{A}-\mathrm{B}}{\alpha^2-\beta^2} + \left(\beta^2\mathbf{m} + m_j \mathrm{D}_j \operatorname{grad}_{\mathbf{X}}\right)\frac{e^{i\beta \mathrm{R}}}{2i\beta}\right\}. \qquad (6.59)\end{aligned}$$

### 6.5.3 Two more identities

We outline two observations which will prove to be useful for discarding terms unimportant in further calculation. First, it will be demonstrated that if $\boldsymbol{K}$ is a vector with constant components, then the terms with the multiplier $e^{i\alpha \mathrm{R}}$ do not occur in the solution $\boldsymbol{\xi}$ of the equation

$$\text{Л}\boldsymbol{\xi} = \operatorname{rot}(\mathrm{X}_l \boldsymbol{K} \mathrm{B}), \qquad (6.60)$$

$l = 1, 2, 3$. We write this assertion as:

$$\boldsymbol{\xi} = (...)e^{i\beta \mathrm{R}} + 0 \cdot e^{i\alpha \mathrm{R}}. \qquad (6.61)$$

Henceforth, the dots (...) stand for a coefficient (being a vector with components polynomial in $X_1$, $X_2$, $X_3$, R, and $\frac{1}{\mathrm{R}}$), in the value of which we are not interested. Indeed, seeking $\boldsymbol{\xi}$ in the form

$$\boldsymbol{\xi} = \operatorname{rot}_{\mathbf{X}} \boldsymbol{\Psi} + \operatorname{grad}_{\mathbf{X}} \Phi,$$

we immediately obtain

$$\Phi \equiv 0, \quad \rho^{\circ}(\nabla_{\mathbf{X}}^2 + \beta^2)\boldsymbol{\Psi} = \mathrm{X}_l \boldsymbol{K} \mathrm{B}.$$

Now it is easy to find that

$$\boldsymbol{\Psi} = \frac{\boldsymbol{K}}{\rho^{\circ}}(\nabla_{\mathbf{X}}^2 + \beta^2)^{-1}\mathrm{X}_l \mathrm{B} = \frac{\boldsymbol{K}}{i\beta\rho^{\circ}}\mathrm{D}_l(\nabla_{\mathbf{X}}^2 + \beta^2)^{-1} e^{i\beta \mathrm{R}},$$

whence the required relation immediately follows.

By a similar manipulation one easily checks that if

$$\text{Л}\boldsymbol{\eta} = \operatorname{grad}(\mathrm{X}_l \mathrm{A}), \qquad (6.62)$$

then

$$\boldsymbol{\eta} = (...)e^{i\alpha \mathrm{R}} + 0 \cdot e^{i\beta \mathrm{R}}. \qquad (6.63)$$

We completed the derivation of auxiliary formulas and can proceed to direct calculation of the diffraction coefficients.

## 6.6 Leading nonzero diffraction coefficient for the wave $S$ from a center of rotation

### 6.6.1 Rearrangement of the expression $\text{Л}^1\mathbf{V}^0$

We start by rearranging, for the case of (6.34), the right-hand side of the equation (6.26) (which is defined by the expression (6.22) at $m = 1$ and $\boldsymbol{u} = \mathbf{V}^0$) (see (6.37))

$$(\text{Л}^1\mathbf{V}^0)_p = 2\text{D}_j(\mu^1(\mathbf{X})\text{D}^2_{pj}\Phi) + \text{D}_p(\lambda^1(\mathbf{X})\nabla^2_{\mathbf{X}}\Phi) + \rho^1(\mathbf{X})\text{D}_p\Phi,$$
$$\Phi = \frac{\alpha^2}{\rho^{\text{o}}}\text{A}. \tag{6.64}$$

*1*. Consider the last term on the right-hand side of (6.64),

$$\rho^1(\mathbf{X})\text{D}_p\Phi = \text{D}_p\left(\rho^1(\mathbf{X})\Phi\right) - \rho_p\Phi\,. \tag{6.65}$$

The linear terms in the Taylor expansions of the volume density and Lamé parameters will be denoted by:

$$\rho^1(\mathbf{X}) = \rho_j X_j, \quad \lambda^1(\mathbf{X}) = \lambda_j X_j, \quad \mu^1(\mathbf{X}) = \mu_j X_j, \tag{6.66}$$

where $\rho_j = \left.\frac{\partial\rho}{\partial x_j}\right|_{\mathbf{x}=\mathbf{x}^{\text{o}}}$, $\lambda_j = \left.\frac{\partial\lambda}{\partial x_j}\right|_{\mathbf{x}=\mathbf{x}^{\text{o}}}$, $\mu_j = \left.\frac{\partial\mu}{\partial x_j}\right|_{\mathbf{x}=\mathbf{x}^{\text{o}}}$.

*2*. The second term on the right-hand side of the formula (6.64), with (6.53) taken into account, becomes

$$\nabla^2_{\mathbf{X}}\Phi = \frac{\alpha^2}{\rho^{\text{o}}}\left(-\alpha^2\text{A} - 4\pi\delta(\mathbf{X})\right). \tag{6.67}$$

Applying the identity

$$X_l\delta(\mathbf{X}) = 0, \quad l = 1, 2, 3, \tag{6.68}$$

which follows from (3.19), we find

$$\text{D}_p\left(\lambda^1(\mathbf{X})\nabla^2_{\mathbf{X}}\Phi\right) = -\frac{\alpha^4}{\rho^{\text{o}}}\text{D}_p\left(\lambda^1(\mathbf{X})\text{A}\right). \tag{6.69}$$

*3*. Address now the first term on the right-hand side of (6.64):

$$\begin{aligned}2\text{D}_j(\mu^1(\mathbf{X})\text{D}^2_{pj}\Phi) &= 2\text{D}_p\left(\mu^1(\mathbf{X})\nabla^2_{\mathbf{X}}\Phi\right) - 2\mu_p\nabla^2_{\mathbf{X}}\Phi + 2\mu_j\text{D}_{pj}\Phi\\ &= 2\frac{\alpha^2}{\rho^{\text{o}}}\left\{-\text{D}_p\left(\alpha^2\mu^1(\mathbf{X})\text{A}\right) + \mu_p\left(\alpha^2\text{A} + 4\pi\delta(\mathbf{X})\right) + \mu_j\text{D}^2_{pj}A\right\}.\end{aligned} \tag{6.70}$$

*4*. As a result, the right-hand side of the equation (6.26) can be rewritten in the form

$$\text{Л}^1\mathbf{V}^0 = \text{grad}_{\mathbf{X}}(k^1(\mathbf{X})\Phi) + \frac{\alpha^2}{\rho^{\text{o}}}\left(\psi\text{A} + 4\pi\varphi\delta(\mathbf{X})\right) + 2\,\text{grad}_{\mathbf{X}}(\mu_j V_j^0), \quad (6.71)$$

where $k^1(\mathbf{X})$ is a homogeneous linear function,

$$k^1(\mathbf{X}) = \rho^1(\mathbf{X}) - \alpha^2\left(\lambda^1(\mathbf{X}) + 2\mu^1(\mathbf{X})\right), \quad (6.72)$$

while $\psi$ and $\varphi$ are vectors with components

$$\psi_j = 2\alpha^2\mu_j - \rho_j, \quad \varphi_j = 2\mu_j. \quad (6.73)$$

### 6.6.2 Discarding terms unimportant for finding $\chi^1$ in $\text{Л}^1\mathbf{V}^0$

*1*. First, we note that the last term on the right-hand side of (6.71) has no effect on generation of the wave $S$. Indeed, consider the corresponding equation

$$\text{Л}\mathbf{P} = 2\,\text{grad}_{\mathbf{X}}(\mu_j V_j^0) = -2\,\text{grad}_{\mathbf{X}}\left(\mu_j \text{D}_j \Phi\right). \quad (6.74)$$

The solution of the equation (6.74) can easily be found (see Section 6.5.3) in the form

$$\mathbf{P} = \text{grad}_{\mathbf{X}}\, p,$$

whence

$$\frac{\rho^{\text{o}}}{\alpha^2}\left(\nabla_{\mathbf{X}}^2 + \alpha^2\right)p = -2\frac{\alpha^2}{\rho^{\text{o}}}\mu_j \text{D}_j \text{A}.$$

Further, (6.56) gives

$$p = -2\left(\frac{\alpha^2}{\rho^{\text{o}}}\right)^2 \mu_j \text{D}_j \frac{e^{i\alpha\text{R}}}{2i\alpha},$$

and thus

$$\mathbf{P} = (...)e^{i\alpha\text{R}} + 0\cdot e^{i\beta\text{R}}, \quad (6.75)$$

where the dots (...) denote the expression which is of no interest.

*2*. Second, thanks to the results of Section 6.5.3, the first term on the right-hand side of (6.71) can also be neglected when calculating the wave $S$. Therefore

$$\mathbf{V}^1 = \widetilde{\mathbf{V}^1} + (...)e^{i\alpha\text{R}}, \quad (6.76)$$

where $\widetilde{\mathbf{V}^1}$ is a solution of the equation

$$\text{Л}\widetilde{\mathbf{V}^1} = -\frac{\alpha^2}{\rho^{\text{o}}}\left(\psi\text{A} + 4\pi\varphi\delta(\mathbf{X})\right). \quad (6.77)$$

### 6.6.3 Final result

Applying the formulas (6.57) and (6.58), we find

$$\widetilde{\mathbf{V}^1} = -\left(\frac{\alpha}{\rho^o}\right)^2 \left\{\frac{1}{\alpha^2-\beta^2}\left[\beta^2\boldsymbol{\psi} + \psi_j \mathrm{D}_j \,\mathrm{grad}_{\mathbf{x}}\right] - \left[\beta^2\boldsymbol{\varphi} + \varphi_j \mathrm{D}_j \,\mathrm{grad}_{\mathbf{x}}\right]\right\}\mathrm{B} + (...)e^{i\alpha \mathrm{R}}. \quad (6.78)$$

Recall now the reasoning in Section 6.3.2. The expression (6.78) proves to have the structure (6.40), (6.41) and to be the spherical wave $S$. Performing the same calculation as in Chapter 3, we observe that the function $\mathrm{D}^2_{mn}\mathrm{B}$ has the far-field asymptotics

$$\mathrm{D}_{mn}\mathrm{B} = -\beta^2 s_m s_n \mathrm{B} + O\left(\frac{1}{\mathrm{R}^2}\right) = e^{i\beta \mathrm{R}}\left\{\frac{\widetilde{Z}_1(\mathbf{s})}{\mathrm{R}} + O\left(\frac{1}{\mathrm{R}^2}\right)\right\}, \quad \mathrm{R}\to\infty, \quad (6.79)$$

where

$$\widetilde{Z}_1(\mathbf{s}) = -\beta^2 s_m s_n. \quad (6.80)$$

With the help of formulas (6.73), one can find the asymptotics of the components of $\widetilde{\mathbf{V}^1}$:

$$\widetilde{V^1}_m = \left\{\left(\frac{\alpha\beta}{\rho^o}\right)^2 \frac{\rho_l - 2\beta^2\mu_l}{\alpha^2-\beta^2}\left(\delta_{lm} - s_l s_m\right)\frac{1}{\mathrm{R}} + O\left(\frac{1}{\mathrm{R}^2}\right)\right\} e^{i\beta\mathrm{R}} + (...)e^{i\alpha\mathrm{R}}. \quad (6.81)$$

Noting that the functions $\left(\frac{\alpha\beta}{\rho^o}\right)^2 \frac{\rho_l - 2\beta^2\mu_l}{\alpha^2-\beta^2}$ are the components of the vector

$$\begin{aligned}\mathbf{q} &= -\frac{1}{\rho^2(a^2-b^2)}\left(\mathrm{grad}_{\mathbf{x}}\,\rho - \frac{2}{b^2}\mathrm{grad}_{\mathbf{x}}\,\mu\right)\Big|_{\mathbf{x}=\mathbf{x}^o} \\ &= \frac{1}{\rho(a^2-b^2)}\left(\frac{\mathrm{grad}_{\mathbf{x}}\,\rho}{\rho} + 4\frac{\mathrm{grad}_{\mathbf{x}}\,b}{b}\right)\Big|_{\mathbf{x}=\mathbf{x}^o},\end{aligned} \quad (6.82)$$

the coefficient of $\mathrm{R}^{-1}$ in (6.81) can be written as

$$\mathbf{Z}_1 = \{\mathbf{q} - (\mathbf{q},\mathbf{s})\mathbf{s}\}. \quad (6.83)$$

Hence, in accordance with (6.42), we obtain

$$\chi^1(\mathbf{s}) = \sqrt{\rho b}\Big|_{\mathbf{x}=\mathbf{x}^o}\{\mathbf{q} - (\mathbf{q},\mathbf{s})\mathbf{s}\}. \quad (6.84)$$

## 6.7 Leading nonzero diffraction coefficient for the wave $P$ from a center of rotation

### 6.7.1 Rearrangement of the expression $\mathrm{Л}^1\mathbf{V}^0$

Let us rearrange the $j$-th component of the right-hand side of the equation (6.26) in the case where $\mathbf{V}^0$ is given by the expression (6.44):

$$(\mathrm{Л}^1\mathbf{V}^0)_j = \mathrm{D}_k\left[\mu^1\left(\mathrm{D}_j\mathrm{V}^0_k + \mathrm{D}_k\mathrm{V}^0_j\right)\right] + \mathrm{D}_j\left(\lambda^1 \operatorname{div}_{\mathbf{X}}\mathbf{V}^0\right) + \rho^1\mathrm{V}^0_j. \tag{6.85}$$

In order to simplify the calculation, in (6.44) we assume that

$$\boldsymbol{m} = \mathbf{e}_1 \quad\Rightarrow\quad \mathbf{V}^0 = \frac{\beta^2}{\rho^\circ}\operatorname{rot}_{\mathbf{X}}(\mathbf{e}_1\mathrm{B}) = \frac{\beta^2}{\rho^\circ}\left(\mathbf{e}_2\mathrm{D}_3 - \mathbf{e}_3\mathrm{D}_2\right)\mathrm{B}. \tag{6.86}$$

The last term in (6.85), with the help of identity

$$\operatorname{rot}_{\mathbf{X}}(f\mathbf{Y}) = f\operatorname{rot}_{\mathbf{X}}\mathbf{Y} + \left[\operatorname{grad}_{\mathbf{X}} f \times \mathbf{Y}\right],$$

takes the form

$$\rho^1\mathbf{V}^0 = \rho^1\operatorname{rot}_{\mathbf{X}}\boldsymbol{\psi} = \operatorname{rot}_{\mathbf{X}}\left(\rho^1\boldsymbol{\psi}\right) - \left[\operatorname{grad}_{\mathbf{X}}\rho^1 \times \boldsymbol{\psi}\right]. \tag{6.87}$$

The first term on the right-hand side of (6.85) is

$$\mathrm{D}_k\left[\mu^1\left(\mathrm{D}_j\mathrm{V}^0_k + \mathrm{D}_k\mathrm{V}^0_j\right)\right] = \mu^1\left(\mathrm{D}_j\operatorname{div}_{\mathbf{X}}\mathbf{V}^0 + \nabla^2_{\mathbf{X}}\mathrm{V}^0_j\right) + \mathrm{D}_j(\mu_p\mathrm{V}^0_p) + \mu_k\mathrm{D}_k\mathrm{V}^0_j. \tag{6.88}$$

Since $\mathbf{V}^0$ is a solenoidal vector, $\operatorname{div}_{\mathbf{X}}\mathbf{V}^0 = 0$. Next,

$$\mu^1\nabla^2_{\mathbf{X}}\mathbf{V}^0 = \mu^1\nabla^2_{\mathbf{X}}\operatorname{rot}_{\mathbf{X}}\boldsymbol{\psi} = \operatorname{rot}_{\mathbf{X}}\left(\mu^1\nabla^2_{\mathbf{X}}\boldsymbol{\psi}\right) - \left[\operatorname{grad}_{\mathbf{X}}\mu^1 \times \nabla^2_{\mathbf{X}}\boldsymbol{\psi}\right].$$

Employing the formula (6.53), the definition (6.44) of the vector $\boldsymbol{\psi}$, and the identity (6.68), we find

$$\mu^1\nabla^2_{\mathbf{X}}\mathbf{V}^0 = \operatorname{rot}_{\mathbf{X}}\left(-\beta^2\mu^1\boldsymbol{\psi}\right) + \left[\beta^2\operatorname{grad}_{\mathbf{X}}\mu^1 \times \boldsymbol{\psi}\right] + 4\pi\frac{\beta^2}{\rho^\circ}\boldsymbol{\Psi}\delta(\mathbf{X}), \tag{6.89}$$

where $\boldsymbol{\Psi}$ stands for the following vector with constant components:

$$\boldsymbol{\Psi} = \left[\operatorname{grad}_{\mathbf{X}}\mu^1 \times \mathbf{e}_1\right] = \mu_3\mathbf{e}_2 - \mu_2\mathbf{e}_3. \tag{6.90}$$

As a result, from the formulas (6.85)–(6.90) we get

$$\mathrm{Л}^1\mathbf{V}^0 = \operatorname{rot}_{\mathbf{X}}\left(\boldsymbol{k}^1\mathrm{B}\right) + \frac{\beta^2}{\rho^\circ}\left\{4\pi\boldsymbol{\Psi}\delta(\mathbf{X}) - \boldsymbol{\Phi}\mathrm{B}\right\} + \operatorname{grad}_{\mathbf{X}}\left(\mu_l V^0_l\right) + \mu_l\mathrm{D}_l\mathbf{V}^0, \tag{6.91}$$

where

$$\boldsymbol{k}^1 = \boldsymbol{k}^1(\mathbf{X}) = \rho^1(\mathbf{X}) - \beta^2\mu^1(\mathbf{X}) \tag{6.92}$$

is a homogeneous linear function of the components of $\mathbf{X}$, and $\boldsymbol{\Phi}$ is the vector with constant components defined by the relation

$$\boldsymbol{\Phi} = \left[\operatorname{grad}_{\mathbf{X}}\boldsymbol{k}^1 \times \mathbf{e}_1\right] = (\rho_3 - \beta^2\mu_3)\mathbf{e}_2 - (\rho_2 - \beta^2\mu_2)\mathbf{e}_3. \tag{6.93}$$

### 6.7.2 Terms in $\mathbf{Л}^1\mathbf{V}^0$ important for calculation of $\psi^1$

The first term on the right-hand side of equation (6.91), in view of relations (6.60)–(6.61), gives no contribution to the diffraction coefficient of the wave $P$, and we can therefore refrain from considering it. Because of the relation $\mu_l \mathrm{D}_l \mathbf{V}^0 = \frac{\beta^2}{\rho^\mathrm{o}} \mu_l \mathrm{D}_l \operatorname{rot}_{\mathbf{X}}(\mathbf{e}_1 \mathrm{B})$, the same is true for the last term in (6.91). Therefore, instead of solving equation (6.26), (6.91), we can deal with the vector function $\widetilde{\mathbf{V}}^1$ defined by

$$\widetilde{\mathbf{V}}^1 = \boldsymbol{f} + \boldsymbol{g} + \boldsymbol{h}, \tag{6.94}$$

where

$$\mathbf{Л}(\boldsymbol{f}) = -4\pi \frac{\beta^2}{\rho^\mathrm{o}} \boldsymbol{\Psi} \delta(\mathbf{X}), \tag{6.95}$$

$$\mathbf{Л}(\boldsymbol{g}) = \frac{\beta^2}{\rho^\mathrm{o}} \boldsymbol{\Phi} \mathrm{B}, \tag{6.96}$$

$$\mathbf{Л}(\boldsymbol{h}) = -\operatorname{grad}_{\mathbf{X}} (\mu_l V_l^0). \tag{6.97}$$

The solution of the concentrated-force problem (6.95) is easily obtained from (6.57), and its components are

$$f_m = -\left(\frac{\beta}{\rho^\mathrm{o}}\right)^2 \Psi_l \mathrm{D}_{lm}^2 \mathrm{A} + (\ldots) e^{i\beta \mathrm{R}}. \tag{6.98}$$

The equation (6.96) is solved with the use of the formula (6.59), whence

$$g_m = \left(\frac{\beta}{\rho^\mathrm{o}}\right)^2 \Phi_l \mathrm{D}_{lm}^2 \frac{\mathrm{A}}{\alpha^2 - \beta^2} + (\ldots) e^{i\beta \mathrm{R}}. \tag{6.99}$$

The solution of the equation (6.97) will be sought in the form

$$\boldsymbol{h} = \operatorname{grad}_{\mathbf{X}} \chi,$$

whence $\mathbf{Л}^1(\boldsymbol{h}) = \frac{\rho^\mathrm{o}}{\alpha^2} Л_\alpha \operatorname{grad}_{\mathbf{X}} \chi$, and the equation for $\chi$ becomes

$$\frac{\rho^\mathrm{o}}{\alpha^2} (\nabla_{\mathbf{X}}^2 + \alpha^2) \chi = -\mu_l V_l^0 = -\frac{\beta^2}{\rho^\mathrm{o}} (\mu_2 \mathrm{D}_3 - \mu_3 \mathrm{D}_2) \mathrm{B}.$$

Applying the formula (6.55), we get

$$\chi = \left(\frac{\alpha\beta}{\rho^\mathrm{o}}\right)^2 (\mu_2 \mathrm{D}_3 - \mu_3 \mathrm{D}_2) \frac{\mathrm{A} - \mathrm{B}}{\alpha^2 - \beta^2},$$

and thus

$$h_m = \left(\frac{\alpha\beta}{\rho^\mathrm{o}}\right)^2 (\mu_2 \mathrm{D}_{3m}^2 - \mu_3 \mathrm{D}_{2m}^2) \frac{\mathrm{A}}{\alpha^2 - \beta^2} + (\ldots) e^{i\beta \mathrm{R}}. \tag{6.100}$$

Combining together formulas (6.94), (6.98)–(6.100) and omitting the terms unimportant for finding the diffraction coefficient $\psi^1$, we arrive at

$$V_m^1 = \left(\frac{\beta}{\rho^\circ}\right)^2 \left\{ \left(\frac{\Phi_l}{\alpha^2-\beta^2} - \Psi_l\right) \mathrm{D}_{lm}^2 + \frac{\alpha^2}{\alpha^2-\beta^2}\left(\mu_2 \mathrm{D}_{3m}^2 - \mu_3 \mathrm{D}_{2m}^2\right)\right\} \mathrm{A} + (...)e^{i\beta \mathrm{R}}. \quad (6.101)$$

By a simple calculation, similar to that of the previous section, we find

$$\mathbf{V}^1 \approx -\left(\frac{\alpha\beta}{\rho^\circ}\right)^2 \frac{(\rho_3 - 2\alpha^2\mu_3)s_2 - (\rho_2 - 2\alpha^2\mu_2)s_3}{\alpha^2-\beta^2} \mathbf{s}\mathrm{A} + (...)e^{i\beta\mathrm{R}}, \quad (6.102)$$

whence it follows that the radiation pattern of the spherical wave $P$ is

$$(\mathbf{Z}_1(\mathbf{s}), \mathbf{s}) = ([\mathbf{s} \times \mathbf{p}], \mathbf{e}_1) \quad (6.103)$$

with

$$\mathbf{p} = \frac{1}{\rho^2(b^2-a^2)} \left(\mathrm{grad}_{\mathbf{x}}\, \rho - \frac{2}{a^2}\, \mathrm{grad}_{\mathbf{x}}\, \mu\right)\Bigg|_{\mathbf{x}=\mathbf{x}^\circ}. \quad (6.104)$$

### 6.7.3 Final result

For the concentrated force in (6.43) acting in the direction of the vector $\boldsymbol{m}$, we must replace $\mathbf{e}_1$ by $\boldsymbol{m}$ in the formula (6.103).

Matching with the ray field, we get $\psi^0 = 0$, as in the case of a homogeneous medium, see (6.10), and

$$\psi^1 = \psi^1(\mathbf{s}) = \sqrt{\rho a}\big|_{\mathbf{x}=\mathbf{x}^\circ} ([\mathbf{s}\times\mathbf{p}], \boldsymbol{m}). \quad (6.105)$$

---

## 6.8 ⋆ Comments to Chapter 6

Investigations of scalar wavefields of point sources in inhomogeneous media were started in the 1890s by Hadamard (see Hadamard 1923 [9] and references therein). Hadamard presented series expansions convergent both near the source point and at a finite distance from it, as long as no caustics are met. Hadamard's theory attained true elegance in view of the theory of analytic parameter continuation of generalized functions, which is traceable to M. Riesz. The Fourier transform with respect to time of Hadamard's formulas gives a specialized modification of the ray expansion, which is valid both close to a source point and at a limited distance therefrom; see Babich 1965, 1992 [5, 6], Qian, Yuan, Liu, Luo, and Burridge 2016 [21]. Hadamard's expansion is available only for scalar second-order equations.[5] For general hyperbolic systems,

[5]The only exception is a construction of two leading terms of such an expansion for the equations of isotropic electrodynamics recently described in Lu, Qian, and Burridge 2016 [20].

a construction of uniform asymptotics of fundamental solutions that generalizes the expansion in plane waves is known; see Lax 1957 [19] and Babich 1961 [4] (with the role of plane waves played by ray solutions). Deriving the correction-terms diffraction coefficient (such as we have found in the current chapter) within this approach seems a too tricky task.

The construction of the current chapter has a clearly pronounced boundary layer nature. The boundary layer method has long been extensively exploited in mathematical physics; see Cole 1972 [8], Babič and Kirpičnikova 1979 [7], Ilin 1992 [10], among others. The simple locality approach built upon freezing the coefficients at the source point and providing the leading diffraction coefficients for point sources has been used since the ray method came into being (see, e.g., Alekseev and Tsepelev 1956 [2]). Where further advance is required, the boundary layer approach suggests itself. The boundary layer approach was first applied to a scalar point source problem by Avila and Keller 1963 [3] (who were, however, mistaken in constructing higher-order terms), and by Babich and Kirpichnikova 1979 [7]. The analytic structure of the eikonal and the geometrical spreading for a central field of rays near the source point in a smooth medium has been described by Hadamard 1923 [9] and Sobolev 1930 [23][6]; see also Babič and Kirpičnikova 1979 [7].

The explicit high-frequency asymptotic expressions for the wave $S$ from a center of expansion and for the wave $P$ from a center of rotation were found by Kiselev 1974, 1975 [11, 12]. Kiselev 1975 [12] gave a detailed description of a scheme of constructing higher-order terms near a source point and their matching with the ray formulas, based upon the techniques of "uniqueness lemmas" (see Babič and Kirpičnikova 1979 [7]). The combinations of gradients of the density and the velocities occurring in the formulas (6.82) and (6.104) appear also in other issues concerning higher-order asymptotic terms, e.g., in the calculation of dilatation (i.e., the divergence) of the wave $S$ and the rotation of the wave $P$ in smoothly inhomogeneous media (see Kiselev and Rogoff 1998 [18]).

A center of expansion is accepted in seismic exploration as a model of the explosive source. The role played by the wave $S$, resulting from the smooth inhomogeneity of the medium, was discussed in this context by Kiselev and Frolova 1983 [17]. The related numerical research was mentioned in Alekseev and Mikhailenko 1982 [1].

The approach developed in this chapter was extended to point sources, which generate modulated waves of the form (4.237) in inhomogeneous elastic media, by Kiselev 1981 [13]. The above formulas for the wave $S$ from a center of expansion and the wave $P$ from a center of rotation remain valid under the assumption of viscoelasticity of a medium (see Kiselev 1992 [15]). The technique described in the present chapter proved to be useful for the asymptotic description of the wave $S$ from a center of expansion and the wave $P$ from

[6] A presentation of Sobolev's paper [23] can be found in the textbook by Smirnov 1964 [22].

a center of rotation in a medium with a weak anisotropy (see Kiselev 1988, 2001 [14, 16]).

---

## References to Chapter 6

[1] Alekseev, A. S. and Mikhailenko, B. G. 1982. Nongeometrical phenomena in the theory of propagation of seismic waves. *Doklady Akad. Nauk USSR.* 267:1079–83. *Алексеев А. С., Михайленко Б. Г.* Нелучевые эффекты в теории распространения сейсмических волн. Доклады АН СССР, 1982. Т. 267(5). С. 1079–1083.

[2] Alekseev, A. S. and Tsepelev, N. V. 1956. Intensity of reflected waves in inhomogeneous layered media. *Izv. AN SSSR. Ser. Geophys.* 9:1022–35. *Алексеев А. С., Цепелев Н. В.* Интенсивность отраженных волн в слоисто-неоднородной среде. Изв. АН СССР. Сер. геофиз., 1956. № 9. С. 1021–1035.

[3] Avila, G. S. S. and Keller, J. B. 1963. The high-frequency asymptotic field of a point source in an inhomogeneous medium. *Comm. Pure Appl. Math.* 16(4):363–81.

[4] Babich, V. M. 1961. Fundamental solutions of the dynamical equations of elasticity for nonhomogeneous media. *J. Appl. Math. Mech.* 25:49–60. *Бабич В. М.* Фундаментальные решения уравнений теории упругости для неоднородной среды. Прикладная математика и механика, 1961. Т. 25(1). С. 38–45.

[5] Babich, V. M. 1965. The short wave asymptotic form of the solution for the problem of a point source in an inhomogeneous medium. *USSR Comput. Math. Math. Phys.* 5:247–51. *Бабич В. М.* О коротковолновой асимптотике решения задачи о точечном источнике в неоднородной среде. Ж. вычисл. матем. и матем. физ., 1965. Т. 5(5). С. 949–951.

[6] Babich, V. M. 1992. The Hadamard ansatz, its analogues, generalizations, and applications. *St. Petersburg Math. Journal.* 3:937–72. *Бабич В. М.* Анзац Адамара, его аналоги, обобщения и приложения. Алгебра и анализ, 1991. Т. 3(5). С. 1–36.

[7] Babič, V. M. and Kirpičnikova, N. Ya. 1979. *The boundary-layer method in diffraction problems.* Berlin: Springer-Verlag. *Бабич В. М., Кирпичникова Н. Я.* Метод пограничного слоя в задачах дифракции. Л.: Изд-во ЛГУ, 1975.

[8] Cole, J. D. 1972. *Perturbation methods in applied mathematics.* Waltham, Mass.: Blaisdell. *Коул Дж.* Методы возмущений в прикладной математике. М.: Мир, 1972.

[9] Hadamard, J. 1923. *Lectures on Cauchy's problem in linear partial differential equations.* New Haven: Yale University Press. *Адамар Ж.* Задача Коши для уравнений в частных производных гиперболического типа. М.: Наука, 1978.

[10] Ilin, A. M. 1992. *Matching of asymptotic expansions of solutions of boundary value problems.* Providence, RI: AMS. *Ильин А. М., Данилин А. Р.* Асимптотические методы в анализе. М.: Физматлит, 2009.

[11] Kiselev, A. P. 1974. High-frequency point sources in inhomogeneous elastic medium. *Sov. Phys. Doklady.* 19(4): 855–6. *Киселев А. П.* О высокочастотных точечных источниках в неоднородных изотропных упругих средах. Докл. АН СССР, 1974. Т. 219(4). С. 829–831.

[12] Kiselev, A. P. 1975. On initial data for ray formulas describing point sources in inhomogeneous elastic media. *Voprosy Dynamicheskoi Teorii Rasprostraneniya Seismicheskikh Voln [Problems in the Dynamic Theory of Propagation of Seismic Waves]*, ed. G. I. Petrashen'. Leningrad: Leningrad Univ. Press, Leningrad, Vol. 15: 6–27 [in Russian]. *Киселев А. П.* О начальных данных для лучевых формул, описывающих поля точечных источников в неоднородных упругих средах. Вопр. динам. теории распростр. сейсм. волн, 1975. Вып. 15. С. 6–26.

[13] Kiselev, A. P. 1982. Excitation of modulated oscillations in inhomogeneous media. *J. Math. Sci.* 20:1118–25. *Киселев А. П.* Возбуждение модулированных колебаний в неоднородных средах. Зап. научн. семин. ЛОМИ АН СССР, 1981. Т. 104. С. 111–122.

[14] Kiselev A. P. 1988. Point sources of vibrations in a weakly anisotropic elastic media. *Sov. Phys. Doklady.* 33:466–7. *Киселев А. П.* Точечные источники колебаний в слабо анизотропной упругой среде. Докл. АН СССР, 1988. Т. 300(4). С. 824–826.

[15] Kiselev, A. P. 1992. Higher-order terms of the ray theory and "non-geometrical phenomena" in inhomogeneous viscoelastic media. *Phys. Solid Earth.* 28:946–8. *Киселев А. П.* Высшие приближения лучевого метода и "нелучевые явления" в неоднородных вязкоупругих средах. Изв. РАН. Физика Земли, 1992. № 11. С. 35–38.

[16] Kiselev, A. P. 2001. Body waves in a weakly anisotropic medium — II. $S$-waves from a centre of expansion and $P$-waves from a centre of rotation. *Geophys. J. Int.* 145:714–20.

[17] Kiselev, A. P. and Frolova, E. N. 1983. Transverse wave from a nondirectional source in an inhomogeneous elastic media. *Izv. Acad. Sci. USSR. Phys. Solid Earth.* 18:323–8. *Киселев А. П., Фролова Е. Н.* Поперечная волна от ненаправленного источника в неоднородной упругой среде. Изв. АН СССР. Физика Земли, 1982. № 5. С. 3–8.

[18] Kiselev, A. P. and Rogoff, Z. M. 1998. Dilatation of $S$-waves in smoothly inhomogeneous isotropic elastic media. *J. Acoust. Soc. Amer.* 104:2592–5.

[19] Lax P. D. 1957. Asymptotic solutions of oscillatory initial value problems. *Duke Math. J.* 24:627–46.

[20] Lu, W., Qian, J. and Burridge, R. 2016. Babich-like ansatz for three-dimensional point-source Maxwell's equations in an inhomogeneous medium at high frequencies. *Multiscale Model. Simul.* 14:1089–122.

[21] Qian, J., Yuan, L., Liu Y., Luo S. and Burridge, R. 2016. Babich's expansion and high-order eulerian asymptotics for point-source Helmholtz equations. *J. Sci. Comput.* 67:883–908.

[22] Smirnov, V. I. 1964. *A course of higher mathematics, Volume 4.* Reading MA: Addison-Wesley. *Смирнов В. И.* Курс высшей математики. Т. 4. Ч. 2. М.: Наука, 1981.

[23] Sobolev, S. L. 1930. Wave equation for inhomogeneous medium. *Trudy Seismolog. Institute.* 6:1–57 [in Russian]. *Соболев С. Л.* Волновое уравнение для неоднородной среды. Труды Сейсмологического института, 1930. № 6. С. 1–57.

# Chapter 7

## The "Nongeometrical" Wave $S^*$

For several decades it has been undisputable that any wave described by the elastodynamics equations can be associated with related rays. For this reason much attention was drawn in the 1980s to numerical investigations of a wave named $S^*$ that showed an astonishing property: it did not seem to be related to any ray path and to be described by the ray method. The wave $S^*$ is generated by a point source placed closely to a plane boundary or interface. It is allied with the total internal reflection. In the present chapter, we show that it can be described within the framework of the ray method, that is, it can be represented in the form (4.2), though, with a complex eikonal. The specific feature of the wave $S^*$ is that the corresponding eikonal has a small imaginary part. We confine our consideration to the case of a plane traction-free boundary of an isotropic elastic body.

In the approach that we develop in the sequel, the reciprocity principle is of fundamental importance. It is not essential (in contrast to the construction of an explicit solution, e.g., by the Fourier method) that the boundary is plane and the medium is homogeneous. The issues we are concerned with have simpler counterparts in the scalar problem of two half-spaces in contact, with a source placed in a faster medium.

## 7.1 Statement of the problem and qualitative discussion

### 7.1.1 Boundary-value problem

Let a point source, namely, a center of expansion, act at an interior point $\mathbf{y} = (x^{\circ}, y^{\circ}, h)$, $h > 0$, of a homogeneous isotropic half-space $x_3 = z > 0$. The point $\mathbf{y}$ is referred to as the *source point* and $h$ is its *depth.* A variable *observation point* will be denoted by $\mathbf{x} = (\xi, \eta, \zeta)$. The related elastodynamics equation (3.47), (3.49) for a time-harmonic displacement $\boldsymbol{u} = \boldsymbol{u}(\mathbf{x}; \mathbf{y})$ reads

$$l_{\mathbf{x}} \boldsymbol{u} = -4\pi \, \mathrm{grad}_{\mathbf{x}} \, \delta(\mathbf{x} - \mathbf{y}), \quad \zeta > 0. \tag{7.1}$$

A boundary condition on the plane $z = 0$ must be added; let it be the traction-free condition

$$\mathbf{t}_{\mathbf{x}}^{\mathbf{e}_3} \boldsymbol{u}\big|_{\zeta=0} = 0. \tag{7.2}$$

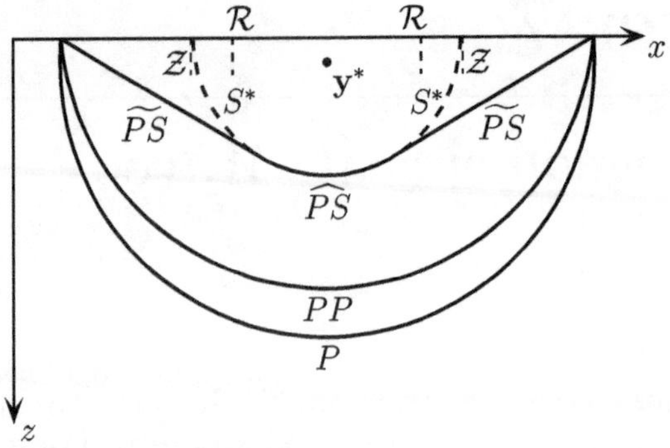


**Figure 7.1**: Wavefronts in the problem (7.1)–(7.2)

Also we assume that the limiting absorption principle holds, to ensure the uniqueness of a solution (which can be immediately constructed by the Fourier method).

We will see that in this relatively simple problem, a "nongeometrical" wave $S^*$ appears, and it may have a considerable amplitude as $\omega$ is large and $h$ is small. Qualitative arguments to prove its existence are given in Section 7.1.2.

### 7.1.2 Qualitative discussion of arising waves

#### What waves do we expect to arise?

Let us consider what waves are generated by this source. Figure 7.1 shows the sections of wavefronts by the plane $\eta = y^{\circ}$. The wavefronts are marked in correspondence with the names of waves. First, an *incident* spherical wave $P$ appears traveling directly from the source point to the observation point. Further, it gives rise to reflected waves: a monotype $PP$ and a converted $PS$. We find it convenient to split the front of the wave $PS$ into the parts, marked by $\widehat{PS}$ and $\widetilde{PS}$, that correspond to subcritical and supercritical angles, respectively. Consideration based on the ray approach seems to predict no other waves.

However, numerical simulations of the wavefield (e.g., by means of a finite differences approach) show the presence of an extra wave in a neighborhood of the bold dashed line $S^*$. Its polarization is predominantly transverse, and for a sufficiently shallow source point this wave is strong. No related rays exist in the sense in which they were introduced in Chapters 2 and 4. Researchers who found this wave numerically called it "nongeometrical" (without quotation marks, however), see Hron and Mikhailenko 1981 [18]. This chapter is concerned with its asymptotic description.

Further on, in the course of traveling along the boundary, the wave $S^*$ also generates a *Zaitsev wave* $\mathcal{Z}$ rapidly attenuating with depth, shown in Fig. 7.1 with a dashed line. This wave will be discussed in Section 7.2.5. Furthermore,

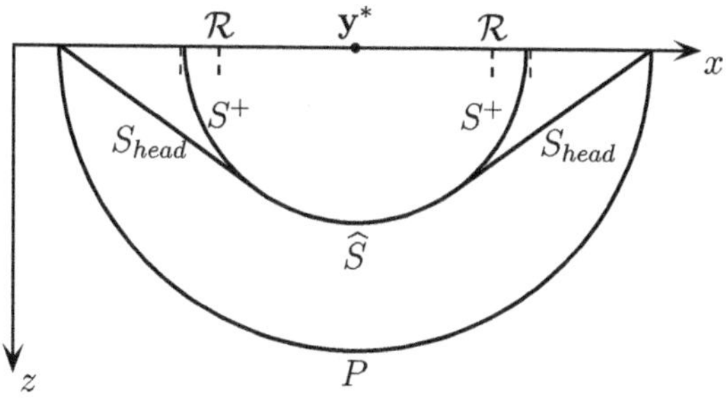


**Figure 7.2**: Section of wavefronts by the plane $\eta = y^\circ$ for the source point $\mathbf{y}^*$

a surface Rayleigh wave arises (shown in Fig. 7.1 with a dotted line and marked by $\mathcal{R}$). Its excitation is beyond the scope of our book.[1]

### Why does the wave $S^*$ arise?

Let the point $\mathbf{y}$ normally approach the boundary of the half-space. In the limit, the wavefield grades into that of a certain point source placed at the projection $\mathbf{y}^* = (x^\circ, y^\circ, 0)$ of $\mathbf{y}$ onto the boundary. The wavefronts corresponding to the limiting source are well known, and they are presented in Fig. 7.2. Here, $P$ is the front of the longitudinal wave, $\widehat{S}$ and $S^+$ are parts of the front of the shear wave, and $S_{head}$ is the front of the *head wave.*

When slightly shifting the source into the half-space, we cannot expect a too abrupt change in the wavefield, and the wave marked by $S^+$ cannot immediately disappear. The resulting wave is expected to be of a comparable strength; we mark it by $S^*$.

### ⋆ Head wave

Digressing for a while from our main subject, we will clarify now the nature of a head wave generated by a point source placed at the surface of a homogeneous half-space. Thorough discussion of head waves is given, e.g., by Brekhovskikh 1960 [12] and Landau and Lifshitz 1987 [25] (by the way, these authors call it a *lateral wave*). In contrast to the wave $S^*$, it is "quite a ray wave." The related rays are polygonal lines connecting the surface source point $\mathbf{y}^*$ and the observation point $\mathbf{x}$ on its wavefront. These rays (as well as any other ones) can be described on the basis of the Fermat principle, which has here some specific features.

Let us take interest in the minimal time required to come from $\mathbf{y}^*$ to some point $\check{\mathbf{y}}$, traveling with velocity $a$ along the surface, and after, up to the inner

[1] This Rayleigh wave (which is so strong for shallow sources and the values of parameters typical of seismology that it is believed to be the cause of the main damage by earthquakes) is described by the exact solution and also in the course of numerical simulations. Description of its excitation, not based upon the exact solution, was called by Babich, Kiselev, Lawry, and Starkov 2001 [11] "a challenge to diffraction community."

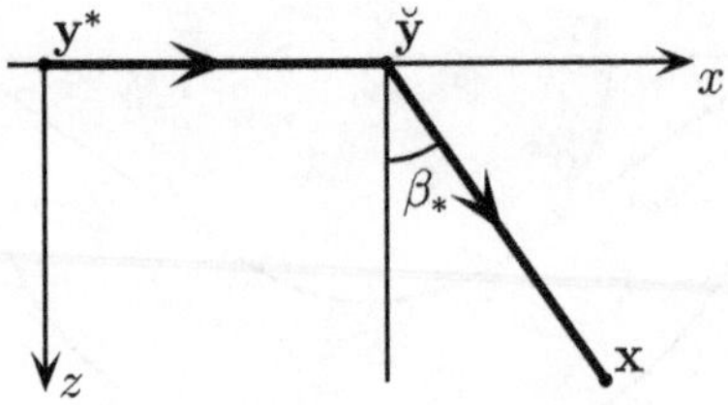


**Figure 7.3**: Ray path of the head wave

point $\mathbf{x}$ of the half-space, with velocity $b$. The related Fermat functional has the form

$$\int_{\mathbf{y}^*}^{\check{\mathbf{y}}} \frac{ds}{a} + \int_{\check{\mathbf{y}}}^{\mathbf{x}} \frac{ds}{b}, \tag{7.3}$$

with $ds$ standing as always for the differential of the arc length. The comparison curves are sums of two smooth segments. One of them, $\mathbf{y}^*\check{\mathbf{y}}$, belongs to the boundary and connects $\mathbf{y}^*$ and $\check{\mathbf{y}}$; the other, $\check{\mathbf{y}}\mathbf{x}$, lies inside the half-space. If the horizontal distance between $\mathbf{x}$ and $\mathbf{y}^*$ is sufficiently large, an easy exercise is to show that the minimum is achieved as $\mathbf{y}^*\check{\mathbf{y}}$ and $\check{\mathbf{y}}\mathbf{x}$ are straight segments lying in the same plane and the angle between $\check{\mathbf{y}}\mathbf{x}$ and the normal to the boundary coincides with the critical angle $\beta_* = \arcsin\frac{b}{a}$ introduced by (2.168) (see Fig. 7.3). The polygonal two-segment line $\mathbf{y}^*\check{\mathbf{y}}\mathbf{x}$ is the ray of the head wave. The corresponding eikonal equals

$$\tau(\mathbf{x}) = \frac{|\mathbf{y}^* - \check{\mathbf{y}}|}{a} + \frac{|\check{\mathbf{y}} - \mathbf{x}|}{b}. \tag{7.4}$$

The head wave is a wave $S$ with the eikonal (7.4). Its wavefront is part of the surface of a circular cone.[2] Its amplitude can be described by the ray method on which we do not dwell. Such a calculation for the scalar case of a contact of two homogeneous isotropic half-spaces can be found in Friedrichs and Keller 1955 [16], and for isotropic elastodynamics in Alekseev and Gel'chinsky 1961 [3].

---

## 7.2 Derivation of formulas

At first glance, the derivation of an asymptotic formula for the wave $S^*$ by the ray method is an easy matter. Let us explain what difficulties are encountered in doing this. Essential is that for the source point close to a boundary, the wavefield does not have a ray-type structure and finding reflected waves

[2] The rays of the head waves generated by the source at $\mathbf{y}^*$ fill the domain called *supercritical*. For points inside this domain, the segment $\check{\mathbf{y}}\mathbf{x}$ is nondegenerate (see Fig. 7.3).

appears to us as a radically novel problem. Its relatively simple solution can be found with the help of the reciprocity principle.

By swapping the source and observation points, we arrive at the consideration of incidence on the boundary of a spherical wave from a remote source, which is not "dangerously close" to the boundary and to which the standard ray method is therefore applicable. The proximity of the observation point to the boundary implies no complication. Such is the idea of the ray calculation of formulas for the wave $S^*$, to which we proceed.

The wave $S^*$ will be sought in the form of the ray series ansatz

$$\boldsymbol{u}^{S^*}(\mathbf{x}) = e^{i\omega\tau}\left\{\boldsymbol{u}^0(\mathbf{x}) + \frac{\boldsymbol{u}^1(\mathbf{x})}{-i\omega} + \ldots\right\}, \quad \omega \to \infty, \tag{7.5}$$

where the formal large parameter $\omega$ corresponds to the dimensionless parameter

$$\frac{\omega r_*}{b} \gg 1, \tag{7.6}$$

where

$$r_* = |\mathbf{x} - \mathbf{y}^*| = \sqrt{(\xi - x^\circ)^2 + (\eta - y^\circ)^2 + \zeta^2} \tag{7.7}$$

is the distance between the observation point and the projection of the source point onto the boundary $\mathbf{y}^* = (x^\circ, y^\circ, 0)$. Therewith, the eikonal and amplitude functions are expansions in powers of the depth $h$:

$$\tau = \tau_0 + h\tau_1 + h^2\tau_2 + \ldots, \tag{7.8}$$

$$\boldsymbol{u}^j = \boldsymbol{u}^{j0} + h\boldsymbol{u}^{j0} + \ldots, \tag{7.9}$$

$j = 0, 1, \ldots$, see in this connection the Section 4.2.5 devoted to the complex eikonal.

### 7.2.1 Auxiliary problem and reciprocity principle

In order to apply the ray method rigorously, we will deal with a source distant from the boundary. Let us introduce an auxiliary problem with a certain point source, in which the roles of the points $\mathbf{x}$ and $\mathbf{y}$ are reversed. We denote by $\boldsymbol{w} = \boldsymbol{w}(\mathbf{y}'; \mathbf{x})$ the solution for the concentrated vertical force acting at the point $\mathbf{x}$, which is not "dangerously close" to the boundary

$$\boldsymbol{l}_{\mathbf{y}'}\boldsymbol{w} = -4\pi\mathbf{e}_3\delta(\mathbf{y}' - \mathbf{x}), \quad z' > 0, \tag{7.10}$$

with the same boundary condition

$$\mathbf{t}^{\mathbf{e}_3}_{\mathbf{y}'}\boldsymbol{w}\Big|_{\mathbf{y}'=(x',y',0)} = 0. \tag{7.11}$$

Apply now the formula (1.93) expressing the reciprocity principle to inner points to $\boldsymbol{u}(\mathbf{y}'; \mathbf{y})$ and $\boldsymbol{w}(\mathbf{y}'; \mathbf{x})$ of the half-space $\mathbb{R}^3_+ = \{z > 0\}$:

$$\int_{\mathbb{R}^3_+} \{\boldsymbol{u}(\mathbf{y}'; \mathbf{y}) \cdot \mathbf{e}_3\delta(\mathbf{y}' - \mathbf{x}) - \boldsymbol{w}(\mathbf{y}'; \mathbf{x}) \cdot \mathrm{grad}_{\mathbf{y}'}\,\delta(\mathbf{y} - \mathbf{y}')\}\, d^3\mathbf{y}' = 0.$$

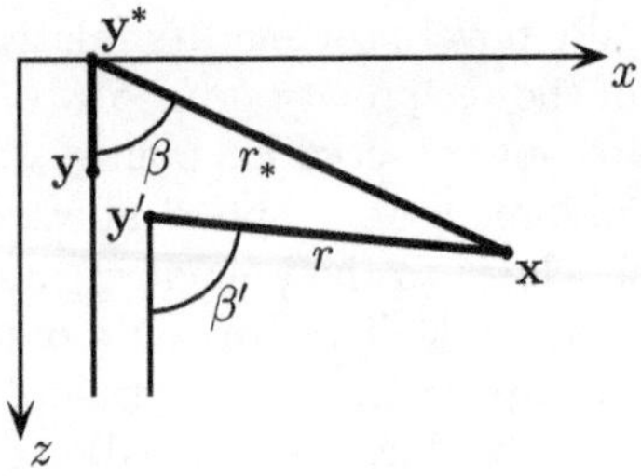


**Figure 7.4**: Supercritical reflection in the auxiliary problem

After integration by parts of the second term in curly brackets and making use of the boundary conditions (7.2) and (7.11) and the property of the delta function (3.17), we get

$$u_3(\mathbf{x};\mathbf{y}) \equiv (\boldsymbol{u}, \mathbf{e}_3) = -\operatorname{div}_{\mathbf{y}} \boldsymbol{w}(\mathbf{y};\mathbf{x}). \tag{7.12}$$

Similarly,

$$u_\gamma(\mathbf{x};\mathbf{y}) \equiv (\boldsymbol{u}, \mathbf{e}_\gamma) = -\operatorname{div}_{\mathbf{y}} \boldsymbol{w}^{(\gamma)}(\mathbf{y};\mathbf{x}), \tag{7.13}$$

where the $\boldsymbol{w}^{(\gamma)}$ are solutions for concentrated horizontal forces at the point $\mathbf{x}$ and acting along the axes $\mathbf{e}_1$ and $\mathbf{e}_2$,

$$\begin{aligned} \boldsymbol{l}_{\mathbf{y}'}\boldsymbol{w}^{(\gamma)} &= -4\pi\mathbf{e}_\gamma\delta(\mathbf{y}'-\mathbf{x}), \quad z'>0, \\ \mathbf{t}^{\mathbf{e}_3}_{\mathbf{y}'}\boldsymbol{w}^{(\gamma)}\big|_{\mathbf{y}'=(x',y',0)} &= 0, \quad \gamma=1,2. \end{aligned} \tag{7.14}$$

Formulas (7.12) and (7.13) will be the basis for obtaining the ray asymptotics of $u_3^{S^*}$, $u_1^{S^*}$, and $u_2^{S^*}$ from solutions of problems with point sources distant from the boundary. Formulas (7.12) and (7.13) are specialized cases of the "reciprocal formulas for point sources," presented by Babich 1962 [5].

### 7.2.2 What is required to find the wave $S^*$?

We start with the note that the full solution $\boldsymbol{w}$ of the auxiliary problem, which is somewhat cumbersome, will not be required.

A concentrated vertical force acting at the point $\mathbf{x}$ generates waves $P$ and $S$, that are incident on the boundary, and each generates reflected waves of both types. They all have different phases. Since waves $S$ are solenoidal, waves $P$ solely give a contribution to the divergence of $\boldsymbol{w}$. Their number is three. The direct wave $P$ going from $\mathbf{x}$ to $\mathbf{y}'$, as well as the monotype reflected wave $PP$, are associated with the corresponding rays, can be described by the standard techniques, and are of no interest in the subsequent calculation.

Quite another matter is the converted wave $SP$, generated at $\mathbf{x}$ as the wave $S$, for which the total internal reflection is possible, provided that the incidence angle $\beta$ is large enough, i.e., $\mathbf{x}$ is such that

$$\sin\beta > \beta_* = \frac{b}{a}. \tag{7.15}$$

In this domain termed supercritical, see Fig. 7.4, the converted wave $SP$ has a complex eikonal, which will be constructed in accordance with the scheme presented in Section 4.2.5. A similar expansion will be considered for vector amplitudes.

### 7.2.3 Solution of the auxiliary problem for $\boldsymbol{w}$

We seek a solution of the auxiliary problem in the form

$$\boldsymbol{w} = \boldsymbol{w}^{\mathrm{i}} + \boldsymbol{w}^{\mathrm{r}}, \tag{7.16}$$

where $\boldsymbol{w}^{\mathrm{i}}$ is the incident wavefield, i.e., the solution of the equation (7.10) in the whole space, and $\boldsymbol{w}^{\mathrm{r}}$ is the reflected wavefield. The sum $\boldsymbol{w}^{\mathrm{i}} + \boldsymbol{w}^{\mathrm{r}}$ satisfies the boundary condition (7.11).

#### Incident wavefield

As follows from the formula (3.86) for the far-field area, the wavefield of a concentrated force $\boldsymbol{w}^{\mathrm{i}}(\mathbf{y}';\mathbf{x})$ at a variable point $\mathbf{y}' = (x', y', z')$ can be represented as the sum of waves $P$ and $S$:

$$\boldsymbol{w}^{\mathrm{i}}(\mathbf{y}';\mathbf{x}) = \boldsymbol{w}^{\mathrm{i}a}(\mathbf{y}';\mathbf{x}) + \boldsymbol{w}^{\mathrm{i}b}(\mathbf{y}';\mathbf{x}). \tag{7.17}$$

As is apparent from Section 7.2.2, of interest is only the term $\boldsymbol{w}^{\mathrm{i}b}$. At a point $\mathbf{y}'$ distant from $\mathbf{x}$, we apply the far-field formula (3.86). For the incident wave $S$, we have

$$\boldsymbol{w}^{\mathrm{i}b}(\mathbf{y}';\mathbf{x}) = e^{i\frac{\omega}{b}r}\boldsymbol{w}^{ib0}\left(1 + O\left(\frac{b}{\omega r}\right)\right), \tag{7.18}$$

where

$$\boldsymbol{w}^{ib0} = \frac{\mathbf{e}_3 - s_3{'}\mathbf{s}'}{\rho b^2 r} = \frac{\sin\beta'(\mathbf{e}_1\cos\beta' + \mathbf{e}_3\sin\beta')}{\rho b^2 r} \tag{7.19}$$

and

$$r = |\mathbf{y}' - \mathbf{x}| = \sqrt{(x'-\xi)^2 + (y'-\eta)^2 + (z'-\zeta)^2}. \tag{7.20}$$

Here, $\mathbf{s}' = \mathbf{e}_1\sin\beta' - \mathbf{e}_3\cos\beta'$ is a unit vector directed from the source point $\mathbf{x}$ to the point $\mathbf{y}'$, $s_3'$ is its projection onto the $z$-axis, and $\beta'$ stands for the angle between $\mathbf{s}$ and the direction of the $z$-axis.

#### Reflected wavefield. Elementary derivation of the leading-order term

We seek the converted wave $SP$ in the form of the ray ansatz

$$\boldsymbol{w}^{SP} = e^{i\omega\tau_{SP}}\left\{\boldsymbol{W}^{SP0} + \frac{\boldsymbol{W}^{SP1}}{-i\omega} + \dots\right\}. \tag{7.21}$$

The eikonal $\tau_{SP} = \tau_{SP}(\mathbf{y}';\mathbf{x})$ and the amplitudes $\boldsymbol{W}^{SPj}$ are considered as series in the depth $z'$ of the observation point $\mathbf{y}'$.

Assume now that the point $\mathbf{y}'$ is close to the point $\mathbf{y}^*$, and let the distance $|\mathbf{y}' - \mathbf{y}^*|$ be of order $h$. Expand the phase of the incident wave, $|\mathbf{y}' - \mathbf{y}^*|$ up to linear terms, inclusively, under the assumption that the points $\mathbf{x}$, $\mathbf{y}^*$, and $\mathbf{y}'$ lie in the same vertical plane, $\eta = y' = y^0$. Then

$$r = |\mathbf{x} - \mathbf{y}'| = r_* + x \sin\beta' - z\cos\beta' + O(h^2/r_*), \tag{7.22}$$

where, for convenience of the comparison with the formulas of Chapter 2, we put $z = z'$, $x = x' - \xi$. Consequently,

$$\begin{aligned} \boldsymbol{w}^{\mathrm{ib}}(\mathbf{y}';\mathbf{x}) &= \mathfrak{A} e^{i\text{æ}(x\sin\beta' - z\cos\beta')}\left(\frac{\cos\beta'}{b}\mathbf{e}_1 + \frac{\sin\beta'}{b}\mathbf{e}_3\right) \\ &\times \left(1 + O\left(\frac{h}{r_*}\right) + O\left(\frac{b}{\omega r_*}\right) + O\left(\frac{\omega h^2}{b r_*}\right)\right), \end{aligned} \tag{7.23}$$

where $\text{æ} = \frac{\omega}{b}$ and

$$\mathfrak{A} = -\frac{\sin\beta'}{\rho b r_*} e^{i\text{æ} r_*} \tag{7.24}$$

is treated (in this approximation) as a constant. According to Chapter 2, the leading term in the expression (7.23) is an upgoing plane wave $P-SV$. From the formula (2.160), the expression for the corresponding converted plane wave $SP$ is

$$\boldsymbol{w}^{SP0} \approx \mathfrak{A} B^a e^{i\frac{\omega}{a}(x\sin\alpha' + z\cos\alpha')}\left(\frac{\sin\alpha'}{a}\mathbf{e}_1 + \frac{\cos\alpha'}{a}\mathbf{e}_3\right). \tag{7.25}$$

Here, $\alpha'$ is the angle connected with the angle $\beta'$ by the Snell law (2.167) which takes the form $\frac{\sin\beta'}{b} = \frac{\sin\alpha'}{a}$ in the notation of this chapter. In the supercritical domain $\alpha'$ is complex. By $B^a = B^a(\beta')$ we denote the related conversion coefficient corresponding to the wave $S$ incident under the angle $\beta'$ with the normal (see Section 2.5.4).

**Calculation of the divergence of the wave $SP$**

To find the leading term of the divergence, we just differentiate the phase. From (7.25) we have

$$\begin{aligned} \operatorname{div}_{\mathbf{y}'} \boldsymbol{w}^{SP0} &\approx \mathfrak{A} B^a \left(\frac{\sin\alpha'}{a}\frac{\partial}{\partial x} + \frac{\cos\alpha'}{a}\frac{\partial}{\partial z}\right) e^{i\frac{\omega}{a}(x\sin\alpha' + z\cos\alpha')} \\ &= \frac{i\omega}{a^2}\mathfrak{A} B^a e^{i\frac{\omega}{a}(x\sin\alpha' + z\cos\alpha')}. \end{aligned}$$

### 7.2.4 Leading term of the asymptotics of the wave $S^*$

**Usage of the reciprocity principle**

Thus,

$$u_3^{S^*}(\mathbf{x};\mathbf{y}) = -\operatorname{div}_{\mathbf{y}} \boldsymbol{w}(\mathbf{y};\mathbf{x}) \approx -\frac{i\omega}{a^2}\mathfrak{A} B^a\Big|_{\beta'=\beta} e^{i\frac{\omega}{a}(x\sin\alpha + z\cos\alpha)}\Big|_{x=0,z=h}, \tag{7.26}$$

or

$$u_3^{S^*}(\mathbf{x};\mathbf{y}) = \sin\beta \frac{e^{iær_*}}{r_*} g(\beta)\left(1 + O\left(\frac{h}{r_*}\right) + O\left(\frac{b}{\omega r_*}\right) + O\left(\frac{\omega h^2}{br_*}\right)\right) \tag{7.27}$$

with $g(\beta)$ dependent solely on $\beta$:

$$g(\beta) = \frac{i\omega}{\rho a^2 b} e^{i\frac{\omega}{a}h\cos\alpha} B^a(\beta). \tag{7.28}$$

Here,

$$\cos\alpha = \cos(\alpha(\beta)) = \sqrt{1 - \sin^2(\alpha(\beta))} = i\sqrt{\frac{b^2}{a^2}\sin^2\beta - 1}.$$

In a similar manner, from the identity (7.13) for $\gamma = 1$ it follows that

$$u_1^{S^*}(\mathbf{x};\mathbf{y}) = \cos\beta \frac{e^{iær_*}}{r_*} g(\beta)\left(1 + O\left(\frac{h}{r_*}\right) + O\left(\frac{b}{\omega r_*}\right) + O\left(\frac{\omega h^2}{br_*}\right)\right). \tag{7.29}$$

From (7.13) for $\gamma = 2$ we obtain $u_1^{S^*}(\mathbf{x};\mathbf{y}) = 0$.

### Final formula for the leading term

Consequently,

$$\begin{gathered} \boldsymbol{u}^{S^*} = \frac{e^{iær_*}}{r_*} g(\beta)\mathsf{s}\left(1 + O\left(\frac{h}{r_*}\right) + O\left(\frac{b}{\omega r_*}\right) + O\left(\frac{\omega h^2}{br_*}\right)\right), \\ \beta > \beta_* = \arcsin\frac{b}{a}, \end{gathered} \tag{7.30}$$

where $\mathsf{s} = \mathsf{s}(\beta) = \mathbf{e}_1\cos\beta + \mathbf{e}_3\sin\beta$ stands for the unit vector orthogonal to the ray lying in the plane of incidence and connecting $\mathbf{y}^*$ with $\mathbf{x}$.

In this approximation, the wave $S^*$ shows itself as a spherical wave emitted from the projection of the source point onto the boundary. It exists only for supercritical observation angles. Its radiation pattern differs by a constant factor from the function

$$B^a(\beta)e^{iæh\cos\alpha(\beta)} = B^a(\beta)e^{-\frac{\omega|\cos\alpha(\beta)|}{a}h}, \quad \beta < \beta_*, \tag{7.31}$$

which strongly depends on the *dimensionless source depth*

$$æh = \frac{\omega h}{b} \tag{7.32}$$

and rapidly tends to zero as $æh$ grows.

### 7.2.5 ⋆ On higher-order terms and other refinements

**Eikonal of the wave $SP$**

Consider the expansion of the eikonal $\tau_{SP}$ in powers of $z'$

$$\tau_{SP} = \tau_0 + z'\tau_1 + z'^2\tau_2 + \dots \tag{7.33}$$

in the auxiliary problem. Following the line of argument in Section 4.2.5, we arrive at the equation

$$\frac{\partial \tau_{SP}}{\partial z'} = i\sqrt{\left(\frac{\partial \tau_{PS}}{\partial x'}\right)^2 + \left(\frac{\partial \tau_{PS}}{\partial y'}\right)^2 - \frac{1}{a^2}} \tag{7.34}$$

with the condition having the form

$$\begin{aligned} \tau_{SP}\big|_{z'=0} &= \frac{r}{b}\bigg|_{z'=0} = |\mathbf{y}' - \mathbf{x}|_{z'=0} \\ &= \frac{1}{b}\sqrt{(x'-\xi)^2 + (y'-\eta)^2 + (z'-\zeta)^2}\bigg|_{z'=0} \end{aligned} \tag{7.35}$$

in the notation (7.20). Calculations (rather easy) show that the leading term of the expansion (7.33) is real, while the next one is imaginary:

$$\tau_0 = \frac{r_*}{b}, \quad \tau_1 = \frac{\cos\alpha'}{a} = \frac{i|\cos\alpha'|}{a}, \tag{7.36}$$

which perfectly agrees with the expression obtained above with the help of the theory of plane waves. After some more algebra (the detailed calculation is given in Babich, Kiselev, Lawry, and Starkov 2001 [11]) the next term in the expansion (7.33) can be found, and it proves to be real:

$$\tau_2 = \tau_2(\beta') = -\frac{\sin^2\beta'}{2br_*}\left(\frac{a\cos\beta'}{b|\cos\alpha'|}\right)^2. \tag{7.37}$$

The above calculations result in the following refinement of the expression (7.31):

$$\boldsymbol{u}^{S^*} = \frac{e^{i\text{æ}r_*}}{r_*}\widetilde{g}(\beta)\mathsf{s}\left(1 + O\left(\frac{h}{r_*}\right) + O\left(\frac{b}{\omega r_*}\right) + O\left(\frac{\omega h^3}{br_*^2}\right)\right), \tag{7.38}$$

where

$$\widetilde{g}(\beta) = \frac{i\omega}{\rho a^2 b} e^{i\omega\left[h\frac{\cos\alpha(\beta)}{a} - h^2\tau_2(\beta)\right]} B^a(\beta). \tag{7.39}$$

The expression of order $h^2$ in the exponential describes the "time advance as compared to the geometric travel time," which was observed in a detailed computer modelling by Tsvankin and Kalinin 1984 [28].

### Higher-order terms for amplitudes

The construction of scalar counterparts of the amplitude vectors $\boldsymbol{u}^0$, $\boldsymbol{u}^1, \ldots$ in the form of expansions with respect to powers of $h$ (7.9) is described in the case of a scalar problem of contact between two homogeneous half-spaces by Babich, Kiselev, Lawry, and Starkov 2001 [11].

### Zaitsev wave and a neighborhood of the limiting ray

The wave with wavefront shown in Fig. 7.1 with a dashed line marked as $\mathcal{Z}$, is an inhomogeneous wave $P$ arising in propagating the wave $S^*$ along the boundary. The wave $\mathcal{Z}$ is needed in order to satisfy, in the sum with the wave $S^*$, the boundary condition. A sketch of calculations of its amplitude in the context of the ray method was presented by Babich, Kiselev, Lawry, and Starkov 2001 [11]. There the wavefield is described in the vicinity of the limiting ray $\beta = \beta_*$ (where the ray expansions of the waves $S^*$ and $PS$ are not applicable, and the wavefield is expressed in terms of the *Pearcey integral*). That paper concerns the scalar case.

---

## 7.3 ⋆ Comments to Chapter 7

The oldest known, though unnoticed over a period of decades, mentioning of the wave $S^*$ can be found in Lapwood 1949 [24], where an exact solution of the two-dimensional problem represented by the Fourier integral was studied. The qualitative consideration of Section 7.1.2 is traceable to this paper. Wide attention has been attracted to the wave $S^*$ due to numerical simulations by Hron and Mikhailenko 1981 [18]. Shortly thereafter, it was identified in the explicit solution of the three-dimensional problem by Daley and Hron 1983 [14] and observed in a physical modeling by Kim and Behrens 1986 [21]. Researchers were unable to associate it with any rays and thus called it a "nongeometrical" wave. The seismic community was impressed by the wave $S^*$ so much that its properties were berhymed by Gutowsky, Hron, Wagner, and Treifel 1984 [17]. Contemporaneously, numerical simulations revealed some other *"nongeometrical phenomena."* A striking example was a wave propagating with velocity of waves $S$ but having the longitudinal polarization on a certain ray (see, e.g., Alekseev and Mikhailenko 1982 [4], Tsvankin and Kalinin 1984 [28], Daley and Hron 1987 [15]). It was the converted wave $PS$ resulted from the reflection of an incident spherical wave $P$ from a plane interface, near the vertical (that is the ray passing across the source point). Later, it was demonstrated that all those "nongeometrical phenomena" perfectly agree with the ray theory (see the review by Babich and Kiselev [10]). So, the anomalous polarization is perfectly described by taking into account the corresponding additional component (see Kiselev and Tsvankin 1989a, 1989b [22, 23] and others).

The above approach to the construction of asymptotic formulas for the wave $S^*$, based on the reciprocity principle given above, has been developed by Babich and Kiselev 1987, 1988 [8, 9]. This approach does not encounter serious difficulties as applied to the case of an interface, not necessarily a plane one. Neither homogeneity nor isotropy of the medium is indispensable. A sketch of the reciprocity-based ray construction of higher-order terms for the wave $S^*$ and of other "nongeometrical waves" for a point source close to an interface is given (for the scalar case) by Babich, Kiselev, Lawry, and Starkov 2001 [11]. We note that many authors used approaches based upon reciprocity principles in finding (mainly in a heuristic manner) asymptotic high-frequency formulas (see, e.g., the derivation of expressions for creeping waves by Keller 1956 [20] and for head waves of interference type by Buldyrev 1967 [13]). Several other applications of reciprocity can be found in Achenbach 2003 [1]. It should be mentioned that a ray approach to the wave $S^*$ is developed by using *complex rays* and without addressing the reciprocity. Its short scalar version is given by Babich 1995a [6]; for a more detailed presentation of a theory of complex rays in the application to elastodynamics, see Babich 1995b [7].

The literature devoted to the study of exact solutions for point sources in layered media is enormous. We confine ourselves to mentioning the textbooks due to Aki and Richards 1980 [2], Brekhovskikh 1960 [12], Hudson 1980 [19] and Petrashen', Molotkov, and Krauklis 1982 [27], together with two remarkably interesting, in our opinion, papers by Petrashen', Marchuk, and Ogurtsov 1950 [26] and Zavorokhin 2012 [30], concerning non-time-harmonic case. The Zaitsev wave was first mentioned by Zaitsev 1959 [29], in the course of analyzing an exact solution.

---

## References to Chapter 7

[1] Achenbach, J. D. 2003. *Reciprocity in elastodynamics.* Cambridge: Cambridge University Press.

[2] Aki, K. and Richards, P. G. 1980. *Quantitative seismology: Theory and methods.* San Francisco: W. H. Freeman and Company. *Аки К., Ричардс П.* Количественная сейсмология. Т. 1, 2. М.: Мир, 1983.

[3] Alekseev, A. S. and Gel'chinsky, B. Ya. 1961. Ray method of calculation of intensity of headv waves. In *Voprosy Dynamicheskoi Teorii Rasprostraneniya Seismicheskikh Voln [Problems in the Dynamic Theory of Propagation of Seismic Waves]*, ed. G. I. Petrashen'. Leningrad: Leningrad Univ. Press, Leningrad, Vol. 5: 54–72 [in Russian]. *Алексеев А. С., Гель-*

*чинский Б. Я.* Лучевой метод вычисления интенсивности головных волн. Вопр. динам. теории распростр. сейсм. волн, 1961. Вып. 5. С. 54–72.

[4] Alekseev, A. S. and Mikhailenko, B. G. 1982. Nongeometrical phenomena in the theory of propagation of seismic waves. *Doklady Akad. Nauk USSR.* 267:1079–83. *Алексеев А. С., Михайленко Б. Г.* Нелучевые эффекты в теории распространения сейсмических волн. Докл. АН СССР, 1982. Т. 267(5). С. 1079–1083.

[5] Babich, V. M. 1962. Reciprocity principle for dynamic equations of theory of elasticity. In *Voprosy Dynamicheskoi Teorii Rasprostraneniya Seismicheskikh Voln [Problems in the Dynamic Theory of Propagation of Seismic Waves]*, ed. G. I. Petrashen'. Leningrad: Leningrad Univ. Press, Leningrad, Vol. 6: 66–74 [in Russian]. *Бабич В. М.* Принцип взаимности для динамических уравнений теории упругости. Вопр. динам. теории распростр. сейсм. волн, 1962. Вып. 6. С. 66–74.

[6] Babich, V. M. 1995a. On applying complex rays to calculation of scalar "nongeometric" waves. *J. Math. Sci.* 73:308–16. *Бабич В. М.* Применение комплексных лучей к расчету "негеометрических" волн в скалярном случае. Зап. научн. семин. ЛОМИ АН СССР, 1990. Т. 186. С. 20–32.

[7] Babich, V. M. 1995b. Calculation of a "nongeometric" $S^*$-wave by using complex rays. *J. Math. Sci.* 77:3146–52. *Бабич В. М.* Расчет "нелучевой" волны $SH$ с помощью комплексных лучей. Зап. научн. семин. ЛОМИ АН СССР, 1992. Т. 200. С. 17–26.

[8] Babich, V. M. and Kiselev, A. P. 1987. Geometro-seismic description of "nongeometrical" waves $P^*$, $S^*$, ... *Preprint P-11-87. Leningrad Branch of Steklov Math. Institute* [in Russian]. *Бабич В. М., Киселев А. П.* Геометросейсмическое описание "нелучевых" волн $P^*$, $S^*$, ... Препринт ЛОМИ АН СССР, 1987.

[9] Babich, V. M. and Kiselev, A. P. 1988. Ray description of the "nongeometrical" $S^*$ wave. *Izv. Acad. Sci. USSR. Phys. Solid Earth.* 24(10):817–20. *Бабич В. М., Киселев А. П.* Геометросейсмическое описание "нелучевой" волны $S^*$. Изв. АН СССР. Физика Земли, 1988. № 10. С. 67–71.

[10] Babich, V. M. and Kiselev A. P. 1989. Nongeometrical waves — are there any? An asymptotic description of some "nongeometrical" phenomena in seismic wave propagation. *Geophys. J. Intern.* 99:415–20.

[11] Babich, V. M., Kiselev, A. P., Lawry, J. M. H. and Starkov, A. S. 2001. A ray description of all wavefields generated by a high-frequency point source near an interface. *SIAM J. Appl. Math.* 62:21–40.

[12] Brekhovskikh, L. M. 1960. *Waves in layered media.* New York: Academic Press. *Бреховских Л. М.* Волны в слоистых средах. М.: Изд-во АН СССР, 1957.

[13] Buldyrev, V. S. 1967. Short-wave interference in diffraction by a nonuniform cylinder of arbitrary cross section. *Radiophys. Quant. Electron.* 10:699–711.

[14] Daley, P. F. and Hron F. 1983. High-frequency approximation to the nongeometrical $S^*$ arrival. *Bull. Seismol. Soc. Amer.* Vol. 73(1). P. 109–23.

[15] Daley, P. F. and Hron, F. 1987. Reflection of an incident spherical $P$ wave on a free surface (near-vertical incidence). *Bull. Seismol. Soc. Amer.* 77:1057–70.

[16] Friedrichs, K. O. and Keller, J. B. 1955. Geometrical acoustics. II. Diffraction, reflection, and refraction of a weak spherical or cylindrical shock at a plane interface. *J. Appl. Phys.* 26:961–6.

[17] Gutowsky, R. R., Hron, F., Wagner, D. E. and Treifel, S. 1984. $S^*$. *Bull. Seismol. Soc. Amer.* 74:61–78.

[18] Hron, F. and Mikhailenko, B. G. 1981. Numerical modelling of nongeometrical effects by Alekseev – Mikhailenko method. *Bull. Seismol. Soc. Amer.* 71:1011–129.

[19] Hudson, J. A. 1980. *Excitation and propagation of elastic waves.* Cambridge: Cambridge University Press.

[20] Keller, J. 1956. Diffraction of a convex cylinder. *IRE Trans. Antennas and Propagation.* 4(3):312–21.

[21] Kim, J. Y. and Behrens J. 1986. Experimental evidence of $S^*$ wave. *Geophysical Prospecting.* 34:100–8.

[22] Kiselev, A. P. and Tsvankin, I. D. 1989a. Comparison of numerical and asymptotic estimates of elastic wave field. *Doklady. Earth Sci. Sections.* 304:4–7. *Киселев А. П., Цванкин И. Д.* О сопоставлении асимптотических и численных расчетов упругих волновых полей. Докл. АН СССР, 1989. Т. 304(1). С. 61–65.

[23] Kiselev, A. P. and Tsvankin, I. D. 1989b. A method of comparison of exact and asymptotic wave field computations. *Geophys. J. Intern.* 96:253–8.

[24] Lapwood, E. R. 1949. The disturbance due to a line source in a semi-infinite elastic medium. *Phil. Trans. Roy. Soc. London.* 242(841):63–100.

[25] Landau, L. D. and Lifshitz, E. M. 1987. *Fluid dynamics.* Oxford: Pergamon. *Ландау Л. Д., Лифшиц Е. М.* Гидродинамика. М.: Наука, 2006.

[26] Petrashen', G. I., Marchuk, G. I. and Ogurtsov, K. I. 1950. On Lamb's problem in the case of a half-space. *Uchenye Zapiski LGU.* 35(21): 71–118 [in Russian]. *Петрашень Г. И., Марчук Г. И., Огурцов К. И.* О задаче Лэмба в случае полупространства. Ученые записки ЛГУ, 1950. № 35. Вып. 21. С. 71–118.

[27] Petrashen', G. I., Molotkov, L. A. and Krauklis, P. V. 1982. *Waves in piecewise homogeneous layered isotropic elasic media. Method of contour integrals in elastodynamics.* Leningrad: Nauka. *Петрашень Г. И., Молотков Л. А., Крауклис П. В.* Волны в слоисто-однородных изотропных упругих средах. Метод контурных интегралов в нестационарных задачах динамики. Л.: Наука, 1982.

[28] Tsvankin, I. D. and Kalinin A. V. 1984. Nongeometrical effects in the generation of converted waves. *Izv. Akad. Nauk SSSR, Fizika Zemli.* 2:34–40 [in Russian]. *Цванкин И. Д., Калинин А. В.* Нелучевые эффекты при образовании обменных сейсмических волн. Изв. АН СССР. Физика Земли, 1984. № 2. С. 34–40.

[29] Zaitsev, L. P. 1959. On a head wave of a surface type. *Voprosy Dynamicheskoi Teorii Rasprostraneniya Seismicheskikh Voln [Problems in the Dynamic Theory of Propagation of Seismic Waves]*, ed. G. I. Petrashen'. Leningrad: Leningrad Univ. Press, Leningrad, Vol. 3: 378–83. [in Russian]. *Зайцев Л. П.* О головной волне поверхностного типа. Вопр. динам. теории распростр. сейсм. волн, 1959. Вып. 3. 378–383.

[30] Zavorokhin, G. L. 2012. The wave field of a point source that acts on the open boundary of a Biot half-plane. *J. Math. Sci.* 185:567–72. *Заворохин Г. Л.* Волновое поле от точечного источника, действующего на открытой границе полуплоскости Био. Зап. научн. семин. ПОМИ РАН, 2011. Т. 393. С. 101–110.

# Chapter 8

## Ray Method for Rayleigh Waves

In Chapters 4 and 5 we dealt with solutions of elastodynamics equations that locally describe, up to higher-order terms, plane volume waves with smoothly varying directions of propagation, amplitudes, and phases. In the present chapter, we describe ray solutions, which locally, near the surface of the body, approximately coincide with the Rayleigh wave known from the theory of plane waves in a homogeneous half-space. By the fact that the Rayleigh wave is localized near the surface of the body, the corresponding ansatz simultaneously demonstrates both ray and boundary layer features. Here, as well as in the classical ray method, the related eikonal equation and transport equations, which can be integrated, arise.

We start with the most general case of a smoothly inhomogeneous and, generally speaking, anisotropic body.[1] First we deal with the general anisotropy and find it convenient to consider isotropy as its particular case. As before, the absolute value of the amplitude is described via the approximate conservation of the related analog of the flow of energy along a ray strip (which is the "surface counterpart" of a ray tube). However, we meet here novelty as compared to the case of volume waves: an additional phase increment arises as the wave propagates. This phenomenon is known nowadays as the *Berry phase*. The term became common after the paper by Berry 1984 [10], in which the phenomenon (together with certain effects in quantum mechanics) got its clear and brilliant description. Analogs of such a phase increment found earlier by other researchers had not attracted much attention in the scientific community.

Formally, the nature of the Berry phase is associated with the complexity (in the sense of the complex analysis) of consistency conditions for the corresponding first-order transport equation, which decouples into two separate equations for the wave amplitude and phase.[2] The one is associated with the conservation of energy, and the other is related to the Berry phase.

In this chapter, we follow the ideas of a relatively fresh paper by Babich and Kirpichnikova 2004 [7], based on the boundary layer approach. Here, elastodynamics equations and boundary conditions on a free, in general, nonplanar surface are written in nonorthogonal curvilinear coordinates, which requires addressing the techniques of tensor calculus. We need just its basics, a short

[1] Recall that an elastic body and an elastic medium are regarded as synonyms.

[2] A discussion of the mathematical nature of the Berry phase for a comparatively simple case is given by Simon 1983 [32].

account of which can be found in the Appendix. The reader with a modest knowledge of the subject is kindly invited to look there beforehand.

---

## 8.1 Equations, boundary conditions, and a recurrent system

We consider an elastic body with smooth volume density and elastic moduli, bounded by a smooth surface $\mathscr{S}$. Assume that regular coordinates $(q^1, q^2)$ are defined on $\mathscr{S}$. Points of the body (which are close enough to $\mathscr{S}$) are characterized by the coordinates $(q^1, q^2, q^3) = (q^1, q^2, n)$, where $n$ is the distance from $\mathscr{S}$; see Section A.4. We assume that $n > 0$ inside the body.

### Strain and stress

We will need expressions for components of strain and stress tensors in the coordinates $(q^1, q^2, q^3)$, which are, generally speaking, non-Cartesian. In a non-Cartesian system, the covariant and contravariant components of a tensor are not the same. The covariant components of the stress tensor are

$$\varepsilon_{ij} = \varepsilon_{ij}(\boldsymbol{u}) = \frac{1}{2}\left(\nabla_i u_j + \nabla_j u_i\right), \tag{8.1}$$

where $u_1, u_2, u_3$ are covariant components of the displacement vector, and $\nabla_l$ stands for the covariant derivative. Indeed, in Cartesian coordinates, $\nabla_j = \frac{\partial}{\partial x_j}$; see Section A.5. Tensors equal in one system of coordinates, are equal in any other. Both sides of the relation (8.1) are tensors, and it holds in Cartesian coordinates. Therefore (8.1) is valid in an arbitrary coordinate system $(q^1, q^2, n)$.

With little difficulty, one can find that in any coordinate system,

$$\varepsilon_{kl} = -\Gamma^r_{kl} u_r + \frac{1}{2}\left(\frac{\partial u_k}{\partial q^l} + \frac{\partial u_l}{\partial q^k}\right), \tag{8.2}$$

where $\Gamma^r_{kl}$ is the Christoffel symbol (A.24) (which vanishes for Cartesian coordinates). Hooke's law takes the form

$$\sigma^{ij} = c^{ijkl}\varepsilon_{kl} = c^{ijkl}\left\{-\Gamma^r_{kl} u_r + \frac{1}{2}\left(\frac{\partial u_k}{\partial q^l} + \frac{\partial u_l}{\partial q^k}\right)\right\}. \tag{8.3}$$

Symmetries analogous to (1.5.2) and (1.5.6) hold true:

$$c^{ijkl} = c^{klij} = c^{ijlk}, \tag{8.4}$$

as well as the condition of positive definiteness of the potential energy

$$c^{ijkl} e_{ij} e_{kl} \geqslant \text{const} \sum_{i,j=1}^{i,j=3} (e_{ij})^2 \tag{8.5}$$

for arbitrary $e_{ij} = e_{ji}$ and some const $> 0$.

**Equations of elastodynamics**

Instead of straightforward recalculation of the first and second derivatives, we prefer to consider the elastodynamics equations in the coordinates $q^1$, $q^2$, $q^3$ in the form of Euler equations for the Lagrangian represented in proper variables. Here,

$$\mathcal{K} = \frac{\rho}{2}(\dot{\mathbf{u}}, \dot{\mathbf{u}}) = \frac{\rho}{2} g^{jl} \dot{u}_j \dot{u}_l, \quad \text{and} \quad \mathcal{W} = \frac{1}{2} \sigma^{ps} \varepsilon_{ps}. \tag{8.6}$$

The expression for the volume element $\sqrt{g} dq^1 dq^2 dq^3$ is given in (A.16). Equating to zero the variation of the integral $\int \mathcal{L} d\mathbf{x} dt = \int \mathcal{L} \sqrt{g} dq^1 dq^2 dq^3 dt$, where $\mathcal{L} = \mathcal{K} - \mathcal{W}$, implies the Euler equations

$$-\frac{\partial}{\partial t} \frac{\partial \left(\sqrt{g}\mathcal{L}\right)}{\partial \dot{u}_j} - \frac{\partial}{\partial q^i} \frac{\partial \left(\sqrt{g}\mathcal{L}\right)}{\partial \varepsilon_{ij}} + \sqrt{g} \frac{\partial \mathcal{L}}{\partial u_j} = 0, \quad j = 1, 2, 3.$$

With account of the relations (8.2) and (8.3), the elastodynamics equations (1.14) take the form

$$\sqrt{g}\left(-\rho g^{lj} \ddot{u}_j + \Gamma^l_{kp} \sigma^{kp}\right) + \frac{\partial \left(\sqrt{g} \sigma^{li}\right)}{\partial q^i} = 0, \quad l = 1, 2, 3. \tag{8.7}$$

**Boundary conditions**

Traction-free conditions on the boundary $\mathscr{S}$ (1.41) can be written componentwise as

$$\sigma^{i3}|_{\mathscr{S}} = c^{i3kl} \left\{ -\Gamma^r_{kl} u_r + \frac{1}{2} \left( \frac{\partial u_k}{\partial q^l} + \frac{\partial u_l}{\partial q^k} \right) \right\} \bigg|_{\mathscr{S}} = 0, \quad i = 1, 2, 3. \tag{8.8}$$

**The ansatz**

An adequate choice of the ansatz is a keystone for subsequent constructions. Roughly speaking, when the ansatz is properly chosen, the rest of the job is just a machinery, which may, however, be tiresome, as in the present case. We seek the Rayleigh wave in the form of the ray ansatz

$$\begin{gathered} \mathbf{u} = e^{i\omega(\tau(q^1, q^2) - t)} \left\{ \boldsymbol{U}^O(q^1, q^2, \nu) + \frac{\boldsymbol{U}^I(q^1, q^2, \nu)}{-i\omega} + \dots \right\}, \\ \nu = \omega n, \quad \omega \to \infty. \end{gathered} \tag{8.9}$$

As usual, the frequency $\omega$ plays the role of a large parameter. The upper indices marking the amplitude vectors $\boldsymbol{U}$ are numbers (of approximations) and not contravariant components.

We require that

$$\boldsymbol{U}^O \to 0, \quad \boldsymbol{U}^I \to 0, \; \dots \quad \text{as} \quad \nu \to \infty. \tag{8.10}$$

Conditions (8.10) guarantee the near-surface localization of the wave (8.9). Expansion (8.9) under the condition (8.10) is the desired ansatz.

As is obvious, the classical Rayleigh wave considered in Chapter 2 (in both isotropic and anisotropic cases) can be represented in the form (8.9)–(8.10). The roles of $q^1$, $q^2$, and $n$ are played there by the Cartesian coordinates $x$, $y$, and $z$, and the sum in (8.9) reduces to a single term.

### Recurrent system

Expand the coefficients in the equations (8.7) and (8.8) in the powers of $n$, replace $n$ by $\frac{\nu}{\omega}$, substitute there the representation (8.9), and equate the coefficients of the powers of $-i\omega$ to zero, as we have already done several times. We arrive at recurrent equations for the vectors $\boldsymbol{U}^O$, $\boldsymbol{U}^I$, ...:

$$\begin{aligned} \boldsymbol{N}\boldsymbol{U}^O &= 0, \\ \boldsymbol{N}\boldsymbol{U}^I &= \boldsymbol{M}\boldsymbol{U}^O, \\ &\dots \end{aligned} \tag{8.11}$$

with the boundary conditions

$$\boldsymbol{\mathcal{T}}\boldsymbol{U}^O|_{\nu=0} = 0, \quad \boldsymbol{\mathcal{T}}\boldsymbol{U}^I|_{\nu=0} = \boldsymbol{\mathcal{S}}\boldsymbol{U}^O|_{\nu=0}, \; \dots \tag{8.12}$$

In (8.11), $\boldsymbol{N}$, $\boldsymbol{M}$, ... are ordinary differential operators with respect to $\nu$ of order no larger than 2, with the coefficients dependent, in general, on $q^1$, $q^2$, and $\nu$. Their dependence on $\nu$ is polynomial. The operator $\boldsymbol{\mathcal{S}}$ in (8.12) is of the first order.

We shall analyze the problems that define $\boldsymbol{U}^O$ and $\boldsymbol{U}^I$, under the assumption that the Barnett – Lothe nondegeneracy condition (see Section 2.10) for the corresponding plane surface wave holds. Also, we shall outline a way of constructing higher-order terms.

---

## 8.2 Boundary value problem for $\boldsymbol{U}^O$

The problem defining $\boldsymbol{U}^O$,

$$\begin{gathered} \boldsymbol{N}\boldsymbol{U}^O = 0, \quad \boldsymbol{\mathcal{T}}\boldsymbol{U}^O\big|_{\nu=0} = 0, \\ \boldsymbol{U}^O \to 0, \quad \nu \to \infty, \end{gathered} \tag{8.13}$$

is identical, up to notation, to that for the Rayleigh wave in a homogeneous elastic half-space, considered in Chapter 2. Elastic stiffnesses and volume density are "frozen" at a point $\mathbf{x} \in \mathscr{S}$, and the aforementioned Rayleigh wave is a local plane Rayleigh wave analogous to local plane volume waves, which we dealt with in Chapters 4 and 5.

### ⋆ About the mathematical background

The coordinates $(q^1, q^2, n)$ are, in general, regular only for small $n$. We assign a half-line $0 \leqslant n < \infty$ to each point $(q^1, q^2) \in \mathscr{S}$, defining thus the topological product $\mathscr{S} \times [0, \infty)$. We thus arrive at a mathematical object known as a normal bundle with base $\mathscr{S}$ and the fiber $[0, \infty)$. In other cases, e.g., for the Stoneley wave, it is natural to consider the fiber $(-\infty, \infty)$. Normal bundles play an important role in contemporary mathematics; see Hirsch [16], Treve [33].

A normal bundle with the fiber $[0, \infty)$ is a space, where the recurrent system of equations (8.11) will be analyzed. The problem (8.13) defines the local plane Rayleigh wave on the direct product of the tangent plane to the surface $\mathscr{S}$ and the half-line $0 \leqslant n < +\infty$.

### 8.2.1 Explicit forms of equation and a boundary condition for $\boldsymbol{U}^O$

The first vector equation (that is, the system of three scalar equations) of (8.11) for the covariant vector $\boldsymbol{U}^O$ is

$$(\boldsymbol{N}\,\boldsymbol{U}^O)^m = \sqrt{\mathring{g}}(-\mathring{\rho}\mathring{g}^{lm}U_l^O + \mathring{c}^{mjqs}p_j p_q U_s^O) = 0, \qquad \nu > 0, \quad m = 1, 2, 3. \tag{8.14}$$

Here, $\mathring{g}^{lm} = g^{lm}|_{\mathscr{S}}$, $\mathring{\rho} = \rho|_{\mathscr{S}}$, $\mathring{c}^{mjqs} = c^{mjqs}|_{\mathscr{S}}$, and $\mathring{g} = g|_{\mathscr{S}}$ stand for the boundary values of the correspondent quantities, and

$$p_1 := \frac{\partial \tau}{\partial q^1}, \quad p_2 := \frac{\partial \tau}{\partial q^2}, \quad p_3 := -i\frac{d}{d\nu}. \tag{8.15}$$

The operators $p_{1,2}$ are operators of multiplication by real factors $\frac{\partial \tau}{\partial q^{1,2}}$ that are the components of the slowness vector. Since $p_3$, up to constant factor, is differentiation with respect to $\nu$, (8.14) is a system of ODE on the half-line $\nu > 0$ with coefficients independent of $\nu$. Boundary condition (8.8) reads

$$\mathring{c}^{l3rs}p_r U_s^O\big|_{\mathscr{S}} = 0, \quad l = 1, 2, 3. \tag{8.16}$$

Also, the condition at infinity (8.10) should hold.

We arrived at a homogeneous problem, one-dimensional with respect to $\nu$, $0 \leqslant \nu < \infty$, and dependent on the parameters $q^1$ and $q^2$. From here, as in the ray theory of volume waves, descriptions for the eikonal and for the zero-order polarization will be derived.

### 8.2.2 The eikonal equation

To derive the eikonal equation and to find the polarization of the leading-term vector amplitude, we consider the following auxiliary problem:

$$\mathring{c}^{mjks}p_j p_k U_s^O \sqrt{\mathring{g}} = H^2 \sqrt{\mathring{g}}\mathring{\rho}\mathring{g}^{lm}U_l^O, \quad m = 1, 2, 3, \tag{8.17}$$

under the boundary conditions (8.10), (8.16). Here, the eigenvalue $H^2$ depends on the components $p_1$ and $p_2$ of the slowness vector. In Chapters 2 and 5, we had already met analogs of such $H^2$.

The eigenvalues of the spectral problem under consideration (8.17) (as well as their analogs in the theory of volume waves) are denoted by $H^2$, because they prove to be positive, which will be easily established. Indeed, dividing both sides of (8.17) by $\sqrt{\mathring{g}}$, multiplying by $U_m^{O*}$, and integrating by parts with the use of the boundary condition, we obtain

$$\int_0^\infty \mathring{c}^{mjks}(p_j U_m^O)^*(p_k U_s^O)d\nu$$

$$= H^2 \int_0^\infty \mathring{\rho}\mathring{g}^{lm} U_l^O U_m^{O*} d\nu = H^2 \int_0^\infty \mathring{\rho}\left(\boldsymbol{U}^O, \boldsymbol{U}^O\right) d\nu. \quad (8.18)$$

The positivity of $H^2$ immediately follows from the positive definiteness of the matrix $\|\mathring{g}^{lm}\|$.

Let us assume that $H^2$ is uniquely determined by the values of $p_1 = \frac{\partial \tau}{\partial q^1}$ and $p_2 = \frac{\partial \tau}{\partial q^2}$. Under the requirement that

$$H^2(p_1, p_2; q^1, q^2) = 1, \quad (8.19)$$

equations (8.17) and (8.13) are equivalent. In our approach, the equation (8.19) plays the role of the eikonal equation.[3]

The function $H(p_1, p_2; q^1, q^2)$ is obviously first-degree homogeneous with respect to $p_1$ and $p_2$, that is, $H(\mathrm{C}p_1, \mathrm{C}p_2; q^1, q^2) = \mathrm{C}H(p_1, p_2; q^1, q^2)$ for any $\mathrm{C} > 0$. Putting $\mathrm{C} = \omega$, we obtain

$$\omega = H(k_1, k_2; q^1, q^2), \quad (8.20)$$

where, by definition, $k_{1,2} := \omega p_{1,2}$. The equation (8.20) where $H$ is homogeneous with respect to $k_{1,2}$ demonstrates the lack of dispersion.

The existence of an eigenvalue of the spectral problem (8.13) follows from the assumption that the Barnet – Lothe nondegeneracy condition is satisfied; see Section 2.10.

### 8.2.3 The Fermat principle, rays, and the group velocity theorem

The equation (8.19) for the eikonal $\tau$ is a nonlinear PDE of the first order. We solve it by the method of characteristics, repeating the argument of Chapter 5 next to word-for-word.

We start by writing the equation (8.19) as $\frac{H^2}{2} = \frac{1}{2}$. An argument similar to that of Section 5.2 immediately yields the equations

$$\frac{dq^\alpha}{d\sigma} = \frac{\partial}{\partial p_\alpha}\left(\frac{H^2}{2}\right), \quad \frac{dp_\alpha}{d\sigma} = -\frac{\partial}{\partial q^\alpha}\left(\frac{H^2}{2}\right), \quad \alpha = 1, 2, \quad (8.21)$$

[3]Note that under the condition $H = 1$, the equation (8.18) is equivalent to the virial theorem for the local plane Rayleigh wave in a half-space with "frozen" density and stiffnesses; see Section 2.9.2. In fact, the left-hand side is proportional to integrated time-averaged potential energy, while the right-hand side is the same for kinetic energy. As was shown at the end of Section 2.9.2, $\overline{\overline{\mathcal{E}}} = 2\overline{\overline{\mathcal{K}}} = 2\frac{1}{4}\mathring{\rho}\omega^2 \int_0^\infty (\boldsymbol{U}^O, \boldsymbol{U}^O)d\nu$.

where $\sigma$ is a parameter along the ray, interpretable as time. Further on, applying the Legendre transformation, we find that the characteristic curves are extremals of the functional $\int \frac{\mathscr{L}^2}{2} d\sigma$, where $\frac{\mathscr{L}^2}{2} = \frac{\mathscr{L}^2(\dot{q}^1, \dot{q}^2; q^1, q^2,)}{2}$ is the Legendre transform of the function $\frac{H^2(p_1, p_2; q^1, q^2)}{2}$. Here, we temporarily employ the notation $\dot{q}^{1,2} := \frac{dq^{1,2}}{d\sigma}$. The function $\mathscr{L}$ is first-degree homogeneous with respect to $\dot{q}^1$, $\dot{q}^2$, whence the integral $\int \mathscr{L} d\sigma$ is independent of the parameterization of the curve $q^\alpha = q^\alpha(\sigma)$. The functional $\int \mathscr{L}^2 d\sigma$ has the same extremals as $\int \mathscr{L} d\sigma$. Arguments closely analogous to those of Section 5.2 lead to conclusions that the equation $\delta \int \mathscr{L} d\sigma = 0$ corresponds to the Fermat principle that $\frac{dq^{1,2}}{d\sigma}$ are contravariant components of the group velocity (which is tangent to $\mathscr{S}$), and also that the extremals of the functional $\int \mathscr{L} d\sigma$ are rays of the Rayleigh wave. Naturally, these rays are curves on the surface $\mathscr{S}$, which is a little surprise as we deal with a surface wave (in this connection, see, e.g., Section 5.2).

Let us consider the Cauchy problem for equation (8.19). We fix the value of $\tau|_\varsigma$ on some curve $\varsigma \subset \mathscr{S}$ and the direction in which $\tau$ grows. These data, with the use of (8.19), uniquely determine $p_{1,2}|_\varsigma$, which, together with (8.21), uniquely determine $\frac{dq^{1,2}}{d\sigma}\big|_\varsigma$. Consider a surface ray launched from a point $M_0 \in \varsigma$ in the direction of the group velocity $\left(\frac{dq^1}{d\sigma}\big|_\varsigma, \frac{dq^2}{d\sigma}\big|_\varsigma, 0\right)$. The value of the integral $\int_{M_0}^{M} \mathscr{L} d\sigma$ along the segment of the ray $M_0M$, connecting $M_0$ and $M \in \mathscr{S}$, defines, as in Chapter 5, the value of $\tau(M)$, because $\tau(M_0)$ is considered to be a known quantity, while $\int_{M_0}^{M} \mathscr{L} d\sigma = \tau(M) - \tau(M_0)$. In the same way as in Chapter 5, we introduce the ray coordinates $(\tau, \eta)$, where $\eta$ characterizes a ray, and $\tau$ characterizes a point on the ray.

At the end of the current section, we derive a formula that relates the group velocity of the Rayleigh wave and the leading term of the expansion for the average density of energy flow of the corresponding local plane Rayleigh wave. Essentially, this will prove the group velocity theorem for the local plane wave.

Differentiating (8.18) with respect to $p_\alpha = \frac{\partial \tau}{\partial q^\alpha}$ and putting $H = 1$, we obtain

$$2\sqrt{\mathring{g}} \int_0^\infty \mathring{c}^{\alpha jms} \operatorname{Re}(U_j^{O*} p_s U_m^O) d\nu = 2\frac{\partial H}{\partial p_\alpha} \int_0^\infty \mathring{\rho} \mathring{g}^{kl} U_k^O U_l^{O*} \sqrt{\mathring{g}} d\nu. \quad (8.22)$$

Multiplication by $\frac{\omega}{4}$ gives the equation

$$\int_0^\infty \overline{\mathbf{S}} d\nu = \boldsymbol{v}^{gr} \int_0^\infty \overline{\mathcal{E}} d\nu, \quad (8.23)$$

where $\boldsymbol{v}^{gr}$ is the group velocity vector, and $\overline{\mathbf{S}}$ is the time-averaged density of the energy flow of the local plane surface wave. Peculiar features of surface waves are the presence of integrals over the depth (as in Section 2.9) and the fact that $\int_0^\infty \overline{\mathbf{S}} d\nu$ and $\boldsymbol{v}^{gr}$ both lie in the tangent plane to $\mathscr{S}$.

### 8.2.4 Consistency condition for the boundary value problem for $\boldsymbol{U}^I$

**Initial formulas**

The boundary value problem (8.13) describes the vector $\boldsymbol{U}^O$, up to scalar factor $\phi(q^1, q^2)$ independent of $\nu$,

$$\boldsymbol{U}^O(q^1, q^2, \nu) = \phi(q^1, q^2)\, \boldsymbol{V}(q^1, q^2, \nu), \tag{8.24}$$

where $\boldsymbol{V}(q^1, q^2, \nu)$ is a somehow fixed solution of the problem (8.13). Consistency condition for the problem for $\boldsymbol{U}^I$ will allow $\phi(q^1, q^2)$ to be uniquely found, up to a respective diffraction coefficient.

The vector $\boldsymbol{U}^I$ satisfies the second equation in (8.11), where the operator $\boldsymbol{N}$ is defined in (8.14). Under the agreement that repeated upper and lower Greek indices imply summation over from 1 to 2, the operator $\boldsymbol{M}$ can be written as

$$\begin{aligned}(\boldsymbol{M}\,\boldsymbol{U})^l = i\nu \left\{ \frac{\partial\left(\rho\sqrt{g}g^{jl}\right)}{\partial n}\bigg|_{n=0} U_j - \frac{\partial\left(c^{ljmr}\sqrt{g}\right)}{\partial n}\bigg|_{n=0} p_j p_m U_r \right\} \\ - \sqrt{\mathring{g}}\left(\mathring{c}^{sjmr}\mathring{\Gamma}^l_{sj} p_m U_r - \mathring{c}^{ljmr}\mathring{\Gamma}^s_{mr} p_j U_s\right) - \frac{\partial\left(c^{ljmr}\sqrt{g}\right)}{\partial q^j}\bigg|_{n=0} p_m U_r \\ - \sqrt{\mathring{g}}\left(\mathring{c}^{lj\alpha m} p_j \frac{\partial U_m}{\partial q^\alpha} + \mathring{c}^{lj\alpha r} p_j \frac{\partial U_r}{\partial q^\alpha} + \frac{\partial^2 \tau}{\partial q^\alpha \partial q^\beta}\mathring{c}^{l\alpha\beta m} U_m\right), \quad l = 1, 2, 3.\end{aligned} \tag{8.25}$$

The transport equation for $\boldsymbol{U}^I$, see (8.11), should be solved under the boundary condition (8.12), which reads as

$$\left\{ \mathring{c}^{l3mr}\left(p_r U^I_m - \mathring{\Gamma}^s_{mr} U^O_s\right) + \mathring{c}^{l3\alpha m}\frac{\partial U^O_m}{\partial q^\alpha} \right\}\bigg|_{\nu=0} = 0, \quad l = 1, 2, 3, \tag{8.26}$$

and the assumption that $\boldsymbol{U}^I$ vanishes at infinity.

This problem, in general, has no solution, because the corresponding homogeneous problem possesses a nonzero solution (which was denoted by $\boldsymbol{V}$ in (8.24)). To derive the related consistency condition, we use a standard procedure. Scalar multiplication of the equation (8.11) for $\boldsymbol{U}^I$ by $\boldsymbol{V}$ and integration with respect to $\nu$ from 0 to $\infty$ give

$$\int_0^\infty \left(\boldsymbol{N}\,\boldsymbol{U}^I - \boldsymbol{M}\,\boldsymbol{U}^O, \boldsymbol{V}\right) d\nu = 0. \tag{8.27}$$

The scalar product is understood here in accordance with (A.12)–(A.13).

We will transform the integral (8.27) by means of integration by parts, moving the derivatives from $\boldsymbol{U}^I$ to $\boldsymbol{V}$ and employing the boundary conditions

(8.26) and conditions at infinity. The equations (8.27) take the form

$$\int_0^\infty \left(\boldsymbol{M}\,\boldsymbol{U}^O, \boldsymbol{V}\right) d\nu - i\sqrt{\mathring{g}}\left\{\mathring{c}^{l3mr}\mathring{\Gamma}^s_{mr}U^O_s V^*_l - \mathring{c}^{l3\alpha m}\left.\frac{\partial U^O_m}{\partial q^\alpha}\right|_{\nu=0} V^*_l\right\}\Bigg|_{\nu=0} = 0,$$

$$m, j, l, r = 1, 2, 3; \quad \alpha = 1, 2, \tag{8.28}$$

in which the components of $\boldsymbol{U}^I$ do not occur. We thus got a linear relation (8.28) between $\phi$ and $\frac{\partial\phi}{\partial q^{1,2}}$. This is the desired transport equation. Further, we will see that it is a linear ODE along the surface rays of the Rayleigh wave. Turn now to analytic manipulations.

## Transformation of equation (8.28)

As seen from the relation (8.25), on the left-hand side of (8.28) a term of the form

$$-\int_0^\infty i\nu \left.\frac{\partial\left(c^{ljmr}\sqrt{g}\right)}{\partial n}\right|_{n=0} V^*_l p_j p_m U_r d\nu \tag{8.29}$$

is present. We integrate it by parts, employing the vanishing of integrated terms. Integration in (8.29) by parts proceeds with the use of the boundary conditions. Let us transform this integral, moving the operators $p_j$ to the components of the vector $\boldsymbol{V}^*$ and taking into account the boundary conditions and the fact that the integrated terms vanish. We arrive at the expression

$$\int_0^\infty \left.\frac{\partial\left(c^{l3mr}\sqrt{g}\right)}{\partial n}\right|_{n=0} V^*_l p_m U^O_r d\nu - i\int_0^\infty \nu \left.\frac{\partial\left(c^{ljmr}\sqrt{g}\right)}{\partial n}\right|_{n=0} (p_m U^O_r)(p_j V_l)^* d\nu. \tag{8.30}$$

The first term cancels with

$$-\int_0^\infty (p_s U^O_m)\frac{\partial(c^{lrms}\sqrt{g})}{\partial q^r} V^*_l d\nu$$

at $r = 3$. Another simplification arises from moving $p_j$ from $U^O_j$ to $V^*_l$ in the expression

$$-\sqrt{\mathring{g}}\int_0^\infty \mathring{c}^{ljmr}\mathring{\Gamma}^s_{mr} V^*_l p_j U^O_s d\nu.$$

The integrated term in the case $j = 3$ cancels with the second term in (8.28).

The final simplifying observation is that, after moving the operator $p_j$ to $V^*_l$ in the integral $-\int \sqrt{\mathring{g}}\mathring{c}^{lj\alpha m}\, V^*_l p_j \frac{\partial U^O_m}{\partial q^\alpha} d\nu$ and integrating it by parts, the integrated term cancels with the third term in formula (8.28).

### 8.2.5 Preliminary analysis of the transport equation

The above transformation results in the equation

$$\int_0^\infty \left\{ i\nu \frac{\partial(\rho\sqrt{g}g^{jl})}{\partial n}\bigg|_{n=0} U_j^O V_l^* - i\nu \frac{\partial(c^{ljmr}\sqrt{g})}{\partial n}\bigg|_{n=0} (p_m U_r^O)(p_j V_l)^* \right.$$
$$- \sqrt{\mathring{g}}\left[\mathring{c}^{sjmr}(p_m U_r^O)\mathring{\Gamma}^l_{sj} V_l^* - \mathring{c}^{ljmr} U_s^O \mathring{\Gamma}^s_{mr}(p_j V_l)^*\right]$$
$$- \frac{\partial(\mathring{c}^{l\alpha mr}\sqrt{\mathring{g}})}{\partial q^\alpha} V_l^*(p_m U_r^O) - \sqrt{\mathring{g}}\mathring{c}^{lj\alpha m}\frac{\partial U_m^O}{\partial q^\alpha}(p_j V_l)^*$$
$$\left. - \sqrt{\mathring{g}}\left[\mathring{c}^{l\alpha mr} V_l^* p_m \frac{\partial U_r^O}{\partial q^\alpha} + \frac{\partial^2 \tau}{\partial q^\alpha \partial q^\beta}\mathring{c}^{l\alpha\beta m} U_m V_l^*\right]\right\} d\nu = 0. \quad (8.31)$$

Substitute here the expression (8.24) for $\boldsymbol{U}^O$. First, select the terms containing derivatives of the amplitude $\phi$, which are

$$- \int_0^\infty \sqrt{\mathring{g}}\left\{\mathring{c}^{lj\alpha m} V_m (p_j V_l)^* + \mathring{c}^{l\alpha mr} V_l^* p_m V_r\right\} \frac{\partial \phi}{\partial q^\alpha} d\nu$$
$$= -\sqrt{\mathring{g}}\mathring{c}^{lj\alpha m} \int_0^\infty \left\{V_m (p_j V_l)^* + V_m^*(p_l V_j)\right\} d\nu \frac{\partial \phi}{\partial q^\alpha}. \quad (8.32)$$

Next, address the formula (8.22) and take into consideration the fact that $\frac{\partial H}{\partial p^\alpha}$ are the components of the group velocity vector, which is tangent to $\mathscr{S}$. An easy calculation shows that the expression (8.32) equals

$$-\frac{d\phi}{d\tau}\int_0^\infty \sqrt{\mathring{g}}\mathring{\rho}\mathring{g}^{kl} V_k V_l^* d\nu = -\frac{d\phi}{d\tau}\int_0^\infty \sqrt{\mathring{g}}\mathring{\rho}(\boldsymbol{V}, \boldsymbol{V}) d\nu, \quad (8.33)$$

where $\frac{d}{d\tau}$ stands for derivation along the surface ray. The coefficient of $\frac{d\phi}{d\tau}$ in (8.33) is nonzero, whence, dividing (8.32) by this coefficient, we arrive at a relation of the form

$$\frac{d\phi}{d\tau} + \Xi\phi = 0, \quad (8.34)$$

with $\Xi$ independent of $\phi$. We thus arrived at an ODE along a ray. Consider the resulting equation in more detail.

### 8.2.6 The Umov equation and an expression for $|\phi|$

We start by putting in (8.24)

$$\phi = |\phi| e^{iv} \quad \Rightarrow \quad \boldsymbol{U}^O = |\phi| e^{iv} \boldsymbol{V}. \quad (8.35)$$

To derive the equation for $|\phi|$, we replace the vector $\boldsymbol{V}$ in (8.31) by the vector $\boldsymbol{U}^O$ and equate to zero the real part of the resulting relation. As if by a wave of wizard's wand, this reduces to

$$\frac{1}{\sqrt{\mathring{g}}}\frac{\partial}{\partial q^\alpha}\int_0^\infty \sqrt{\mathring{g}}\mathring{c}^{l\alpha mj}\,\mathrm{Re}\left[U_l^O (p_m U_j^O)^*\right] d\nu = 0. \quad (8.36)$$

Since no sorcery happens in mathematical physics, some in-depth reason must exist for a dramatic simplification of an expression. In the case under consideration, the "sorcery" is based on the "surface" version of energy balance. The matter is that, by virtue of the energy conservation relation (8.22)–(8.23), see also Section 2.9, the equation (8.36) can be written as

$$\operatorname{div}\int_0^\infty \overline{\overline{\mathcal{E}}}(\boldsymbol{U}^O)\boldsymbol{v}^{gr}d\nu = \operatorname{div}(E_{\boldsymbol{U}^O}\boldsymbol{v}^{gr}) = \frac{1}{\sqrt{\mathring{g}}}\frac{\partial}{\partial q^\alpha}\left(\sqrt{\mathring{g}}E_{\boldsymbol{U}^O}(\boldsymbol{v}^{gr})^\alpha\right) = 0, \tag{8.37}$$

with the notation

$$E_{\boldsymbol{U}^O}(M) := \int_0^\infty \mathring{\rho}(\boldsymbol{U}^O, \boldsymbol{U}^O)d\nu = \int_0^\infty \mathring{\rho}\mathring{g}^{lm}U_l^O U_m^{O*}d\nu, \tag{8.38}$$

and

$$\boldsymbol{v}^{gr} = \left(\frac{dq^1}{ds}, \frac{dq^2}{ds}, 0\right). \tag{8.39}$$

Here, as well as in Section 2.9, $\boldsymbol{v}^{gr}$ stands for the group velocity that essentially is a two-dimensional vector, which we treat, however, as a three-dimensional vector with the third component taken identically zero. The quantity denoted by $\overline{\overline{\mathcal{E}}}_{\boldsymbol{U}^O}$ in (8.37) is the reduced (that is, divided by $\frac{\omega}{2}$) density of energy in a half-space with "frozen" parameters, averaged over the period and integrated over the depth $n$. In other words, $\frac{\omega^2}{2}\overline{\overline{\mathcal{E}}}_{\boldsymbol{U}^O}$ is the reduced integrated averaged density of energy of the local plane surface wave.

Equation (8.37) has the form of the classical continuity equation for the energy fluid (cf. (4.67) and (5.39)). Its physical interpretation is the conservation of mass (the role of which is played by $\overline{\overline{\mathcal{E}}}_{\boldsymbol{U}^O}$) in the course of motion of the fluid. Equation (8.37) is the Umov equation for surface waves; it is analogous to (4.67) and (5.37).

We introduce the ray coordinates $(\tau, \eta)$ on the surface $\mathscr{S}$. In arbitrary surface coordinates $(q^1, q^2)$, the contravariant components of the group velocity are $\left(\frac{dq^1}{d\tau}, \frac{dq^2}{d\tau}, 0\right)$. In our case, they are $(1, 0, 0)$, and the relation (8.37) reduces to

$$\frac{1}{\sqrt{\mathring{g}}}\frac{\partial}{\partial \tau}\left(\sqrt{\mathring{g}}E_{\boldsymbol{U}^O}\right) = 0. \tag{8.40}$$

It immediately follows from (8.40) that

$$\sqrt{\mathring{g}}E_{\boldsymbol{U}^O} = \text{const} \tag{8.41}$$

along a ray. Now we dwell on a hydrodynamic interpretation of this result.

Let the boundary of our elastic body be described in parametric form by $\mathbf{x} = \mathring{\mathbf{x}}(\tau, \eta)$. Consider a *ray strip*, formed by surface rays, corresponding to the values of the parameter $\eta$ in the interval $\eta \leqslant \eta' \leqslant \eta + d\eta$. It is a "surface" analog of a ray tube. Calculate, in the leading approximation, the amount of energy that passes through the section $\tau = \text{const}$ of the ray strip. Let $AB$ be

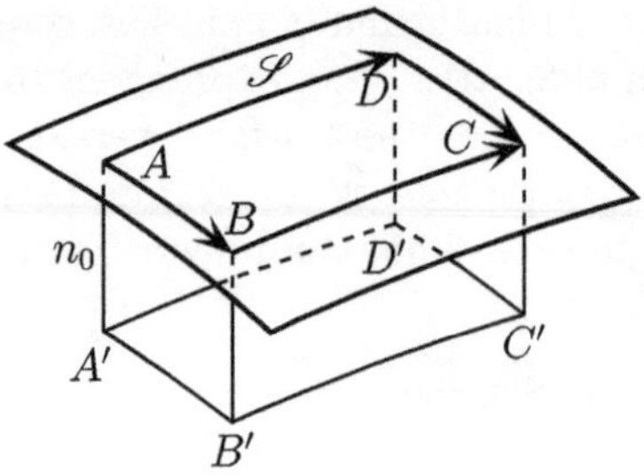


**Figure 8.1**: The domain that appears in the derivation of the formula (8.43)

this section; see Fig. 8.1. Obviously, $\overrightarrow{AB} \approx \frac{\partial \mathring{\mathbf{x}}}{\partial \eta} d\eta$. The energy fluid flows with the velocity $\boldsymbol{v}^{gr}$.

Moving along with the fluid for the time interval from $\tau$ to $\tau + d\tau$, the vector $\overrightarrow{AB}$ takes the position $\overrightarrow{DC}$; see Fig. 8.1, and $\overrightarrow{AD} \approx \overrightarrow{BC} \approx \boldsymbol{v}^{gr} d\tau = \frac{\partial \mathring{\mathbf{x}}}{\partial \tau} d\tau$. Consider the parallelepiped $ABCDA'B'C'D'$, the upper face of which is a parallelogram $ABCD$, the edges $AA'$, $BB'$, $CC'$, $DD'$ are orthogonal to the surface $\mathscr{S}$, and $|AA'| = |BB'| = |CC'| = |DD'| = n_0$; see Fig. 8.1. Denote by $n_0$, $n_0 > 0$, an independent of $\omega$ value of the coordinate $n = q^3$, although small but such that the energy of the Rayleigh wave localized near the surface for $n > n_0$ is small. Then the amount of energy flowing through the section $AB$ of the ray strip, in the leading approximation, is

$$\frac{\omega^2}{2} \int_0^{n_0} \overline{\overline{\mathcal{E}}} dn \cdot (\text{area of } ABCD) \approx \frac{\omega^2}{2} \int_0^{n_0} \overline{\overline{\mathcal{E}}} dn |\mathring{\mathbf{x}}_\eta \times \mathring{\mathbf{x}}_\tau| d\eta d\tau$$
$$\approx \frac{1}{\omega} \frac{\omega^2}{2} \int_0^{\infty} \overline{\overline{\mathcal{E}}} d\nu |\mathring{\mathbf{x}}_\eta \times \mathring{\mathbf{x}}_\tau| d\eta d\tau. \qquad (8.42)$$

Recall that $\overline{\overline{\mathcal{E}}}$ is a function of the variables $\eta$, $\tau$, and $n = \frac{\nu}{\omega}$. The expression (8.42), up to a factor $\frac{\omega}{2} d\eta d\tau$, coincides with the expression (8.41). Thus, the hydrodynamic meaning of the relation (8.41) is the constancy, in the first approximation, of the averaged over the period amount of energy passing per unit time through the section $\tau = \text{const}$ of the ray strip ($\eta \leqslant \eta' \leqslant \eta + d\eta$, $0 \leqslant \nu < \infty$). Here, $\frac{\omega^2}{2} \overline{\overline{\mathcal{E}}}$ is the time-averaged density of the energy of the local plane Rayleigh wave, related to the point $(\eta, \tau) \in \mathscr{S}$. This wave is defined in a half-space given as a topological product of the tangent plane to $\mathscr{S}$ at the point $(\eta, \tau)$ and the half-line $0 \leqslant n \leqslant \infty$.

Let $\mathbf{x}^{\circ}$ and $\mathbf{x}$ be points on the same surface ray. Then (8.36) implies that

$$E_{\boldsymbol{U}^{\circ}}(\mathbf{x}) = E_{\boldsymbol{U}^{\circ}}(\mathbf{x}^{\circ}) \frac{\sqrt{\mathring{g}(\mathbf{x}^{\circ})}}{\sqrt{\mathring{g}(\mathbf{x})}}. \qquad (8.43)$$

Here

$$E_{\boldsymbol{U}^{\circ}}(\mathbf{x}) = |\phi(\mathbf{x})|^2 E_{\boldsymbol{V}}(\mathbf{x}),$$

where

$$E_{\boldsymbol{V}}(M) := \int_0^\infty \mathring{\rho}(\boldsymbol{V},\boldsymbol{V})d\nu = \int_0^\infty \mathring{\rho}\mathring{g}^{lm}V_l^O V_m^{O*}d\nu \tag{8.44}$$

is the reduced integrated over the depth energy of the local plane wave. Then formula (8.43) implies the following expression for $|\phi|$:

$$|\phi(\mathbf{x})| = |\phi(\mathbf{x}^o)|\frac{\sqrt{E_{\boldsymbol{V}}(\mathbf{x}^o)}}{\sqrt{E_{\boldsymbol{V}}(\mathbf{x})}}\frac{\sqrt[4]{\mathring{g}(\mathbf{x}^o)}}{\sqrt[4]{\mathring{g}(\mathbf{x})}}. \tag{8.45}$$

The formula (8.45) is a two-dimensional analog of the formulas (4.83) and (5.41) for absolute values of the amplitudes of volume waves. We rewrite these formulas in a form even more akin to (5.41). For this purpose, we address the metric tensor describing the squared distance between two closely spaced points in $\mathbb{R}^3$. In the coordinates $q^1 = \eta_1$, $q^2 = \eta_2$, $q^3 = \tau$ employed in Chapter 4, we had $ds^2 = g_{ij}dq^i dq^j$, and $\mathscr{D} = \sqrt{\det\|g_{ij}\|} = \sqrt{g}$, where $g = \mathscr{D}^2$. Therefore, (4.83) can be represented as

$$\left|\boldsymbol{u}^{P0}(\mathbf{x})\right| = \left|\boldsymbol{u}^{P0}(\mathbf{x}^o)\right| \frac{\sqrt{\rho(\mathbf{x}^o)\sqrt[4]{g(\mathbf{x}^o)}}}{\sqrt{\rho(\mathbf{x}))\sqrt[4]{g(\mathbf{x}^o)}}}. \tag{8.46}$$

The formula (5.41) can be rewritten in a similar way.

### 8.2.7 The Berry phase

The argument $\upsilon$ of the amplitude $\phi$ (the term "argument" is understood here in the sense of complex numbers) is analogous to numerous Berry phases in quantum mechanics and in the theory of wave phenomena. The function $\upsilon$ satisfies the respective transport equation, which is a first-order ODE along a ray. However, the expression for its solution does not look simple.

A simple and transparent formula (8.45) for $|\phi|$ has been derived above with the help of energy considerations, which is not applicable in this case. Now, without energy considerations we will arrive at cumbersome formulas of little mathematical elegance.

In order to find $\upsilon = \arg\phi$, we substitute (8.35) into (8.31) and compare the real and imaginary parts, respectively. We obtain

$$\begin{aligned}
&2\frac{d\upsilon}{d\tau}\int_0^\infty \mathring{\rho}\mathring{g}^{kl}V_k V_l^* \sqrt{\mathring{g}}d\nu = 2\frac{d\upsilon}{d\tau}E_{\boldsymbol{V}}\\
&= \int_0^\infty \nu\left\{\frac{\partial(\rho\sqrt{g}g^{lm})}{\partial n}\bigg|_{n=0} V_l V_m^* - \frac{\partial(c^{lrsm}\sqrt{g}g^{lm})}{\partial n}\bigg|_{n=0}(p_s V_m)(p_r V_l)^*\right\}d\nu\\
&\quad + \operatorname{Im}\int_0^\infty\left\{2\sqrt{\mathring{g}}\mathring{c}^{ijrk}\mathring{\Gamma}^l_{ij}(p_r V_k)V_l^* - \frac{\partial(\sqrt{\mathring{g}}\mathring{c}^{l\alpha km})}{\partial q^\alpha}(p_k V_m)V_l^*\right\}d\nu, \quad (8.47)
\end{aligned}$$

where $E_{\boldsymbol{V}}$ was introduced in (8.44). Here, the coefficient of $\frac{d\upsilon}{d\tau}$ is obviously positive. Equation (8.47) allows finding the Berry phase by integration.

We arrive at the representation for $\phi = |\phi|e^{iv}$ of the form

$$\begin{gathered}\phi(\mathbf{x}) = \frac{\chi(\eta)}{\sqrt{E_{\boldsymbol{V}}(\mathbf{x})}\,\sqrt[4]{\mathring{g}(\mathbf{x})}} e^{iv_1} e^{iv_2} e^{iv_3} e^{iv_4}, \\ v_{1,2,3,4} = \int_{\tau_0}^{\tau} \widehat{v}_{1,2,3,4}\, d\tau.\end{gathered} \tag{8.48}$$

Here, $\chi = \chi(\eta) = |\chi(\eta)|e^{iv(\tau_0,\eta)}$ is a diffraction coefficient, which is an arbitrary smooth function dependent only on the ray, and $\tau_0$ is a somehow fixed value of $\tau$. Formulas (8.47) and (8.48) imply

$$\begin{aligned}\widehat{v}_1 &= \frac{1}{2E_{\boldsymbol{V}}} \int_0^{\infty} \nu \left.\frac{\partial(\rho\sqrt{g}g^{lm})}{\partial n}\right|_{n=0} V_l V_m^* d\nu, \\ \widehat{v}_2 &= -\frac{1}{2E_{\boldsymbol{V}}} \int_0^{\infty} \nu \left.\frac{\partial(c^{lrsm}\sqrt{g}g^{lm})}{\partial n}\right|_{n=0} (p_s V_m)(p_r V_l)^* d\nu, \\ \widehat{v}_3 &= \frac{2}{2E_{\boldsymbol{V}}} \operatorname{Im} \int_0^{\infty} \sqrt{\mathring{g}}\mathring{c}^{ijrk}\mathring{\Gamma}^l_{ij}(p_r V_k)V_l^* d\nu, \\ \widehat{v}_4 &= -\frac{1}{2E_{\boldsymbol{V}}} \operatorname{Im} \int_0^{\infty} \frac{\partial(\sqrt{\mathring{g}}\mathring{c}^{l\alpha km})}{\partial q^{\alpha}} (p_k V_m)V_l^* d\nu.\end{aligned} \tag{8.49}$$

We have split the Berry phase into four factors dependent on their own sets of parameters: $v_1$ depends on the speed of variation of volume density with depth, $v_2$ depends on speeds of variation of elastic moduli with depth, $v_3$ is ruled by the curvature of the surface $\mathscr{S}$, and $v_4$ depends on the variation of elastic moduli in directions tangent to $\mathscr{S}$.

## 8.3 On the construction of higher-order terms

Let us shortly outline the way of constructing higher-order terms in the asymptotic series (8.9). It resembles the corresponding consideration for volume waves. The depth dependence of the local plane Rayleigh wave will be considered known (e.g., found numerically).

The vector $\boldsymbol{V}$ and the function $\phi$, see (8.24), which describe the zeroth-order amplitude are assumed known. The equation (8.11) for $\boldsymbol{U}^I$ is a nonhomogeneous Sturm – Liouville vector problem for an operator with constant coefficients on the half-line $0 \leqslant \nu < \infty$. Its solution exists, because equation (8.28) (which is the consistency condition for the problem for the second equation in (8.11)) holds. The solution is a linear combination of exponentials of $\nu$, with coefficients dependent on $\tau$ and $\eta$. The vector $\boldsymbol{U}^I$ is found, up to addition of a solution of the corresponding homogeneous problem, i.e., it has the form

$$\boldsymbol{U}^I = \boldsymbol{U}^{\perp} + \phi^I \boldsymbol{V}, \tag{8.50}$$

where $\phi^I = \phi^I(\tau, \eta)$ is a scalar function, and $\boldsymbol{U}^\perp = \boldsymbol{U}^\perp(\tau, \eta, \nu)$ is a fixed solution of the homogeneous problem for (8.11). We choose $\boldsymbol{U}^\perp$ as the *additional component*, which is a unique solution of the problem

$$\begin{gathered} \boldsymbol{N}\boldsymbol{U}^{I\perp} = \boldsymbol{M}\boldsymbol{U}^O, \quad \boldsymbol{\mathcal{T}}\boldsymbol{U}^{I\perp}|_{\nu=0} = \boldsymbol{\mathcal{S}}\boldsymbol{U}^O|_{\nu=0}, \\ \boldsymbol{U}^{I\perp} \to 0, \quad \nu \to \infty, \end{gathered} \tag{8.51}$$

that satisfies the condition

$$\int_0^\infty (\boldsymbol{U}^\perp, \boldsymbol{V})d\nu = 0$$

and vanishes for $\nu \to \infty$.

To find $\phi^I$, consider the problem for $\boldsymbol{U}^{II}$ having the form

$$\begin{gathered} \boldsymbol{N}\boldsymbol{U}^{II} = \boldsymbol{M}\boldsymbol{U}^I - \boldsymbol{\mathcal{L}}\boldsymbol{U}^O, \quad \nu > 0, \\ \boldsymbol{\mathcal{T}}\boldsymbol{U}^{II}|_{\nu=0} = \boldsymbol{\mathcal{S}}\boldsymbol{U}^I|_{\nu=0} - \boldsymbol{\mathcal{R}}\boldsymbol{U}^O|_{\nu=0}, \quad \boldsymbol{U}^{II} \to 0, \quad \nu \to \infty. \end{gathered} \tag{8.52}$$

Here, $\boldsymbol{\mathcal{S}}$ and $\boldsymbol{\mathcal{R}}$ denote certain first-order differential operators with respect to $\nu$, given by explicit expressions, which we omit. The scalar multiplication of both sides of the first equation in (8.52) by $\boldsymbol{V}$ and integration with respect to $\nu$ provide

$$\int_0^\infty (\boldsymbol{N}\boldsymbol{U}^{II} - \boldsymbol{M}\boldsymbol{U}^I + \boldsymbol{\mathcal{L}}\boldsymbol{U}^O, \boldsymbol{V})d\nu = 0. \tag{8.53}$$

Releasing, with the help of integration by parts, the components of the vector $\boldsymbol{U}^{II}$ from the derivatives with respect to $\nu$, employing the boundary conditions, as we did in Section 8.2.5, and substituting the expression for $\boldsymbol{U}^I$ (see (8.50)), we obtain, generally speaking, a nonhomogeneous transport equation for $\phi^I$ (compare with (8.34)):

$$\frac{d\phi^I}{d\tau} + \Xi\phi^I = f. \tag{8.54}$$

Finding $\phi^I$ from equation (8.54) completes the construction of $\boldsymbol{U}^I$.

We omit similar consideration of the procedure related to further approximations.

## 8.4 Case of an isotropic body

The case of isotropic but, in general, inhomogeneous media seems to be of particular importance in applications. Since the results obtained earlier in this chapter are applicable in this case, we will just specify the above formulas.

### 8.4.1 Peculiar features of an isotropic case

The local velocity $c(\mathbf{x})$ of the Rayleigh wave is found from equation (2.209) with the parameters $\lambda = \lambda(\mathbf{x})$, $\mu = \mu(\mathbf{x})$, and $\rho = \rho(\mathbf{x})$, $\mathbf{x} \in \mathscr{S}$. The rays are extremals of the Fermat functional $\mathrm{I} = \int \frac{ds}{c}$, where we integrate along the arc on the surface $\mathscr{S}$, $ds^2 = \mathring{g}_{\alpha\beta} dq^\alpha dq^\beta$ ($\mathring{g}_{\alpha\beta}$ is the first fundamental form on $\mathscr{S}$). The theory of such rays and the relevant wavefronts for I, in essence, differs little from that for the case of volume waves, developed in Chapter 5. Here, $q^1$, $q^2$ are still ray coordinates on the surface, ($q^1 = \tau$, $q^2 = \eta$), and $n \geqslant 0$ is the distance along the inner normal to the surface. In these coordinates, the amplitude vector $\boldsymbol{U}^O$ has the covariant components $U^O_\tau$, $U^O_\eta$, and $U^O_n$. The matrix of the metric tensor at $n = 0$ is diagonal:

$$\|\mathring{g}_{ij}\| = \begin{pmatrix} c^2 & 0 & 0 \\ 0 & \mathrm{j}^2 & 0 \\ 0 & 0 & 1 \end{pmatrix}, \quad \|\mathring{g}^{ij}\| = \begin{pmatrix} \frac{1}{c^2} & 0 & 0 \\ 0 & \frac{1}{\mathrm{j}^2} & 0 \\ 0 & 0 & 1 \end{pmatrix}, \quad \mathrm{j}^2 := \mathring{g}_{22}, \tag{8.55}$$

where $\mathrm{j} := \mathrm{j}(\tau, \eta)$ is the *geometrical spreading of surface rays*. Obviously, $\sqrt{\mathring{g}} = \mathring{c}\mathrm{j}$. The contravariant components of the stiffness tensor (8.4) read

$$c^{ijkl} = \mu\left(g^{ik}g^{jl} + g^{il}g^{jk}\right) + \lambda g^{ij}g^{kl} \tag{8.56}$$

with $\lambda = \lambda(\mathbf{x})$ and $\mu = \mu(\mathbf{x})$.

Addressing equations (8.14), (8.15), we first note that the operators $p_j$ transform under a change of variables as covariant vectors. Introducing the covariant vector $\boldsymbol{p} = (p_1, p_2, p_3)$, we write the equation $\boldsymbol{N}\boldsymbol{U}^O = 0$ (8.14) in the form

$$\sqrt{\mathring{g}}\left[-\mathring{\rho}\boldsymbol{U}^O + (\mathring{\lambda} + \mathring{\mu})\boldsymbol{p}(\boldsymbol{p}\cdot\boldsymbol{U}^O) + \mathring{\mu}(\boldsymbol{p}\cdot\boldsymbol{p})\boldsymbol{U}^O\right] = 0, \quad \nu > 0, \tag{8.57}$$

where $\cdot$ denotes contraction: $(\boldsymbol{f}\cdot\boldsymbol{h}) = f_l \mathring{g}^{lm} h_m$ ($\boldsymbol{f}$ and $\boldsymbol{h}$ are real-valued).

By the formula (8.24), $\boldsymbol{U}^0 = \phi\boldsymbol{V}$, where $\boldsymbol{V}$ stands for an arbitrarily normalized solution of the problem (8.13). It is convenient to start by defining $\boldsymbol{V}$ in the local Cartesian frame: we take the direction of $\operatorname{grad}\tau$ as the direction of the $x$-axis, the $z$-axis is directed along the inner normal $\boldsymbol{n}$, and the $y$-axis is directed along the vector $[\boldsymbol{n}\times\operatorname{grad}\tau]$. Expressions for $V_x, V_y$ and $V_z$ are given now by the formulas (2.212)–(2.213), and

$$\boldsymbol{V} = V_x\, c \operatorname{grad}\tau + V_z\boldsymbol{n}. \tag{8.58}$$

The contravariant components of the vector $\boldsymbol{V}$ in the coordinates $\tau$, $\eta$, $n$ are

$$V^1 = \frac{1}{c}V_x, \quad V^2 = 0, \quad V^3 = V_z,$$

respectively.

The equation $\boldsymbol{N}\,\boldsymbol{U}^I = \boldsymbol{M}\,\boldsymbol{U}^O$ takes the form

$$\sqrt{\mathring{g}}\Big[-\mathring{\rho}U^{I\tau} + \frac{\mathring{\lambda}+\mathring{\mu}}{c^2}(U^{I\tau}+p_3U^{I\nu}) + \mathring{\mu}\Big(\frac{1}{c^2}+p_3^2\Big)U^{I\tau}\Big] = (\boldsymbol{M}\,\boldsymbol{U}^O)^\tau,$$
$$\sqrt{\mathring{g}}\Big[-\mathring{\rho}U^{I\tau} + \mathring{\mu}\Big(\frac{1}{c^2}+p_3^2\Big)U^{I\eta}\Big] = (\boldsymbol{M}\,\boldsymbol{U}^O)^\eta,$$
$$\sqrt{\mathring{g}}\Big[-\mathring{\rho}U^{I\nu} + \frac{\mathring{\lambda}+\mathring{\mu}}{c^2}p_3(U^{I\tau}+p_3U^{I\nu}) + \mathring{\mu}\Big(\frac{1}{c^2}+p_3^2\Big)U^{I\nu}\Big] = (\boldsymbol{M}\,\boldsymbol{U}^O)^\nu, \tag{8.59}$$

and the boundary conditions become

$$\mathring{\mu}\Big(p_3U^{I\tau}+\frac{1}{c^2}U^{I\nu}\Big)\Big|_{\nu=0} = (\boldsymbol{\mathcal{T}}\boldsymbol{U}^O)^\tau\Big|_{\nu=0},$$
$$\mathring{\mu}p_3U^{I\eta}\Big|_{\nu=0} = (\boldsymbol{\mathcal{T}}\boldsymbol{U}^O)^\eta\Big|_{\nu=0}, \tag{8.60}$$
$$\Big[\mathring{\lambda}\Big(U^{I\tau}-p_3U^{I\nu}+\frac{1}{c^2}U^{I\nu}\Big)+2\mathring{\mu}p_3U^{I\nu}\Big]\Big|_{\nu=0} = (\boldsymbol{\mathcal{T}}\boldsymbol{U}^O)^\nu\Big|_{\nu=0}.$$

The problem splits in two: one is of the $P-SV$ type for $U^{\tau,\nu}$, and the other is of the $SH$ type for $U^{I\eta}$. The second problem is uniquely solvable, but the first one is not, in general, because the corresponding homogeneous problem has a nontrivial solution. The scalar function $\phi$ of two variables should be found from the condition for its consistency.

### 8.4.2 Explicit formulas for the leading-order term

These formulas follow from the equations (8.24), and (8.48)–(8.49).

**Formula for the modulus of $\phi(\tau,\eta)$**

The resulting expression for $|\phi(\tau,\eta)|$ has the form

$$|\phi(\tau,\eta)| = \frac{|\chi(\eta)|}{\sqrt{\mathring{c}\mathrm{j}E_{\boldsymbol{V}}(\tau)}}, \tag{8.61}$$

where $\chi(\eta) = |\chi|e^{iv}$ is the diffraction coefficient. The expression for $E_{\boldsymbol{V}}(\tau)$ can be taken as the right-hand side of (2.214), multiplied by $\frac{2}{\omega}$, that is (see (2.214) and (8.44))

$$E_{\boldsymbol{V}} = \int_0^\infty \mathring{\rho}\mathring{g}^{lm}V_lV_m^*d\nu = \frac{2\mathring{\rho}c^2}{\sqrt{\frac{1}{c^2}-\frac{1}{\mathring{b}^2}}}\left[\frac{6}{c^2}\left(\frac{1}{\mathring{b}^2}-\frac{1}{\mathring{a}^2}\right)+\frac{4}{\mathring{a}^2\mathring{b}^2}-\frac{6}{\mathring{b}^4}+\frac{c^2}{\mathring{b}^6}\right], \tag{8.62}$$

where $\mathring{a}$ and $\mathring{b}$ stand for the values of the velocities of $P$ and $S$ waves on the surface of the body under consideration. We obtained an explicit expression for the modulus of the amplitude factor $|\phi|$ (which is relatively simple as compared to what follows).

**Formula for the phase $\upsilon(\tau, \eta)$ in (8.35)**

Here, we will employ the notation

$$\mathfrak{p} = \sqrt{\frac{1}{c^2} - \frac{1}{a^2}}, \quad \mathfrak{q} = \sqrt{\frac{1}{c^2} - \frac{1}{b^2}}, \quad l = \frac{2}{c^2} - \frac{1}{b^2} = 2\xi^2 - \frac{1}{b^2}, \quad \xi = \frac{1}{c}, \tag{8.63}$$

see, in this connection, (2.207).

The phase of the amplitude factor can also be found explicitly but looks far more cumbersome. It follows from (8.34) that

$$\upsilon(\tau, \eta) = \upsilon(\tau_0, \eta) + \int_{\tau_0}^{\tau} (-\operatorname{Im}\Xi) d\tau. \tag{8.64}$$

By a remarkably laborious calculation (see Babich and Kirpichnikova 2004 [7]), the following result is found

$$-\operatorname{Im}\Xi = \frac{\mathring{C}}{A}\frac{\mathfrak{b}_{11}}{c^2} + \frac{\mathring{D}}{A}\frac{\mathfrak{b}_{22}}{g_{22}} + \frac{\mathring{F}}{A}, \tag{8.65}$$

$$e^{i\upsilon(\eta,\tau)} = e^{i\upsilon(\eta,\tau_0)} \exp\left\{\int_0^{\tau} \left(\frac{\mathring{C}}{A}\frac{\mathfrak{b}_{11}}{c^2} + \frac{\mathring{D}}{A}\frac{\mathfrak{b}_{22}}{\mathring{g}_{22}} + \frac{\mathring{F}}{A}\right) d\tau\right\}. \tag{8.66}$$

Here, $\mathfrak{b}_{11}$ and $\mathfrak{b}_{22}$ are coefficients of the second fundamental form of the surface $\mathscr{S}$ in the coordinates $\tau$ and $\eta$, and $A$ designates the expression

$$A := \frac{l}{\mathring{\rho}\mathring{\mathfrak{p}}c^4} E_{\mathbf{V}}. \tag{8.67}$$

The positivity of $E_{\mathbf{V}}$, see (8.62), implies that $A > 0$.

The coefficients $\mathring{C}$, $\mathring{D}$, and $\mathring{F}$ are the values, at $n = 0$, of the quantities $C$, $D$, and $F$ defined as

$$C = il\frac{C_1\mathfrak{p} + C_2\mathfrak{q}}{\mathfrak{p}^2\mathfrak{q}^2},$$

where $C_1$ and $C_2$ are given by

$$C_1 = 2\mathfrak{p}^2\left(\frac{1}{c^2b^2} - \frac{1}{b^4} - \frac{1}{c^4}\right), \quad C_2 = l\mathfrak{p}^2\left(\frac{3}{2b^2} - \frac{1}{c^2}\right) - \frac{2\mathfrak{q}^2}{c^2}\left(\frac{3}{a^2} - \frac{2}{c^2}\right).$$

Further,

$$D = 2ib^2 l\left[\mathfrak{p}\left(\frac{1}{b^4} - \frac{1}{c^4}\right) + \frac{\mathfrak{q}}{4}\left(\frac{1}{b^4} - \frac{4}{c^4}\right)\right], \tag{8.68}$$

and

$$F = iF_1 \left.\frac{\partial \ln\rho}{\partial n}\right|_{n=0} + iF_2 \left.\frac{\partial}{\partial n}\left(\frac{1}{a^2}\right)\right|_{n=0} + iF_3 \left.\frac{\partial}{\partial n}\left(\frac{1}{b^2}\right)\right|_{n=0}, \tag{8.69}$$

with

$$F_1 = -\left[\frac{a^2 l^2}{a^2-b^2} - \frac{4}{c^4} + \frac{2}{c^2(a^2-b^2)}\right]\mathrm{p} + \frac{2lb^2}{c^2(a^2-b^2)}\mathrm{q}, \quad F_2 = \frac{l\mathrm{q}}{c^2\mathrm{p}^2},$$
$$F_3 = -\frac{2b^2\mathrm{p}}{a^2-b^2}\left[b^2 l^2 + \frac{2}{c^2}\right] + 2l\mathrm{p} + \frac{l\mathrm{p}}{c^2\mathrm{q}^2} + \frac{4b^4 l\mathrm{q}}{c^2(a^2-b^2)}. \tag{8.70}$$

Formulas (8.65)–(8.70) explicitly describe the Berry phase. Each term in the expressions (8.65) and (8.66) describes the impact of various parameters on the Berry phase:

The first term on the right-hand side of (8.65) is associated with the curvature of the surface $\mathscr{S}$ along the ray.

The second term is associated with the curvature of the surface $\mathscr{S}$ in the direction orthogonal to the ray.

The third term is associated with variation of the parameters of the medium with depth.

**Relation between (8.48)–(8.49) and (8.64)–(8.70)**

It may seem striking that the expressions (8.48)–(8.49) for $\upsilon$ show so little likeness to those in (8.64)–(8.70), which are their specifications. The point is that the use of a particular structure of elastic stiffnesses $c^{ijkl}$ for the isotropic case, together with detailed expressions for the components of $\boldsymbol{V}$, allow (after a depressingly long calculation) a major cancellation. For this reason, the formulas (8.64)–(8.70) look so unexpectedly different from those of Section 8.2.7.

It is worthy of note that no terms with tangent derivatives of $\lambda$, $\mu$, and $\rho$ (respectively, $a$, $b$, and $\rho$) are present in (8.69). This is due to the fact that in the isotropic case, the expression

$$\mathrm{P} := V_l^*(p_k V_m)\frac{\partial\left(\mathring{c}^{l\alpha km}\sqrt{g}\right)}{\partial q^\alpha}$$

in the equation (8.49) proves to be real, which can be established by a straightforward calculation.

### 8.4.3 Final formula for the leading-order term

To summarize, the leading term is described by the relation (8.35) that is:

$$\boldsymbol{U}^O = \phi(\tau,\eta)\,\boldsymbol{V}(\tau,\eta,\nu) = |\phi(\tau,\eta)|e^{i\upsilon}\,\boldsymbol{V}(\tau,\eta,\nu), \quad \nu = \omega n, \tag{8.71}$$

where the expression for $\boldsymbol{V}$ is given by the formulas (8.59) and (2.212)–(2.213), and $c$ stands for the velocity of the local Rayleigh wave, $c|\mathrm{grad}\,\tau| = 1$, $|\phi|$ is described by (8.61), and the Berry phase $\upsilon$ is given by the expressions (8.63)–(8.70).

## 8.5 ⋆ Comments to Chapter 8

The ray theory of Rayleigh waves on a smooth surface of an isotropic elastic body was originally developed by Babich 1961 [3] and Babich and Rusakova (Kirpichnikova) 1963 [9]. They started from the classical formulas for the Rayleigh wave (see Chapter 2), replaced the inhomogeneous plane $P$- and $S$-waves by related ray expansions with complex eikonals, and required that their sum satisfy boundary conditions. The velocity of the resulting surface wave is the local velocity $c$ of the Rayleigh wave, which perfectly matches with results of an old paper by Petrovsky 1945 [31]. There, it was proved that "surface" discontinuities of a nonstationary solution may propagate (under certain assumptions) along the surface of an elastic body only with the Rayleigh velocity. The paper by Petrovsky 1945 [31] anticipated the modern research of discontinuities of surface waves by methods of microlocal analysis; see, e.g., Hansen 2014 [15].

The expressions for a particular case of isotropic media found independently from the general theory (presented in Section 8.2) as far back as the early 1960s (see Babich 1961 [3], Babich and Rusakova (Kirpichnikova) 1963 [9]) were specified and rectified (see Krauklis 1971 [24], Babich and Kirpichnikova [6]). In the above papers, attention was given to finding detailed expressions for the phase increments arising in the course of propagation of a wave along surface rays, that is, those later named the Berry phase.

Further advance of ideas of the papers by Babich 1961, Babich and Rusakova (Kirpichnikova) 1963 [3, 9] is due to Nomofilov 1979, 1982 [29, 30]; operators similar to $p_j$ (8.15) were introduced therein. Gregory 1971 [14] presented a detailed study of the special cases of cylindrical and spherical surfaces. Mozhaev 1984 [28] dealt with Rayleigh waves on curvilinear surfaces by a kind of perturbation approach. The paper by Babich and Kirpichnikova 2004 [7], on which the present chapter is based, combined the ideas of ray methods with boundary layer techniques. Similarly to the case of volume waves, the finding of the leading-order intensity employed the corresponding Umov equation.

A version of the ray theory was presented for approximately layered elastic bodies, whenever the structure changes slightly at distances comparable to the wavelength. In the description of Rayleigh (as well as Love) waves there, the consideration of energy flow is of crucial importance. The development is comparatively simple under the assumptions that the curvature of the boundary is not large and the horizontal variations of properties of the body are slow (or nil). Then no Berry phase appears (see Woodhouse [36], Babich, Chikhachev, and Yanovskaya 1976 [5], Levshin, Yanovskaya, Lander, Bukchin et al. 1989 [27], Tromp and Dahlen [34]). For a sufficiently curved boundary and/or a considerable horizontal variation of parameters of the structure, the Berry phase is nontrivial (see Babič and Čihačev 1980 [4], Tromp and Dahlen 1992, 1993 [34, 35], and others). The aforementioned papers by Tromp and Dahlen (see also Dahlen and Tromp 1998 [12]) are characterized by a wide employment of variational ideas and ideas based on the consideration of energy flow.

Special cases where the anomalously polarized component can be found in explicit form were described by Yanovskaya and Roslov 1989 [37], Kirpichnikova and Kiselev 1990, 1995 [19, 20], Aref'ev and Kiselev 1998, 1999 [1, 2]. The cases where the ray solution reduces to a finite number of terms and is thus an exact solution were considered by Kiselev 2004 [21], Kiselev and Tagirdzhanov 2008 [23], and Kiselev and Rogerson 2009 [22]. A number of researchers (see, e.g., Woodhouse 1974 [36], Babich, Chikhachev, and Yanovskaya 1976 [5], Levshin, Yanovskaya, Lander, Bukchin et al. 1989 [27], Fu, Rogerson, and Wang 2013 [13]) took into account topography. The effects of gravitation and unelasticity, as well as those of the Coriolis force, on the propagation of surface waves were discussed by Dahlen and Tromp 1998 [12]. Keller and Karal 1960, 1964 [17, 18] applied complex rays to the theory of surface waves.

A theory of surface *interference waves*, which are superpositions of whispering-gallery $S$-waves and inhomogeneous $P$-waves, exponentially decreasing with depth was developed by Brekhovsikh 1968 [11], Krauklis and Tsepelev 1976, 1987 [25, 26]. A condition for the existence of such a wave is the positivity of the *effective curvature of the surface*, defined by

$$\frac{1}{\mathfrak{R}} - \frac{1}{b}\frac{\partial b}{\partial n}\bigg|_{\mathscr{S}}.$$

Here, $\frac{1}{\mathfrak{R}}$ stands for the normal section curvature in the direction of propagation of the wave, and $b$ is the velocity of the wave $S$. References concerning the propagation of Gaussian beams along the surface of an elastic body are given in Babich and Kiselev 2004 [8].

## References to Chapter 8

[1] Aref'ev, A. V. and Kiselev, A. P. 1998. Viscoelastic Rayleigh waves in a layered structure with weak lateral inhomogeneity. *J. Math. Sci.* 91:2701–05. *Арефьев А. В., Киселев А. П.* Вязкоупругие рэлеевские волны в слоистой структуре, слабо неоднородной по горизонтали. Зап. научн. сем. ПОМИ, 1995. Т. 230. С. 7–13.

[2] Aref'ev, A. V. and Kiselev, A. P. 1999. Viscoelastic Love waves in a layered structure with weak lateral inhomogeneity. *J. Math. Sci.* 96:3289–91. *Арефьев А. В., Киселев А. П.* Вязкоупругие лявовские волны в слоистой структуре, слабо неоднородной по горизонтали. Зап. научн. сем. ПОМИ, 1997. Т. 239. С. 7–11.

[3] Babich, V. M. 1961. Propagation of Rayleigh waves along the surface of a homogeneous elastic body of arbitrary shape. *Dokl. Akad. Nauk SSSR.* 137:1263–6 [in Russian]. *Бабич В. М.* О распространении волн Рэлея вдоль поверхности неоднородного упругого тела произвольной формы. Докл. АН СССР, 1961. Т. 137(6). С. 1263–1266.

[4] Babič, V. M. and Čihačev, B. A. 1980. Propagation of Love and Rayleigh waves in a weakly nonhomogeneous laminated medium. *Vestnik Leningrad. Univ. Ser. Math.* 30:32–8. *Бабич В. М., Чихачев Б. А.* Распространение волн Лява и Рэлея в слабо неоднородной упругой среде. Вестник ЛГУ. Сер. мат. мех. астр., 1975. № 1. С. 32–38.

[5] Babich, V. M., Chikhachev, B. A. and Yanovskaya, T. B. 1976. Surface waves in a vertically inhomogeneous elastic half-space with a weak horizontal inhomogeneity. *Izv. Ac. Sci. USSR. Phys. Solid Earth.* 12:242–5. *Бабич В. М., Чихачев Б. А., Яновская Т. Б.* Поверхностные волны в вертикально-неоднородном упругом полупространстве со слабой горизонтальной неоднородностью. Изв. АН СССР. Физика Земли, 1976. № 4. С. 24–31.

[6] Babich, V. M. and Kirpichnikova, N. Ya. 1990. On the question of Rayleigh waves propagating along the surface of an inhomogeneous elastic body. *J. Sov. Math.* 50:1693–5. *Бабич В. М., Кирпичникова Н. Я.* К вопросу о волнах Рэлея, распространяющихся вдоль поверхности неоднородного упругого тела. Зап. научн. сем. ЛОМИ, 1986. Т. 156. С. 20–23.

[7] Babich, V. M. and Kirpichnikova, N. Ya. 2004. A new approach to the problem of the Rayleigh wave propagation along the boundary of a nonhomogeneous elastic body. *Wave Motion.* 40:209–23.

[8] Babich, V. M. and Kiselev, A. P. 2004. "Nongeometrical phenomena" in propagation of elastic surface waves. In *Surface waves in anisotropic and laminated bodies and defects detection.* ed. R. V. Goldstein, and G. A. Maugin, 119–129. New York: Kluwer.

[9] Babich, V. M. and Rusakova (Kirpichnikova), N. Ya. 1963. The propagation of Rayleigh waves over the surface of a non-homogeneous elastic body with an arbitrary form. *U.S.S.R. Comput. Math. Math. Phys.* 2:719–35. *Бабич В. М., Русакова Н. Я.* О распространении волн Рэлея по поверхности неоднородного упругого тела произвольной формы. Ж. вычисл. матем. матем. физ., 1962. Т. 2(4). С. 652–665.

[10] Berry, M. 1984. Quantal phase factors accompanying adiabatic changes. *Proc. Roy. Soc. London. Ser. A.* 392(1802):45–57.

[11] Brekhovskikh, L. M. 1968. Surface waves confined to the curvature of the boundary in solid. *Sov. Phys. Acoust.* 13:462–72. *Бреховских Л. М.*

О поверхностных волнах в твердом теле, удерживаемых кривизной границы. Акуст. журн., 1967. Т. 13(4). С. 541–555.

[12] Dahlen, F. A. and Tromp, J. 1998. *Theoretical global seismology.* Princeton University Press.

[13] Fu, Y. B., Rogerson, G. A. and Wang, W. F. 2013. Surface waves guided by topography in an anisotropic elastic half-space. *Proc. Roy. Soc. London. Ser. A.* 469(2149) pap. 20120371.

[14] Gregory, R. D. 1971. The propagation of Rayleigh waves over curved surfaces at high frequency. *Proc. Cambridge Philos. Soc.* 70:103–21.

[15] Hansen, S. 2014. Subsonic free surface waves in linear elasticity. *SIAM J. Math. Anal.* 46:2501–24.

[16] Hirsch, M. W. 1997. Differential topology. Berlin: Springer-Verlag. *Хирш М.* Дифференциальная топология. М.: Мир, 1997.

[17] Keller, J. B. and Karal, F. C. 1960. Surface wave excitation and propagation. *J. Appl. Phys.* 31:1039–46.

[18] Keller, J. B. and Karal, F. C. 1964. Geometrical theory of elastic surface-wave excitation and propagation. *J. Acoust. Soc. Amer.* 36:32–40.

[19] Kirpichnikova, N. Ya. and Kiselev A. P. 1990. Depolarization of elastic surface waves in a vertically inhomogeneous half-space. *Sov. Phys. Acoust.* 36:96–7. *Кирпичникова Н. Я., Киселев А. П.* Деполяризация упругих поверхностных волн в вертикально-неоднородном полупространстве. Акуст. журн., 1990. Т. 36(1). С. 173–175.

[20] Kirpichnikova, N. Ya. and Kiselev, A. P. 1995. On the anomalous polarization of Love and Rayleigh elastic waves in layered medium, *J. Math. Sci.* 73:383–4. *Кирпичникова Н. Я., Киселев А. П.* Об аномальной поляризации упругих волн Лява и Рэлея в слоистой структуре. Зап. научн. семин. ЛОМИ АН СССР, 1990. Т. 186. С. 134–136.

[21] Kiselev, A. P. 2004. Rayleigh wave with a transverse structure. *Proc. Roy. Soc. London. Ser. A.* 460(2050):3059–64.

[22] Kiselev, A. P. and Rogerson, G. A. 2009. Laterally dependent surface waves in an elastic medium with a general depth dependence. *Wave Motion.* 46:539–47.

[23] Kiselev, A. P. and Tagirdzhanov, A. M. 2008. Love waves with a transverse structure, *Vestnik St. Petersburg University. Mathematics.* 41(3):278–81. *Киселев А. П., Тагирджанов А. М.* Лявовские волны с поперечной структурой. Вестн. СПбГУ. Сер. 1. мат. мех. астрон., 2008. № 3. С. 136–139.

[24] Krauklis, P. V. 1971. Estimation of the intensity of Rayleigh and Stonely surface waves on an inhomogeneous path. *Seminars in Mathematics in Steklov Math. Institute.* 15:63–6. *Крауклис П. В.* К оценке интенсивности поверхностных волн Рэлея и Стоунли на неоднородной трассе. Зап. научн. семин. ЛОМИ, 1969. Т. 15. С. 115–121.

[25] Krauklis, P. V. and Tsepelev, N. V. 1976. High-frequency asymptotic representation of a wave field concentrated near the boundary of an elastic medium. *J. Math. Sci.* 34:72–92. *Крауклис П. В., Цепелев Н. В.* О построении высокочастотной асимптотики волнового поля, сосредоточенного вблизи границы упругой среды. Зап. научн. сем. ЛОМИ, 1973. Т. 34. С. 72–92.

[26] Krauklis, P. V. and Tsepelev, N. V. 1987. On question of distant propagation of $P_n$ and $S_n$ waves in the oceanic lithosphere. *Izv. Acad. Sci. USSR. Phys. Solid Earth.* 23(3):217–20. *Крауклис П. В., Цепелев Н. В.* К задаче о распространении волн $P_n$ и $S_n$ в океанической литосфере. Изв. АН СССР. Физика Земли, 1987. № 3. С. 51–55.

[27] Levshin, A. L., Yanovskaya, T. B., Lander, A. V., Bukchin, B. G. et al. 1989. *Seismic surface waves in a laterally inhomogeneous Earth.* Dordrecht: Kluwer. *Левшин А. Л., Яновская Т. Б., Ландер А. В. и др.* Поверхностные волны в горизонтально-неоднородной Земле. М.: Наука, 1986.

[28] Mozhaev, V. G. 1984. Application of the perturbation method for calculating the characteristics of surface-waves in anisotropic and isotropic solids with curved boundaries. *Sov. Phys. Acoustics.* 30:394–400. *Можаев В. Г.* Применение метода возмущений для расчета характеристик поверхностных волн в анизотропных и изотропных твердых телах с искривленными границами. Акуст. журн., 1984. Т. 30(5). С. 673–678.

[29] Nomofilov, V. E. 1979. Quasistationary Rayleigh waves on the surface of an inhomogeneous, anisotropic elastic body. *Sov. Phys. Doklady.* 24:609–11. *Номофилов В. Е.* Квазистационарные волны Рэлея на поверхности неоднородного анизотропного упругого тела. Докл. АН СССР, 1979. Т. 247(5). С. 1107–1111.

[30] Nomofilov, V. E. 1982. Propagation of quasi-stationary Rayleigh waves in a nonuniform anisotropic elastic medium. *J. Sov. Math.* 19:1466–75. *Номофилов В. Е.* О распространении квазистационарных волн Рэлея в неоднородной анизотропной упругой среде. Зап. научн. семин. ЛОМИ, 1979. Т. 89. С. 234–245.

[31] Petrovsky, I. G. 1945. On the propagation velocity of discontinuities of the displacement derivatives on the surface of an inhomogeneous elastic body of arbitrary form. *Dokl. Acad. Sci. URSS*, 47:255–8 [in Russian].

*Петровский И. Г.* О скорости распространения разрывов производных смещения на поверхности неоднородного упругого тела произвольной формы. Докл. АН СССР, 1945. Т. 47(4). С. 252–261.

[32] Simon, B. 1983. Holonomy, the quantum adiabatic theorem and Berry's phase. *Phys. Rev. Lett.* 51:2167–70.

[33] Treves, F. 1980. *Introduction to pseudodifferential and Fourier integral operators: Volume 1.* New York: Springer. *Трев Ф.* Введение в теорию псевдодифференциальных операторов и операторов Фурье. Т. 1. М.: Мир, 1984.

[34] Tromp, J. and Dahlen, F. A. 1992. Variational principles for surface wave propagation on a laterally heterogeneous Earth – II. Frequency-domain JWKB theory. *Geophys. J. Int.* 109:599–619.

[35] Tromp, J. and Dahlen, F. A. 1993. Surface wave propagation in a slowly varying anisotropic waveguide. *Geophys. J. Intern.* 113:239–49.

[36] Woodhouse, J. H. 1974. Surface waves in a laterally varying layered structure. *Geophys. J. Intern.* 37:461–90.

[37] Yanovskaya, T. B. and Roslov, Yu. V. 1989. Peculiarities of surface wave fields in laterally inhomogeneous media in the framework of ray theory. *Geophys. J. Intern.* 99:297–303.

# Appendix: Elements of Tensor Analysis and Differential Geometry

## A.1 Definition of tensor

We have already dealt with tensors in special simple cases. In the early chapters, the stress and strain tensors were treated simply as matrices. However, in Chapter 8, we cannot do without techniques of tensor analysis.

The definition of a tensor does not essentially depend on the dimension of the space. Actually, we are interested in the dimensions $m = 3$ and $m = 2$. We start with the definition related to an arbitrary domain $\Omega \subset \mathbb{R}^m$, $m \geqslant 2$. Let $\mathbf{x} = (x^1, \dots, x^m)$ and $\mathbf{z} = (z^1, \dots, z^m)$ be two arbitrary *regular coordinate systems*[1] in $\Omega$, $x^i = x^i(z^1, \dots, z^m)$, $z^j = z^j(x^1, \dots, x^m)$. Consider a set of numbers $T^{i_1 \dots i_r}_{j_1 \dots j_s}$ described in the coordinate system $(x^1, \dots, x^m)$, all the indices range from 1 to $m$.

This set defines a tensor $\mathbf{T}$ of order $s+r$ of the type $(s, r)$, $s$ times covariant and $r$ contravariant; when passing to a coordinate system $z^j = z^j(x^1, \dots, x^m)$, the components $T^{i_1 \dots i_r}_{j_1 \dots j_s}$ are transformed as follows:

$$\widetilde{T}^{i_1 \dots i_r}_{j_1 \dots j_s} = \frac{\partial z^{i_1}}{\partial x^{k_1}} \cdots \frac{\partial z^{i_r}}{\partial x^{k_r}} \frac{\partial x^{l_1}}{\partial z^{j_1}} \cdots \frac{\partial x^{l_s}}{\partial z^{j_s}} T^{k_1 \dots k_r}_{l_1 \dots l_s}. \tag{A.1}$$

Here, $\widetilde{T}^{i_1 \dots i_r}_{j_1 \dots j_s}$ are the components of $\mathbf{T}$ in the coordinates $\mathbf{z}$ and $T^{k_1 \dots k_r}_{l_1 \dots l_s}$ are its components in the coordinates $\mathbf{x}$. The Einstein summation convention is implied, assuming summation from 1 to $m$ over repeated upper and lower indices.

Tensors can be considered not only in a domain but also on a smooth $m$-dimensional surface $\mathscr{S} \subset \mathbb{R}^{m+1}$. The definition of a tensor remains as before, but $(x^1, \dots, x^m)$ and $(z^1, \dots, z^m)$ become the coordinates on the surface $\mathscr{S} \subset \mathbb{R}^{m+1}$.

[1]This means that all the functions $x^i = x^i(z^1, \dots, z^m)$, $z^i = z^i(x^1, \dots, x^m)$ are smooth, and the Jacobians $\frac{D(x^1, \dots, x^m)}{D(z^1, \dots, z^m)}$ and $\frac{D(z^1, \dots, z^m)}{D(x^1, \dots, x^m)}$ do not vanish.

## A.2 Simple operations with tensors

Tensors can be multiplied by scalar functions. Linear combinations of tensors having the same values of $r$ and $s$ are defined with coefficients that are scalar functions.

The product of tensors is defined as follows. Let $T^{i_1\ldots i_r}_{j_1\ldots j_s}$ be the components of an $(r, s)$ tensor $\mathbf{T}$ and $S^{p_1\ldots p_l}_{q_1\ldots q_t}$ be the components of an $(l, t)$ tensor $\mathbf{S}$. The values

$$W^{i_1\ldots i_r p_1\ldots p_l}_{j_1\ldots j_s q_1\ldots q_t} = T^{i_1\ldots i_r}_{j_1\ldots j_s} S^{p_1\ldots p_l}_{q_1\ldots q_t} \tag{A.2}$$

define an $(r+l, s+t)$ tensor $\mathbf{W}$ referred to as the tensor product of $\mathbf{T}$ and $\mathbf{S}$.

An important operation of contraction of indices (which is needed, e.g., in defining the scalar product of vectors) will be illustrated with an example. Let a pair of indices (one is a subscript and the other is a superscript) of the tensor be set equal, e.g., $T^{i_1\ldots i_{r-1} i_r}_{j_1\ldots j_{s-1} i_r}$. In accordance with the summation rule, the summation over $i_r$ is understood. It can easily be proved that we arrive at an $(r-1, s-1)$ tensor

$$V^{i_1\ldots i_{r-1}}_{j_1\ldots j_{s-1}} = T^{i_1\ldots i_r}_{j_1\ldots i_r}. \tag{A.3}$$

## A.3 Metric tensor. Raising and lowering indices

In an $m$-dimensional domain $\Omega$, consider a smooth, real, symmetric, $(0, 2)$ (that is, twice covariant) tensor $\boldsymbol{g}$

$$g_{ij} = g_{ji} \tag{A.4}$$

with a positive definite matrix $\|g_{ij}\|$. Such a tensor is called a metric tensor. The Riemannian length of a (parametrically defined) smooth curve $x^i = x^i(\sigma)$, $-\infty < \sigma_0 < \sigma < \sigma_1 < +\infty$, $i = 1, \ldots, m$, joining points $M_0$ and $M_1 \in \Omega$ is defined by the integral

$$\int_{\sigma_0}^{\sigma_1} \sqrt{g_{ij}\frac{dx^i}{d\sigma}\frac{dx^j}{d\sigma}}\,d\sigma,$$
$$M_0 = (x^1(\sigma_0), \ldots, x^m(\sigma_0)), \quad M_1 = (x^1(\sigma_1), \ldots, x^m(\sigma_1)). \tag{A.5}$$

The integral (A.5) is independent of the parameterization of the curve. We say that a *Riemannian metric* with *metric tensor* $\boldsymbol{g}$ is defined in $\Omega$. The metric tensor has already been encountered in Chapter 4 and Chapter 8.

The matrix $\|g^{ij}\|$ inverse to $\|g_{ij}\|$ is a twice contravariant (that is, $(2, 0)$) tensor. It is symmetric and positive definite as well. The contraction

$g^{is}g_{sj} = \delta^i_j$ is a once covariant and once contravariant tensor. It is evident that the matrix $\|\delta^i_j\|$ is the identity matrix:

$$\delta^i_j = 1 \text{ if } i = j, \text{ and } \delta^i_j = 0 \text{ if } i \neq j.$$

If $(x^1, \ldots, x^m)$ are Cartesian, then $\|g_{sj}\|$ is the identity matrix $g_{ij} = \delta_{ij}$, and the expression (A.5) is reduced to the classical formula for the length of a curve

$$\int_{M_0}^{M_1} \sqrt{\sum_{j=1}^{j=m} (dx^j)^2} = \int_{\sigma_0}^{\sigma_1} \sqrt{\sum_{j=1}^{j=m} \left(\frac{dx^j}{d\sigma}\right)^2} d\sigma.$$

Let $(q^1, \ldots, q^m)$ be some regular coordinates in $\Omega$, and $x^i = x^i(q^1, \ldots, q^m)$, $i = 1, \ldots, m$. Then

$$\sum_{j=1}^{m} (dx^j)^2 = \frac{\partial x^j}{\partial q^k} dq^k \frac{\partial x^j}{\partial q^l} dq^l = g_{kl} dq^k dq^l, \tag{A.6}$$

where

$$g_{kl} := \frac{\partial x^j}{\partial q^k} \frac{\partial x^j}{\partial q^l} = g_{lk} \tag{A.7}$$

is again a metric tensor.

From the formula (A.7) for $g_{kl}$ it follows that

$$\det \|g_{ij}\| = \left(\frac{D(x^1, \ldots, x^m)}{D(q^1, \ldots, q^m)}\right)^2 =: \mathscr{D}^2. \tag{A.8}$$

The traditional notation is

$$g := \det \|g_{ij}\|. \tag{A.9}$$

The tensors $g^{ps}$ and $g_{kl}$ are useful in raising and lowering indices. For example, let $T^{rl}$ be a twice contravariant tensor, then the second index can be lowered by the contraction of $T^{rl}$ with $g_{kl}$:

$$T^{rl} g_{lp} = T^r_p. \tag{A.10}$$

Tensors of $(1, 0)$ and $(0, 1)$ types are both called *vectors*, namely, *contravariant vectors* and *covariant vectors*, respectively. Let $T_1, \ldots, T_m$ be the components of a covariant vector, then $T^1, \ldots, T^m$, where $T^j = g^{jl}T_l$, are called contravariant components of the same vector $\mathbf{T}$. An important example of a covariant vector is the *gradient* of a function. Let $\psi = \psi(q^1, \ldots, q^m)$ be a smooth function; its gradient is given by

$$\operatorname{grad} \psi = (\psi_1, \ldots, \psi_m), \quad \psi_j = \frac{\partial \psi}{\partial q^j}. \tag{A.11}$$

The *scalar product* of complex vectors $\mathbf{T}$ and $\mathbf{S}$ is defined by the formula

$$(\mathbf{T}, \mathbf{S}) := g^{jl} T_j S_l^*, \tag{A.12}$$

where $^*$ is the complex conjugation. It is not difficult to observe that the formulas

$$(\mathbf{T},\mathbf{S}) = T^j S_j^*, \quad (\mathbf{T},\mathbf{S}) = g_{lj} T^l S^{j*}, \quad (\mathbf{T},\mathbf{S}) = T_k S^{k*} \tag{A.13}$$

are equivalent to (A.12). The scalar product is a *scalar*, i.e., it does not depend on the system of coordinates. The contraction of two vectors $\mathbf{T}$ and $\mathbf{S}$

$$(\mathbf{T}\cdot\mathbf{S}) := T_k S^k = g_{jl} T^j S^l = g^{jl} T_j S_l = T^k S_k \tag{A.14}$$

is also a scalar. Of importance to us is the following example of a scalar:

$$(\operatorname{grad}\psi)^2 := (\operatorname{grad}\psi\cdot\operatorname{grad}\psi) = g^{jl}\frac{\partial\psi}{\partial q^j}\frac{\partial\psi}{\partial q^l}. \tag{A.15}$$

Let a point $(q^i,\ldots,q^m)$ run over a finite domain $\Omega\subset\mathbb{R}^m$ where a Riemannian metric and the corresponding metric tensor are defined. The *Riemannian volume* of $\Omega$ is defined by the integral

$$\int_\Omega \sqrt{g}dq^1\ldots dq^m = \int_\Omega \sqrt{g}dq^1\ldots dq^m, \quad g=\det\|g_{ik}(M)\|, \quad M\in\Omega. \tag{A.16}$$

With a little difficulty, one proves that the Riemannian volume does not depend on the coordinates. If the coordinates $q^i = x^i$ are Cartesian, then $\sqrt{g}=1$, and (A.16) becomes the classical formula for the volume of a domain. The expression $\sqrt{g}dq^1\ldots dq^m$ is called the *(Riemannian) volume element.*

---

## A.4 Coordinates $(q^1, q^2, n)$ associated with a surface in $\mathbb{R}^3$. The first and second fundamental forms

### A.4.1 Coordinates $(q^1, q^2, n)$

Let $q^l$, $l=1,2,\ldots,s$, $s<m$, be coordinates on an $s$-dimensional manifold embedded into $\mathbb{R}^m$, i.e., $x^j = x^j(q^1,\ldots,q^s)$, where $x^j$ are Cartesian coordinates on $\mathbb{R}^m$. We are concerned only with the case of $m=3$ and $s=2$, which is the case of a smooth surface $\mathscr{S}\subset\mathbb{R}^3$.

Let $\mathbf{x} = (x^1,x^2,x^3) = \big(\mathring{x}^1(q^1,q^2),\mathring{x}^2(q^1,q^2),\mathring{x}^3(q^1,q^2)\big)$ be a parametric definition of the surface $\mathscr{S}$ in $\mathbb{R}^3$. Let us draw the normal to $\mathscr{S}$ through a point $\mathring{\mathbf{x}}\in\mathscr{S}$. It is a straight line orthogonal to $\mathscr{S}$ in the Euclidian metric. The points of the line are specified by the distance $n$ from $\mathscr{S}$. The points on the one side from $\mathscr{S}$ are endowed with positive $n>0$, the negative $n<0$ corresponds to the points on the other side of the surface. We thus defined a coordinate system $(q^1,q^2,q^3)$, $q^3=n$, in a neighborhood of $\mathscr{S}$. The Cartesian coordinates $(x^1, x^2, x^3)$ and $(q^1, q^2, q^3)$ are connected by the relation

$$\mathbf{x} = (\mathring{x}^1(q^1,q^2), \mathring{x}^2(q^1,q^2), \mathring{x}^3(q^1,q^2)) + n\boldsymbol{n}, \quad \boldsymbol{n} = \frac{\frac{\partial \mathring{\mathbf{x}}}{\partial q^1} \times \frac{\partial \mathring{\mathbf{x}}}{\partial q^2}}{\left| \frac{\partial \mathring{\mathbf{x}}}{\partial q^1} \times \frac{\partial \mathring{\mathbf{x}}}{\partial q^2} \right|}. \tag{A.17}$$

It can be proved that for a small $n$, there is a smooth bijection between $(x^1, x^2, x^3)$ and $(q^1, q^2, q^3)$. The coordinates $q^1$ and $q^2$ will be chosen in such a way that

$$\mathscr{D} = \frac{D(x^1, x^2, x^3)}{D(q^1, q^2, q^3)} = \left( \left[ \frac{\partial \mathring{\mathbf{x}}}{\partial q^1} \times \frac{\partial \mathring{\mathbf{x}}}{\partial q^2} \right], \boldsymbol{n} \right) > 0. \tag{A.18}$$

### A.4.2 First fundamental form

In what follows, we assume that the Latin indices range from 1 to 3, the Greek indices range from 1 to 2, and the Einstein summation convention is applied to both Latin and Greek indices.

Let $g_{ij}$ be the components of a metric tensor in the coordinates $(q^1, q^2, q^3)$, $q^3 = n$. It is not difficult to establish that

$$g_{3j} = g_{j3} = \begin{cases} 0, & j = 1, 2, \\ 1, & j = 3. \end{cases} \tag{A.19}$$

The quadratic form $\mathring{g}_{\alpha\beta} dq^\alpha dq^\beta = \mathring{g}_{\alpha\beta}(q^1, q^2) dq^\alpha dq^\beta$, $\alpha, \beta = 1, 2$ is called the *first fundamental form* of the surface $\mathscr{S}$. It is obvious that $\mathring{g}_{\alpha\beta}(q^1, q^2)$ is a metric tensor on $\mathscr{S}$. The differential form $\sqrt{\mathring{g}} dq^1 dq^2$, where $\mathring{g} := \det \|\mathring{g}_{\alpha\beta}\|$, is called the *area element* on $\mathscr{S}$.

### A.4.3 Second fundamental form

For $n \to 0$, we have

$$g_{\alpha\beta}(q^1, q^2, n) = \mathring{g}_{\alpha\beta}(q^1, q^2) - 2n\mathfrak{b}_{\alpha\beta}(q^1, q^2) + O\left(n^2\right), \tag{A.20}$$

where $\mathfrak{b}_{\alpha\beta}(q^1, q^2) = \mathfrak{b}_{\alpha\beta}(q^1, q^2)$ are some coefficients. The coefficients $\mathfrak{b}_{\alpha\beta}(q^1, q^2)$, where $\alpha, \beta = 1, 2$, are called the coefficients of the *second fundamental form.* It is obvious that $\mathfrak{b}_{\alpha\beta}(q^1, q^2)$ is a twice covariant tensor on $\mathscr{S}$.

The second fundamental form is associated with the curvatures of the surface $\mathscr{S}$ and curves that lie on $\mathscr{S}$. Let $\ell$ be a smooth curve on $\mathscr{S}$ defined by $q^1 = q^1(t)$, $q^2 = q^2(t)$. The ratio

$$\frac{1}{\mathfrak{R}} = \frac{\mathfrak{b}_{\alpha\beta} \dot{q}^\alpha(t) \dot{q}^\beta(t)}{\mathring{g}_{\alpha\beta} \dot{q}^\alpha(t) \dot{q}^\beta(t)}, \quad \dot{q}^\gamma(t) := \frac{dq^\gamma}{dt}, \tag{A.21}$$

is the *curvature of the normal section* of $\mathscr{S}$ at the point $(q^1 = q^1(t), q^2 = q^2(t))$. The normal cut of $\mathscr{S}$ is the intersection of $\mathscr{S}$ and the plane passing through the point $(q^1(t), q^2(t), 0)$ and containing vectors $\boldsymbol{n}$ and $(\dot{q}^1(t), \dot{q}^2(t), 0)$. The curvature $\frac{1}{\mathfrak{R}}$ may be either positive or negative or vanish.

The scalar

$$\mathfrak{H} := \frac{\det \|\mathfrak{b}_{\alpha\beta}\|}{\det \|\mathring{g}_{\alpha\beta}\|} \tag{A.22}$$

is called the *Gaussian curvature of the surface* $\mathscr{S}$ at the corresponding point.

## A.5 Covariant derivative. Divergence

Let $T_i(x^1, \dots, x^m)$ be a covariant tensor and $(x^1, \dots, x^m)$ be arbitrary coordinates. In the general case, the derivatives $\frac{\partial T_i}{\partial x^p}$ are not tensors. However,

$$\mathfrak{U}_{ip} = \frac{\partial T_i}{\partial x^p} - T_s \Gamma^s_{ip} =: \nabla_p T_i \tag{A.23}$$

is a twice covariant tensor, provided that

$$\Gamma^s_{ip} := \frac{1}{2} g^{ks} \left( \frac{\partial g_{ik}}{\partial x^p} + \frac{\partial g_{kp}}{\partial x^i} - \frac{\partial g_{ip}}{\partial x^k} \right). \tag{A.24}$$

The expression $\Gamma^s_{ip}$ is called the *Christoffel symbol (of the second kind)*, and the expression $\nabla_p T_i$ is called the *covariant derivative* of $T_i$. The covariant derivative of the vector $T^i$ is defined by

$$\nabla_p T^i := \frac{\partial T^i}{\partial x^p} + T^j \Gamma^i_{jp}. \tag{A.25}$$

The expressions (A.24) and (A.25) are $(0, 2)$ and $(1, 1)$ tensors, respectively.

Of much importance in applications is the scalar defined by the formula

$$\operatorname{div} \mathbf{T} := \nabla_j T^j \equiv \frac{1}{\sqrt{g}} \frac{\partial}{\partial q^j} \left( \sqrt{g} T^j \right) \tag{A.26}$$

and called the *divergence of the vector* $\mathbf{T}$. We present a proof of the right-hand side of this formula. From (A.24) and (A.25) it follows that

$$\nabla_j T^j = \frac{\partial T^j}{\partial x^j} + \Gamma^i_{ij} T^j. \tag{A.27}$$

The expression $\Gamma^i_{ij}$ can be simplified as follows:

$$\Gamma^i_{ij} = \frac{1}{2} g^{ki} \left( \frac{\partial g_{ik}}{\partial x^j} + \frac{\partial g_{kj}}{\partial x^i} - \frac{\partial g_{ij}}{\partial x^k} \right) = \frac{1}{2} \left( g^{ki} \frac{\partial g_{ik}}{\partial x^j} + g^{ki} \frac{\partial g_{kj}}{\partial x^i} - g^{ki} \frac{\partial g_{ij}}{\partial x^k} \right).$$

The second and third summands on the right-hand side cancel by virtue of the symmetry of the metric tensor, and consequently

$$\Gamma^i_{ij} = \frac{1}{2} g^{ki} \frac{\partial g_{ik}}{\partial x^j}.$$

Consider the expression $\frac{1}{\sqrt{g}} \frac{\partial \sqrt{g}}{\partial x^j} = \frac{1}{2g} \frac{\partial g}{\partial x^j}$. Differentiating the determinant $g$ defined by the formula (A.9), we get

$$\frac{1}{2g} \frac{\partial g}{\partial x^j} = \frac{1}{2g} \frac{\partial g}{\partial g_{ik}} \frac{\partial g_{ik}}{\partial x^j} = \frac{A_{ik}}{2g} \frac{\partial g_{ik}}{\partial x^j}, \tag{A.28}$$

where $A_{ik}$ is the algebraic complement of $g_{ik}$ of the determinant $g$. It is well known (see, e.g., the textbook by Gantmacher 1959 [1]) that the ratio $\frac{A_{ik}}{g}$ is equal to the element $g^{ik}$ of the matrix inverse to $\|g_{ik}\|$. Therefore,

$$\frac{1}{\sqrt{g}} \frac{\partial \sqrt{g}}{\partial x^j} = \frac{1}{2} g^{ik} \frac{\partial g_{ik}}{\partial x^j} = \Gamma^i_{ij}. \tag{A.29}$$

Now we turn back to the formula (A.27) that takes the form

$$\nabla_j T^j = \frac{\partial T^j}{\partial x^j} + \frac{1}{\sqrt{g}} \frac{\partial \sqrt{g}}{\partial x^j} T^j = \frac{1}{\sqrt{g}} \frac{\partial \left(\sqrt{g} T^j\right)}{\partial x^j}, \tag{A.30}$$

which completes the proof of the relation (A.26).

A detailed presentation of the above material can be found, e.g., in Dubrovin, Fomenko, and Novikov 1984 [2] or in Novikov and Taimanov 2006 [3].

---

## References to Appendix


[1] Gantmacher, F. R. 1959. *The theory of matrices.* New York: Chelsea Publishing Company. *Гантмахер Ф. Р.* Теория матриц. М.: Наука, 1966.

[2] Dubrovin, B. A, Fomenko, A. T. and Novikov, S. P. 1984. *Modern geometry — methods and applications. Part I. The geometry of surfaces, transformation groups, and fields.* New York: Springer. *Дубровин Б. А., Новиков С. П., Фоменко, А. Т.* Современная геометрия. Методы и приложения. Т. 1. М.: Наука, 1998.

[3] Novikov, S. P. and Taimanov, I. A. 2006. *Modern geometric structures and fields.* AMS. *Новиков С. П., Тайманов И. А.* Современные геометрические структуры и поля. М.: МЦНМО, 2005.

# Index